# Association for Women in Mathematics Series

Volume 2



More information about this series at http://www.springer.com/series/13764

# Association for Women in Mathematics Series

Focusing on the groundbreaking work of women in mathematics past, present, and future, Springer's Association for Women in Mathematics Series presents the latest research and proceedings of conferences worldwide organized by the Association for Women in Mathematics (AWM). All works are peer-reviewed to meet the highest standards of scientific literature, while presenting topics at the cutting edge of pure and applied mathematics. Since its inception in 1971, The Association for Women in Mathematics has been a non-profit organization designed to help encourage women and girls to study and pursue active careers in mathematics and the mathematical sciences and to promote equal opportunity and equal treatment of women and girls in the mathematical sciences. Currently, the organization represents more than 3000 members and 200 institutions constituting a broad spectrum of the mathematical community in the United States and around the world.

Marie José Bertin • Alina Bucur • Brooke Feigon
Leila Schneps
Editors

# Women in Numbers Europe

## Research Directions in Number Theory

Springer

*Editors*
Marie José Bertin
Number Theory
Jussieu Institute of Mathematics
Paris, France

Alina Bucur
Department of Mathematics
University of California, San Diego
La Jolla, CA, USA

Brooke Feigon
Department of Mathematics
The City College of New York
New York, NY, USA

Leila Schneps
Algebraic Analysis
Jussieu Institute of Mathematics
Paris, France

Association for Women in Mathematics Series
ISBN 978-3-319-36836-8 ISBN 978-3-319-17987-2 (eBook)
DOI 10.1007/978-3-319-17987-2

Springer Cham Heidelberg New York Dordrecht London

Softcover re-print of the Hardcover 1st edition 2015


Printed on acid-free paper

Springer International Publishing AG Switzerland is part of Springer Science+Business Media (www.springer.com)

# Preface

The Women In Numbers—Europe workshop (WINE) was held on October 13–18, 2013, at the Centre International de Rencontres Mathématiques, in Luminy, France. It was the first workshop in the Women in Numbers (WIN) series held outside North America.

The WIN conferences bring together female researchers in number theory at multiple career stages to conduct research projects during a weeklong workshop. The project leaders are generally senior faculty, while the participants are graduate students, postdocs, and junior and senior faculty. The benefits of these workshops flow in both directions: senior women meet, mentor, and collaborate with young researchers, while junior women encounter important new research problems and develop a network of colleagues, supporters, and mentors. Each of the WIN conferences has given rise to a proceedings volume, and some of the collaborations initiated at the conferences went on to generate publications in research journals. Based on the model of the WIN conferences, the Association for Women in Mathematics (AWM) now conducts similar events in the Research Collaboration Conferences for Women (RCCW) series, in areas such as topology (WIT), algebraic combinatorics (Combinatorixx), computer vision (WISH), and applied math (WHAM).

Due to funding restrictions, the WIN workshops were at first limited to North American researchers; despite strong interest shown by European mathematicians, most of them could not be accommodated. This problem led to the organization of the WINE conference, as an attempt to correct this imbalance and to build a bridge between female number theorists on both sides of the Atlantic.

Prior to the conference, the participants were divided into 10 working groups of 3–7 members each, according to their research interests. Each group contained a mix of senior and junior mathematicians, including advanced graduate students, with leaders drawn from the participating senior mathematicians. The leaders suggested background reading and references before the conference, and arrived there with a specific research project.

During the conference itself, participants attended background lectures given by the group leader of each project, with several hours set aside each day for collaborations of the working groups. Project leaders directed the research efforts of their respective groups, providing specific tasks such as computations or partial proofs as a way of penetrating the problem. By the end of the week, group participants had understood their research problems and grasped where they fit into a larger scheme, and many groups had already taken some concrete first steps in the direction of a solution. The week closed with short presentations given by junior members of each research group on their progress, as well as future directions for the work.

The CIRM mathematics research center in Luminy—after initial hesitation at the hitherto-unheard of request to organize a conference for women only—turned out to be an ideal setting for this project. The seamless and hospitable running of the practical side of things allowed the conference organizers to devote most of their energies to their respective research groups, while the incredible natural beauty of the surroundings, with rocky cliffs plunging down to small coves of turquoise sea, gave inspiration and also much-needed opportunities for some rugged physical activity. Each day saw some groups working outside in the sunshine, thanks to moveable blackboards, and evenings after dinner were devoted to group research—an intense form of socializing.

Nearly all of the groups went on with their collaborations after the conference. Eventually, after more than a year, the papers started coming in, and according to standard practice, were sent to anonymous referees for assessment. This volume contains the final results of the research initiated at the WINE conference. Covering topics in graph theory, $L$-functions, $p$-adic geometry, Galois representations, elliptic fibrations, genus 3 curves and bad reduction, harmonic analysis, symplectic groups and mould combinatorics, the articles span a wide swath of number theory.

The WIN conference series is a remarkable initiative that has produced significant advantages for its participants in terms of research, conference experience, and networking. More WINs are planned for the future, and we hope the trend will spread to include anyone, in any country, who would like to participate in a WIN one day.

**Acknowledgements** We would like to thank the following organizations for providing financial support for WINE: Centre International de Rencontres Mathématiques, Clay Mathematics Institute, Microsoft Research, National Science Foundation, and the Number Theory Foundation.

Paris, France — Marie José Bertin
San Diego, CA, USA — Alina Bucur
New York, NY, USA — Brooke Feigon
Paris, France — Leila Schneps

# Contents

# Contributors

**Sara Arias-de-Reyna** Mathematical Research Unit, University of Luxembourg, Luxembourg, Luxembourg

**Cécile Armana** Laboratory of Mathematics, University of Franche-Comté, Besançon, France

**Christina Ballantine** Department of Mathematics and Computer Science, College of the Holy Cross, Worcester, MA, USA

**Marie José Bertin** Jussieu Institute of Mathematics, Pierre and Marie Curie University, Paris, France

**Irene Bouw** Institute of Pure Mathematics, University of Ulm, Ulm, Germany

**Alina Bucur** Department of Mathematics, University of California San Diego, San Diego, CA, USA

**Jenny Cooley** Mathematics Institute, University of Warwick, Coventry, UK

**Agnés David** Mathematical Research Unit, University of Luxembourg, Luxembourg, Luxembourg

**Anne-Maria Ernvall-Hytönen** Department of Mathematics and Statistics, University of Helsinki, Helsinki, Finland

**Brooke Feigon** Department of Mathematics, The City College of New York, New York, NY, USA

**Sharon M. Frechette** Department of Mathematics and Computer Science, College of the Holy Cross, Worcester, MA, USA

**Radhika Ganapathy** Mathematics Department, The University of British Columbia, Vancouver, BC, Canada

**Alice Garbagnati** Department of Mathematics, University of Milan, Milan, Italy

**Julia Gordon** Mathematics Department, The University of British Columbia, Vancouver, BC, Canada

**Marcela Hanzer** Department of Mathematics, University of Zagreb, Zagreb, Croatia

**Ruthi Hortsch** Department of Mathematics, Massachusetts Institute of Technology, Cambridge, MA, USA

**Valentijn Karemaker** Mathematical Institute, Utrecht University, Utrecht, The Netherlands

**Janne Kool** Max Planck Institute for Mathematics, Bonn, Germany

**Kristin Lauter** Microsoft Research, Redmond, WA, USA

**Odile Lecacheux** Jussieu Institute of Mathematics, Pierre and Marie Curie University, Paris, France

**Elisa Lorenzo García** Mathematics Institute, University of Leiden, Leiden, The Netherlands

**Judith Ludwig** Mathematical Institute, University of Bonn, Bonn, Germany

**Michelle Manes** Department of Mathematics, University of Hawaii, Honolulu, HI, USA

**Makiko Mase** Tokyo Metropolitan University, Tokyo, Japan

**Kathrin Maurischat** Mathematical Institute, University of Heidelberg, Heidelberg, Germany

**Rachel Newton** Max Planck Institute for Mathematics, Bonn, Germany

**Almasa Odžak,** Department of Mathematics, University of Sarajevo, Sarajevo, Bosnia & Herzegovina

**Ekin Ozman** Mathematics Department, University of Texas at Austin, Austin, TX, USA

Department of Mathematics, Bogazici University, Istanbul, Turkey

**Marusia Rebolledo** Laboratory of Mathematics, Blaise Pascal University, Clermont-Ferrand, Aubière, France

**Lance Robson** Mathematics Department, University of British Columbia, Vancouver, BC, Canada

**Edva Roditty-Gershon** Raymond and Beverly Sackler School of Mathematical Sciences, Tel Aviv University, Tel Aviv, Israel

**Adriana Salerno** Mathematics Department, Bates College, Lewiston, ME, USA

**Cecília Salgado** Institute for Mathematics, Federal University of Rio de Janeiro, Rio de Janeiro, RJ, Brazil

**Damaris Schindler** Hausdorff Center for Mathematics, University of Bonn, Bonn, Germany

**Lejla Smajlović** Department of Mathematics, University of Sarajevo, Sarajevo, Bosnia & Herzegovina

**Lara Thomas** Pure and Applied Mathematics Unit, ENS Lyon, Lyon, France

**Amanda Tucker** Department of Mathematics, State University of New York College at Geneseo, Geneseo, NY, USA

**Núria Vila** Department of Algebra and Geometry, University of Barcelona, Barcelona, Spain

**Ursula Whitcher** Department of Mathematics, University of Wisconsin–Eau Claire, Eau Claire, WI, USA

**Amy Wooding** Department of Mathematics and Statistics, McGill University, Montreal, QC, Canada

# Explicit Construction of Ramanujan Bigraphs

**Cristina Ballantine, Brooke Feigon, Radhika Ganapathy, Janne Kool, Kathrin Maurischat, and Amy Wooding**

**Abstract** We construct explicitly an infinite family of Ramanujan graphs which are bipartite and biregular. Our construction starts with the Bruhat–Tits building of an inner form of $SU_3(\mathbb{Q}_p)$. To make the graphs finite, we take successive quotients by infinitely many discrete co-compact subgroups of decreasing size.

**MSC 2010:** 11E39, 11E57, 16W10

C. Ballantine
Department of Mathematics and Computer Science, College of the Holy Cross,
1 College Street, Worcester, MA 01610, USA
e-mail: cballant@holycross.edu

B. Feigon (✉)
Department of Mathematics, The City College of New York, NAC 8/133,
New York, NY 10031, USA
e-mail: bfeigon@ccny.cuny.edu

R. Ganapathy
Department of Mathematics, The University of British Columbia, 1984 Mathematics Road,
Vancouver, BC, Canada V6T 1Z2
e-mail: rganapat@math.ubc.ca

J. Kool
Max Planck Institute for Mathematics, Vivatsgasse 7, 53111 Bonn, Germany
e-mail: kool79@mpim-bonn.mpg.de

K. Maurischat
Mathematisches Institut, Universität Heidelberg, Im Neuenheimer Feld 288,
69120 Heidelberg, Germany
e-mail: maurischat@mathi.uni-heidelberg.de

A. Wooding
The Department of Mathematics and Statistics, McGill University, Burnside Hall, Room 1023,
805 Sherbrooke W., Montreal, QC, Canada H3A 0B9
e-mail: amy.cheung@mail.mcgill.ca

 
M.J. Bertin et al. (eds.), *Women in Numbers Europe*, Association for Women in Mathematics Series 2, DOI 10.1007/978-3-319-17987-2_1

## 1 Introduction

Expander graphs are highly connected yet sparse graphs. By a highly connected graph we mean a graph in which all small sets of vertices have many neighbors. They have wide ranging applications, especially in computer science and coding theory. They also model neural connections in the brain and many other types of networks. One is usually interested in regular or biregular expanders. The expansion property is controlled by the size of the spectral gap of the graph. Asymptotically, Ramanujan graphs are optimal expanders as we will explain below. Infinite families of regular Ramanujan graphs of fixed degree were first constructed in the late 1980s by Lubotzky et al. (1988), and independently by Margulis (1988). Since then, the study of problems related to the existence and construction of Ramanujan graphs has become an active area of research. Until recently, all constructions of families of regular Ramanujan graphs have been obtained using tools from number theory, including deep results from the theory of automorphic forms. As a result, the graphs obtained have degree $q + 1$, where $q$ is a power of a prime. Using similar methods, Ballantine and Ciubotaru (2011) gives a roadmap toward the construction of infinite families of Ramanujan bigraphs, i.e., biregular, bipartite graphs satisfying the Ramanujan condition, of bidegree $(q^3 + 1, q + 1)$, where $q$ is a power of a prime. However, they stop short of providing explicit examples. Very recently, Marcus et al. (2014) used the method of interlacing polynomials to prove the existence of arbitrary degree Ramanujan bigraphs. By making the two degrees equal, this implies the existence of arbitrary degree (regular) Ramanujan graphs. Their proof is non-constructive.

In this article, we follow the roadmap given in Ballantine and Ciubotaru (2011) to explicitly construct an infinite family of Ramanujan bigraphs. We start with a quadratic extension, $E/\mathbb{Q}$, and define a division algebra $D$ which is non-split over $E$, i.e., $D$ is not isomorphic to the matrix algebra $M_3(E)$. We then use this to define a special unitary group $\mathbb{G}$ over $E$ from $D$ by means of an involution of the second kind. We define this involution such that the corresponding local unitary group is isomorphic to $\mathrm{SU}_3(\mathbb{Q}_p)$ at the place $p$, i.e., $\mathbb{G}_p = \mathbb{G}(\mathbb{Q}_p) \cong \mathrm{SU}_3(\mathbb{Q}_p)$, and compact at infinity. We also give a concrete description of an infinite family of discrete co-compact subgroups of $\mathbb{G}_p$ which act without fixed points on $\mathbb{G}_p$.

Since the division algebra $D$ is non-split, Corollary 4.6 of Ballantine and Ciubotaru (2011) guarantees that each quotient of the Bruhat–Tits tree of $\mathbb{G}_p$ by one of the above subgroups satisfies the Ramanujan condition. Therefore, we obtain an infinite family of Ramanujan bigraphs of bidegree $(p^3 + 1, p + 1)$. We note that most of this work could be carried out over a general totally real number field but we often choose to work over $\mathbb{Q}$ to simplify the notation.

# 2 Preliminaries and Notation

In this section we introduce the notation used throughout the article and give a brief review of Ramanujan graphs and bigraphs, unitary groups, and buildings.

## *2.1 Ramanujan Graphs and Bigraphs*

While Ballantine and Ciubotaru (2011) also contains a concise review of this topic, we find it useful for the reader to have an overview within the current article. Lubotzky (2012) gives a review of expander graphs with applications within mathematics. Hoory et al. (2006) provides a review accessible to the nonspecialist with many applications, especially to computer science. For an elementary introduction to regular Ramanujan graphs, we refer the reader to Davidoff et al. (2003).

A graph $X = (V, E)$ consists of a set of vertices $V$ together with a subset of pairs of vertices called edges. In this article, all graphs are undirected. Thus, the pair of vertices forming an edge is unordered. The *degree* of a vertex is the number of edges incident to it. A graph is called *k-regular* if all vertices have degree $k$. A graph is called *$(l,m)$-biregular* if each vertex has degree $l$ or $m$. A *bipartite* graph is a graph that admits a coloring of the vertices with two colors such that no two adjacent vertices have the same color. A *bigraph* is a biregular, bipartite graph.

We denote by $\mathrm{Ad}(X)$ the adjacency matrix of $X$ and by $\mathrm{Spec}(X)$ the spectrum of $X$. Thus, $\mathrm{Spec}(X)$ is the collection of eigenvalues of $\mathrm{Ad}(X)$. Since the adjacency matrix is symmetric, $\mathrm{Spec}(X) \subset \mathbb{R}$. For a $k$-regular graph, we have $k \in \mathrm{Spec}(X)$. For an $(l,m)$-biregular graph, we have $\sqrt{lm} \in \mathrm{Spec}(X)$. Moreover, if we denote by $\lambda_i$ the eigenvalues of a graph, for a connected $k$-regular graph we have

$$k = \lambda_0 > \lambda_1 \geq \lambda_2 \geq \cdots \geq -k.$$

Thus, $k$ is the largest absolute value of an eigenvalue of $X$. We denote by $\lambda(X)$ the next largest absolute value of an eigenvalue. If $X$ is bipartite, the spectrum is symmetric and $-k$ is an eigenvalue. Let $X$ be a finite connected bigraph with bidegree $(l,m)$, $l \geq m$. Suppose $X$ has $n_1$ vertices of degree $l$ and $n_2$ vertices of degree $m$. We must have $n_2 \geq n_1$. Then, $\mathrm{Spec}(X)$ is the multiset

$$\{\pm\lambda_0, \pm\lambda_1, \ldots, \pm\lambda_{n_1}, \underbrace{0, \ldots, 0}_{n_2 - n_1}\},$$

where $\lambda_0 = \sqrt{lm} > \lambda_1 \geq \cdots \geq \lambda_{n_1} \geq 0$. Then, with the above notation, $\lambda(X) = \lambda_1$.

If $W$ is a subset of $V$, the *boundary* of $W$, denoted by $\partial W$, is the set of vertices outside of $W$ which are connected by an edge to a vertex in $W$, i.e.,

$$\partial(W) = \{v \in V \setminus W \mid \{v, w\} \in E, \text{ for some } w \in W\}.$$

The *expansion coefficient* of a graph $X = (V, E)$ is defined as

$$c = \inf \left\{ \frac{|\partial W|}{\min\{|W|, |V \setminus W|\}} \;\middle|\; W \subseteq V : 0 < |W| < \infty \right\}.$$

Note that, if $|V| = n$ is finite, then

$$c = \min \left\{ \frac{|\partial W|}{|W|} \;\middle|\; W \subseteq V : 0 < |W| \leq \frac{n}{2} \right\}.$$

A graph $X = (V, E)$ is called an $(n, k, c)$-*expander* if $X$ is a $k$-regular graph on $n$ vertices with expansion coefficient $c$. The expansion coefficient $c$ of a regular graph is related to $\lambda(X)$, the second largest absolute value of an eigenvalue (Lubotzky et al. 1986, Proposition 1.2) by

$$2c \geq 1 - \frac{\lambda(X)}{k}.$$

Good expanders have large expansion coefficient. Thus, good expanders have small $\lambda(X)$ (or large spectral gap, $k - \lambda(X)$). Alon and Boppana (Alon 1986; Lubotzky et al. 1988) showed that asymptotically $\lambda(X)$ cannot be arbitrarily small. They proved that, if $X_{n,k}$ is a $k$-regular graph with $n$ vertices, then

$$\liminf_{n \to \infty} \lambda(X_{n,k}) \geq 2\sqrt{k-1}.$$

Lubotzky et al. (1986) defined a Ramanujan graph to be a graph that beats the Alon–Boppana bound.

**Definition 2.1.** A $k$-regular graph $X$ is called a Ramanujan graph if $\lambda(X) \leq 2\sqrt{k-1}$.

Feng and Winnie Li (1996) proved the analog to the Alon–Boppana bound for biregular bipartite graphs. They showed that, if $X_{n,l,m}$ is a $(l, m)$-biregular graph with $n$ vertices, then

$$\liminf_{n \to \infty} \lambda(X_{n,l,m}) \geq \sqrt{l-1} + \sqrt{m-1}.$$

Then, Solé (1999) defines Ramanujan bigraphs as the graphs that beat the Feng-Li bound.

**Definition 2.2.** A finite, connected, bigraph $X$ of bidegree $(l, m)$ is a Ramanujan bigraph if

$$|\sqrt{l-1} - \sqrt{m-1}| \le \lambda(X) \le \sqrt{l-1} + \sqrt{m-1}.$$

Solé's definition is equivalent to the following definition given by Hashimoto (1989).

**Definition 2.3.** A finite, connected, bigraph of bidegree $(q_1 + 1, q_2 + 1)$ is a Ramanujan bigraph if

$$|(\lambda(X))^2 - q_1 - q_2| \le 2\sqrt{q_1 q_2}.$$

Our goal is to construct an infinite family of Ramanujan bigraphs of the same bidegree and with the number of vertices growing without bound. In general, it is difficult to check that a large regular or biregular graph is Ramanujan. In this article, the graphs are quotients of the Bruhat–Tits building attached to an inner form of the special unitary group in three variables. We then employ a result of Ballantine and Ciubotaru (2011), which uses the structure of the group, to estimate the spectrum of the building quotient in order to conclude that the graphs constructed are Ramanujan.

## 2.2 *Unitary Groups in Three Variables*

We denote by $F$ a local or global field of characteristic zero. For a detailed discussion on unitary groups, we refer the reader to Rogawski (1990). Let $E/F$ be a quadratic extension and $\phi : E^3 \times E^3 \to E$ be a Hermitian form. Then the *special unitary group* with respect to $\phi$ is an algebraic group over $F$ whose functor of points is given by

$$\mathrm{SU}(\phi, R) = \{g \in \mathrm{SL}_3(E \otimes_F R) \mid \phi(gx, gy) = \phi(x, y)\ \forall x, y \in E^3 \otimes_F R\}$$

for any $F$-algebra $R$. We use $\mathrm{SU}_3$ to denote the standard special unitary group corresponding to the Hermitian form given by the identity matrix; that is,

$$\mathrm{SU}_3(R) = \{g \in \mathrm{SL}_3(E \otimes_F R) \mid {}^t\bar{g}g = \mathtt{Id}_3\}$$

where $\bar{g}$ is conjugation with respect to the extension $E/F$.

Let $D$ be a central simple algebra of degree three over $E$ and $\alpha$ be an involution of the second kind, i.e., an anti-automorphism of $D$ that acts on $E$ by conjugation with respect to $E/F$. By Wedderburn's theorem (Knus et al. 1998, Theorem 19.2), $D$ is a cyclic algebra over $E$. Let $N_D$ denote the reduced norm of $D$. Then, $(D, \alpha)$ defines a special unitary group $\mathbb{G}$ by

$$\mathbb{G}(R) = \{d \in (D \otimes_F R)^\times \mid \alpha(d)d = 1,\ N_{D\otimes_F R}(d) = 1\}.$$

Moreover, all special unitary groups are obtained in this way from $(D, \alpha)$ (Rogawski 1990, section 1.9).

## 2.3 Buildings

Let $F$ be a non-archimedean local field and let $E/F$ be an *unramified* separable quadratic extension. Let $G = \mathrm{SU}_3$ be defined as above. Let $\mathcal{O} = \mathcal{O}_E$ be the ring of integers of $E$ and $\mathfrak{p}$ be the unique maximal ideal in $\mathcal{O}$. Let $k = \mathcal{O}/\mathfrak{p}$ be the residue field. We denote by $B$ the Borel subgroup of upper-triangular matrices and by $B(k)$ the $k$-points of $B$. We denote by $I$ the preimage of $B(k)$ under the reduction mod $\mathfrak{p}$ map $G(\mathcal{O}) \to G(k)$. The group $I$ is an Iwahori subgroup. Then the Weyl group $W$ of $G$ is the infinite dihedral group. Let $s_1$ and $s_2$ be the reflections generating $W$. For $i = 1, 2$, let $U_i = I \cup I s_i I$. These subgroups are the representatives of the $G$-conjugacy classes of maximal compact subgroups of $G$ (Hashimoto and Hori 1989). Moreover, $I = U_1 \cap U_2$.

The Bruhat–Tits building associated with $G$ is a one dimensional simplicial complex defined as follows. The set of 0-dimensional simplices consists of one vertex for each maximal compact subgroup of $G$. If $K_1$ and $K_2$ are two maximal compact subgroups of $G$, we place an edge between the vertices corresponding to $K_1$ and $K_2$ if and only if $K_1 \cap K_2$ is conjugate to $I$ in $G$. The edges form the set of one-dimensional simplices of the building. Since they are the faces of the largest dimension, they are the chambers of the building. The group $G$ acts simplicially on the building in a natural way. The building associated with $\mathrm{SU}_3$ is a $(q^3 + 1, q + 1)$ tree, where $q$ is the cardinality of the residue field $k$. For more details on buildings, we refer the reader to Tits (1979) and Garrett (1997).

## 2.4 Ramanujan Bigraphs from Buildings

Let $G$ be the group $\mathrm{SU}_3$ over $\mathbb{Q}_p$ (or a finite extension of $\mathbb{Q}_p$). Let $\tilde{X}$ be the Bruhat–Tits tree of $G$. Let $E$ be an imaginary quadratic extension of $\mathbb{Q}$ and let $D$ be a central simple algebra of degree 3 over $E$ and $\alpha$ an involution of the second kind on $D$. Let $\mathbb{G}$ be the special unitary group over $\mathbb{Q}$ determined by $(D, \alpha)$. We have the following theorem of Ballantine and Ciubotaru (2011) that motivates our work.

**Theorem 2.4 (Ballantine and Ciubotaru (2011, Theorem 1.2)).** *Let $\Gamma$ be a discrete, co-compact subgroup of $G$ which acts on $G$ without fixed points. Assume that $D \neq M_3(E)$, $\mathbb{G}(\mathbb{Q}_p) = G$ and $\mathbb{G}(\mathbb{R})$ is compact. Then the quotient tree $X = \tilde{X}/\Gamma$ is a Ramanujan bigraph.*

In the rest of this article we give a description of an algebra $D$ together with an involution $\alpha$ fulfilling the assumptions of the above theorem, as well as an infinite collection of discrete, co-compact subgroups of $G$ which act on $G$ without fixed points.

# 3 Choosing the Algebra and the Involution

The goal of this section is to determine *explicitly* a global division algebra $D$ which is central simple of degree three over its center $E$ and is equipped with an involution $\alpha$ of the second kind with fixed field $F$ such that the related special unitary group $\mathbb{G}$,

$$\mathbb{G}(R) = \{d \in (D \otimes_F R)^\times \mid \alpha(d)d = 1 \text{ and } N_{D\otimes_F R}(d) = 1\},$$

yields compactness at infinity. Such an algebra exists by the Hasse principle (see for example Harris and Labesse 2004, p. 657), which actually is much stronger: For any set of local data, there is a global one localizing to it. We note that in Ballantine and Ciubotaru (2011) the authors refer to Clozel et al. (2008) for the existence of the global group (and thus the algebra defining it). The example of central simple algebra with involution given in Ballantine and Ciubotaru (2011) does not necessarily lead to Ramanujan bigraphs. It is not a division algebra and the resulting unitary group has non-tempered representations occurring as local components of automorphic representations. Therefore, Rogawski's Theorem (Rogawski 1990, Theorem 14.6.3) does not apply.

## *3.1 Cyclic Central Simple Algebras of Degree Three*

Let $E$ be a number field. Let $L$ be a cyclic algebra of degree three over $E$, and let $\rho$ be a generator of its automorphism group which is isomorphic to the cyclic group $C_3$. Then, define a cyclic central simple algebra $D$ of degree three over $E$ by

$$D = L \oplus Lz \oplus Lz^2,$$

where $z$ is a generic element satisfying $z^3 = a \in E^\times$ subject to the relation

$$zl = \rho(l)z \text{ for any } l \in L.$$

By a theorem of Wedderburn (Knus et al. 1998, Theorem 19.2), any central simple algebra of degree three is cyclic. From now on we will assume $D$ is in the form given above. As $D$ is a vector space over $L$ with basis $\{1, z, z^2\}$, we write the multiplication by $d \in D$ from the right in terms of matrices to obtain an embedding $D \hookrightarrow M_3(L)$,

$$d = l_0 + l_1 z + l_2 z^2 \mapsto A(l_0, l_1, l_2) := \begin{pmatrix} l_0 & l_1 & l_2 \\ a\rho(l_2) & \rho(l_0) & \rho(l_1) \\ a\rho^2(l_1) & a\rho^2(l_2) & \rho^2(l_0) \end{pmatrix},$$

for $l_0, l_1, l_2 \in L$. Let $N_{L/E}$ denote the norm of $L/E$, and $Tr_{L/E}$ denote the trace of $L/E$. Then for the reduced norm of $D$ we have

$$N_D(d) = \det A(l_0, l_1, l_2) = N_{L/E}(l_0) + aN_{L/E}(l_1) + a^2 N_{L/E}(l_2)$$
$$-aTr_{L/E}(l_0\rho(l_1)\rho^2(l_2)).$$

In order for $D$ to be a division algebra, we have to assume that $L/E$ is a field extension. Since $L$ is a cyclic $C_3$-algebra over $E$, it follows that $L/E$ is $C_3$-Galois. Additionally, $D$ is a division algebra if and only if neither $a$ nor $a^2$ belongs to the norm group $N_{L/E}$ of $L/E$ (Pierce 1982, p. 279).

## 3.2 Involutions of the Second Kind

Let $E/F$ be a quadratic extension of number fields, and let $\langle\tau\rangle \cong C_2$ be its Galois group. In order to equip a division algebra $D$ over $E$ with an involution $\alpha$ of the second kind with fixed field $F$, we need to extend the nontrivial automorphism $\tau$ of $E$ to $D$.

We start by extending $\tau$ to $L$, $\tau : L \to D$. For this we have two possibilities. Either, the image $L' := \tau(L)$ equals $L$ or it does not. If $L'$ does not equal $L$, then $\tau$ gives rise to an isomorphism of $L$ to $L'$ inside some field extension containing both. However, $L$ and $L'$ are not isomorphic as extensions of $E$, otherwise $D$ would not be a division algebra. So $L/F$ is not Galois. Notice that in this case $L$ along with $L'$ generate $D$. In contrast, if we extend $\tau : L \to L$, i.e., $\tau(L) = L$, then $\langle\tau, \rho\rangle$ is an automorphism group of $L/\mathbb{Q}$ of order at least six. That is, the degree six extension $L/F$ is Galois with Galois group $\langle\tau, \rho\rangle$, which is isomorphic to the cyclic group $C_6$ or the symmetric group $S_3$.

## 3.3 Compactness at Infinity

We now assume $F$ is totally real. For simplicity, let $F = \mathbb{Q}$.

In order for the unitary group defined by $(D, \alpha)$ to be compact at infinity, we need $E/\mathbb{Q}$ to be imaginary quadratic. To see this, assume $E/\mathbb{Q}$ is real quadratic. Then, $E_\infty = E \otimes_{\mathbb{Q}} \mathbb{R} \cong \mathbb{R} \oplus \mathbb{R}$ would split. Therefore, $L_\infty$ would split as well and we would be able to find an isomorphism $D_\infty \cong M_3(\mathbb{R}) \oplus M_3(\mathbb{R})$, where the involution is given by (see Platonov and Rapinchuk 1994, p. 83)

$$(x, y) \mapsto ({}^t y, {}^t x),$$

and the reduced norm is given by

$$N_D(x, y) = \det(x)\det(y).$$

Thus,

$$\begin{aligned}\mathbb{G}_\infty := \mathbb{G}(\mathbb{R}) = \{(x,y) \in D_\infty \mid ({}^t yx,\, {}^t xy) \\ = (\mathtt{Id}_3, \mathtt{Id}_3) \text{ and } \det(x)\det(y) = 1\} \cong GL_3(\mathbb{R})\end{aligned}$$

is not compact.

Next we remark that in the case when $L/\mathbb{Q}$ is Galois, the Galois group is necessarily $C_6$. To see this, assume $L/E$ is a $C_3 = \langle\rho\rangle$-Galois extension such that $L/\mathbb{Q}$ is $S_3$-Galois. At infinity, we have

$$E_\infty = E \otimes_{\mathbb{Q}} \mathbb{R} \cong \mathbb{C}$$

and $\tau$ acts by complex conjugation. Therefore,

$$L_\infty = L \otimes_{\mathbb{Q}} \mathbb{R} \cong L \otimes_E \mathbb{C} \cong \mathbb{C} \oplus \mathbb{C} \oplus \mathbb{C},$$

with the isomorphism given by

$$l \otimes s \mapsto (\rho^0(l)s, \rho^1(l)s, \rho^2(l)s) \text{ for } l \in L \text{ and } s \in E_\infty.$$

Notice that $[L : E] = 3$, so there is always a real primitive element, and thus there is an $E$-basis for $L$ which is $\tau$-invariant. Here multiplication in $L_\infty$ is defined coordinate wise. The $S_3$-action is given by

$$\rho(l \otimes s) = \rho(l) \otimes s \mapsto (\rho^1(l)s, \rho^2(l)s, \rho^0(l)s)$$

and

$$\tau(l \otimes s) \mapsto (\tau(l)\tau(s), \rho^2\tau(l)\tau(s), \rho\tau(l)\tau(s)).$$

Thus, for any $(t_0, t_1, t_2) \in L_\infty$ we have

$$\begin{aligned}\rho(t_0, t_1, t_2) &= (t_1, t_2, t_0),\\ \tau(t_0, t_1, t_2)) &= (\bar{t}_0, \bar{t}_2, \bar{t}_1),\end{aligned}$$

with the usual complex conjugation. Without specifying the algebra $(D, \alpha)$ containing $L$ any further, we read off that $D_\infty$ is isomorphic to the matrix algebra $M_3(\mathbb{C})$ with $L_\infty$ embedded diagonally. This leads to the following result.

**Proposition 3.1.** *Let $E$,$L$, and $(D, \alpha)$ be as above, and assume $L/\mathbb{Q}$ is $S_3$-Galois. Then the split torus*

$$T_\infty = \{(\bar{t}t^{-1}, t, \bar{t}^{-1}) \mid t \in \mathbb{C}^\times\} \subset L_\infty$$

*is contained in $\mathbb{G}_\infty$. In particular, $\mathbb{G}_\infty$ is non-compact.*

*Proof of Proposition 3.1.* We check the definition of $\mathbb{G}$ for elements of $T_\infty$. We have

$$N_D((\bar{t}t^{-1}, t, \bar{t}^{-1})) = N_{L_\infty/E_\infty}((\bar{t}t^{-1}, t, \bar{t}^{-1})) = \bar{t}t^{-1} \cdot t \cdot \bar{t}^{-1} = 1,$$

as well as

$$\begin{aligned}\alpha((\bar{t}t^{-1}, t, \bar{t}^{-1})) \cdot (\bar{t}t^{-1}, t, \bar{t}^{-1}) &= \tau((\bar{t}t^{-1}, t, \bar{t}^{-1})) \cdot (\bar{t}t^{-1}, t, \bar{t}^{-1}) \\ &= (t\bar{t}^{-1}, t^{-1}, \bar{t}) \cdot (\bar{t}t^{-1}, t, \bar{t}^{-1}) = 1.\end{aligned}$$

Therefore, $T_\infty$ defines a non-compact torus of $\mathbb{G}_\infty$. □

In the case when $L/\mathbb{Q}$ is Galois, there is an obvious (but not unique) choice of an involution of the second kind. As $\tau$ extends to an automorphism of $L$, it is defined on any coefficient of $A(l_0, l_1, l_2)$. Thus, the map

$$\alpha(A(l_0, l_1, l_2)) := {}^t\tau(A(l_0, l_1, l_2)) = \begin{pmatrix} \tau(l_0) & \tau(a\rho(l_2)) & \tau(a\rho^2(l_1)) \\ \tau(l_1) & \tau\rho(l_0) & \tau(a\rho^2(l_2)) \\ \tau(l_2) & \tau\rho(l_1) & \tau\rho^2(l_0) \end{pmatrix}$$

clearly satisfies the conditions

$$\begin{aligned}\alpha^2 &= id, \\ \alpha(A \cdot B) &= \alpha(B) \cdot \alpha(A), \\ \alpha\,|_E &= \tau.\end{aligned}$$

In order for $\alpha$ to be an involution on $D$ of the second kind, we must have that the image $\alpha(D)$ is contained in $D$. Defining

$$\tilde{l}_0 = \tau(l_0), \quad \tilde{l}_1 = \tau(a)\tau\rho(l_2), \quad \tilde{l}_2 = \tau(a)\tau\rho^2(l_1),$$

this condition is equivalent to

$$\alpha(A(l_0, l_1, l_2)) = A(\tilde{l}_0, \tilde{l}_1, \tilde{l}_2)).$$

That is,

$$\begin{pmatrix} \tau(l_0) & \tau(a\rho(l_2)) & \tau(a\rho^2(l_1)) \\ \tau(l_1) & \tau\rho(l_0) & \tau(a\rho^2(l_2)) \\ \tau(l_2) & \tau\rho(l_1) & \tau\rho^2(l_0) \end{pmatrix} = \begin{pmatrix} \tilde{l}_0 & \tilde{l}_1 & \tilde{l}_2 \\ a\rho(\tilde{l}_2) & \rho(\tilde{l}_0) & \rho(\tilde{l}_1) \\ a\rho^2(\tilde{l}_1) & a\rho^2(\tilde{l}_2) & \rho^2(\tilde{l}_0) \end{pmatrix}.$$

This evidently reduces to the following conditions

$$\tau\rho = \rho\tau \text{ on } L$$

and

$$a\tau(a) = 1.$$

We summarize the above discussion in the following theorem.

**Theorem 3.2.** *Assume the extension $L/\mathbb{Q}$ is Galois, and that $\alpha$ is defined by*

$$\alpha(A(l_0, l_1, l_2)) = {}^t\tau(A(l_0, l_1, l_2)).$$

*Then $(D, \alpha)$ is a division algebra which is central simple over $E$ with involution $\alpha$ of the second kind if and only if the following conditions are satisfied:*

(i) $a \in E^\times$, *and* $a, a^2 \notin N_{L/E}$
(ii) $N_{E/\mathbb{Q}}(a) = a\tau(a) = 1$
(iii) $\tau\rho = \rho\tau$ *on $L$, i.e. $L/\mathbb{Q}$ is $C_6$-Galois.*

*Moreover, if these conditions are satisfied, the group $\mathbb{G}_\infty$ is compact.*

*Proof of Theorem 3.2.* The first part of the theorem is proved above. What is left to show is compactness at infinity. The realization of $L$ inside the diagonal subgroup of $M_3(L)$ is chosen such that it is compatible with the isomorphism

$$L_\infty = L \otimes_{\mathbb{Q}} \mathbb{R} \cong L \otimes_E \mathbb{C} \cong \mathbb{C}^3$$

induced by the three embeddings of $L$ into $\mathbb{C}$. Indeed, $D_\infty \cong M_3(\mathbb{C})$ with involution $\alpha : M_3(\mathbb{C}) \to M_3(\mathbb{C})$ given by $\alpha(A) = {}^t\overline{A}$. Therefore,

$$\mathbb{G}_\infty \cong \{A \in M_3(\mathbb{C}) \mid {}^t\overline{A} \cdot A = \mathtt{Id}_3,\ \det A = 1\} = \mathrm{SU}_3(\mathbb{R}),$$

is induced by the standard hermitian form of signature $(3, 0)$. Thus, $\mathbb{G}_\infty$ is compact. $\square$

Notice that it is non-trivial to satisfy condition (iii) of Theorem 3.2, as a quadratic field $E/\mathbb{Q}$ does not necessarily allow an extension $L$ of degree three which is $C_6$-Galois over $\mathbb{Q}$. However, there are situations which allow for the conditions of Theorem 3.2 to be satisfied. Below we provide such an example.

*Example 3.3 (An Example in the Galois-Case).* Let $E = \mathbb{Q}(\sqrt{-3})$. Therefore, $E$ contains a primitive third root of unity, $\zeta_3$, and Kummer theory applies. That is, any cyclic $C_3$-extension $L/E$ can be obtained by adjoining a third root, $L = E(\sqrt[3]{b})$, where $b \in E^\times \backslash (E^\times)^3$. In particular, choose $b = \zeta_3$. Then, $\sqrt[3]{\zeta_3} = \zeta_9$ is a primitive 9th root of unity. Then, $L = E(\zeta_9) = \mathbb{Q}(\zeta_9)$ is a cyclotomic field, which is tautologically cyclic over $\mathbb{Q}$. Its relative Galois group is $Gal(L/E) = \langle\rho\rangle$, where $\rho(\zeta_9) = \zeta_3\zeta_9$. Extending $\tau$ (complex conjugation) from $E$ to $L$ means $\tau(\zeta_9) = \zeta_9^8$. Thus,

$$\rho\tau(\zeta_9) = \rho(\zeta_9)^8 = \zeta_3^8\zeta_9^8 = \bar{\zeta}_3\zeta_9^8 = \tau\rho(\zeta_9).$$

Now choose an element $a \in E^\times$ such that $a, a^2 \notin N_{L/E}$ and $N_{E/\mathbb{Q}}(a) = 1$. One can take for example

$$a = \frac{2+\sqrt{-3}}{2-\sqrt{-3}}.$$

Then, trivially, $N_{E/\mathbb{Q}}(a) = 1$, and we verified using Magma that $a, a^2 \notin N_{L/E}$.

*Example 3.4 (An Example in the Non-Galois Case).* Again, choose $E = \mathbb{Q}(\sqrt{-3})$. But this time, choose a cyclic degree three extension $L = E(\theta)$, $\theta^3 = b$, where $b \in E^\times \backslash (E^\times)^3$ is chosen such that $L/\mathbb{Q}$ is not Galois. For example, one could choose $b = 2\zeta_3$. The automorphism $\rho$ of $L/E$ is given by $\rho(\theta) = \zeta_3\theta$, and the minimal polynomial is given by $X^3 - b$. Let $\theta'$ be a root of $X^3 - \tau(b)$, and let $L' = E(\theta')$. Then (within any field extension containing both) $L$ and $L'$ are non-equal, but there is an isomorphism $\alpha : L \to L'$ extending $\tau$ given by $\tau(\theta) = \theta'$. For the cyclic algebra $(D, \alpha)$ with involution, choose $L$ as above and $a = \tau(b)$, i.e. $z$ may be identified with $\theta'$. Then the above constraint

$$\alpha(\theta) = z$$

determines an involution on $D$ of the second kind, as $\alpha(z) = \alpha^2(\theta) = \theta$. For convenience, let $d = l_0 + l_1 z + l_2 z^2$, $l_j \in L$, be an arbitrary element of $D$, then

$$\alpha(d) = \alpha(l_0) + \theta\alpha(l_1) + \theta^2\alpha(l_2),$$

and one easily checks $\alpha^2(d) = d$. Using the identification of $D$ with a subring of $M_3(L)$ as before, we write down this involution for matrices:

$$z = \begin{pmatrix} 0 & 1 & 0 \\ 0 & 0 & 1 \\ a & 0 & 0 \end{pmatrix} \mapsto \alpha(z) = \begin{pmatrix} \theta & 0 & 0 \\ 0 & \rho(\theta) & 0 \\ 0 & 0 & \rho^2(\theta) \end{pmatrix}.$$

So for an element $e_0 + e_1 z + e_2 z^2 \in L' = E(z) \subset D$, i.e. $e_j \in E$,

$$\alpha(A(e_0, e_1, e_2)) = A(\tau(e_0) + \tau(e_1)\theta + \tau(e_2)\theta^2, 0, 0).$$

As $\alpha^2 = id$, we read off the image of $l = e_0 + e_1\theta + e_2\theta^2 \in L$ under $\alpha$ in matrix form:

$$\alpha(A(l, 0, 0)) = \begin{pmatrix} \tau(e_0) & \tau(e_1) & \tau(e_2) \\ a\tau(e_2) & \tau(e_0) & \tau(e_1) \\ a\tau(e_1) & a\tau(e_2) & \tau(e_0) \end{pmatrix}.$$

Thus, finally

$$\alpha(A(l_0, l_1, l_2)) = \alpha(A(l_0, 0, 0)) + \theta\alpha(A(l_1, 0, 0)) + \theta^2\alpha(A(l_2, 0, 0)),$$

that is, for $l_j = e_{j0} + e_{j1}\theta + e_{j2}\theta^2 \in L$ with $e_{jk} \in E$, we find

$$\alpha(A(l_0, l_1, l_2)) = A(\tilde{l}_0, \tilde{l}_1, \tilde{l}_2),$$

where $\tilde{l}_j = \tau(e_{0j}) + \tau(e_{1j})\theta + \tau(e_{2j})\theta^2$.

## 4 Choosing the Family of Subgroups

Let $\mathbb{G}$ be the global special unitary group constructed from the division algebra and the involution of the second kind given in Example 3.3 of the previous section. Let $p$ be a place where $\mathbb{G}_p := \mathbb{G}(\mathbb{Q}_p)$ is isomorphic to $\mathrm{SU}_3(\mathbb{Q}_p)$. In this section, we will give an explicit infinite family of discrete co-compact subgroups of $\mathbb{G}$ which act without fixed points on the Bruhat–Tits tree of $\mathbb{G}_p$. Before we proceed, we need to describe the place $p$ explicitly. From Rogawski (1990, 14.2) we have that $\mathbb{G}_p$ is isomorphic to $\mathrm{SU}_3(\mathbb{Q}_p)$ if and only if $p$ is inert in $E$. In fact, we can see this directly as shown below. If $p$ does not remain prime (i.e., is not inert), then there are two cases. Either (i) $p$ ramifies in $E$ (i.e., $(p) = \mathfrak{p}^2$, $\mathfrak{p} = \overline{\mathfrak{p}}$) or (ii) $p$ splits into two non-equal prime ideals in $E$ (i.e., $(p) = \mathfrak{p}\overline{\mathfrak{p}}$ with $\mathfrak{p} \neq \overline{\mathfrak{p}}$).

(i) The only prime ramified in $E$ is $(p) = (3) = \mathfrak{p}^2$, where $\mathfrak{p} = (\sqrt{-3}) = (-\sqrt{-3}) = \overline{\mathfrak{p}}$. In this case, $E_\mathfrak{p}/\mathbb{Q}_p$ is a ramified field extension. The involution $\alpha$ is then trivial on the localization $E_\mathfrak{p}$, as $\alpha(\mathfrak{p}) = \overline{\mathfrak{p}} = \mathfrak{p}$. The group $\mathbb{G}_p$ will lead to a $(p+1)$-regular tree, the Bruhat–Tits building on $\mathrm{SL}_2(\mathbb{Q}_p)$. This case has been treated in Lubotzky et al. (1988).

(ii) There are many primes $p$ which are split in $E = \mathbb{Q}(\sqrt{-3})$. The minimal polynomial is $X^2 + 3$. This is reducible modulo $p$ if and only if $p$ is split in $E$. Equivalently, the minimal polynomial is reducible if and only if $-3$ is a square mod $p$. The two localizations $E_\mathfrak{p}$ and $E_{\overline{\mathfrak{p}}}$ here are both equal to $\mathbb{Q}_p$. Therefore, the field extension $E/\mathbb{Q}$ localizes as a split algebra $E_p = E_\mathfrak{p} \oplus E_{\overline{\mathfrak{p}}} = \mathbb{Q}_p \oplus \mathbb{Q}_p$. The involution $\alpha$ is conjugation on $E$, so $\alpha(\mathfrak{p}) = \overline{\mathfrak{p}}$. That is, $\alpha$ exchanges the two summands of $E_p$. Then, $D_p = D_\mathfrak{p} \oplus D_{\overline{\mathfrak{p}}}$, and for an element $(g_1, g_2) \in D_p$ to be in $\mathbb{G}_p$ we need

$$N(g_1, g_2) = (1, 1)$$

and

$$(1, 1) = \alpha((g_1, g_2))(g_1, g_2) = (\alpha(g_2), \alpha(g_1))(g_1, g_2) = (\alpha(\alpha(g_1)g_2), \alpha(g_1)g_2).$$

That is, $g_2 = \alpha(g_1)^{-1}$ for some $g_1$ in the reduced norm one group, $N^1_{D_\mathfrak{p}}$, of $D_\mathfrak{p}$. Thus, $\mathbb{G}_p \cong N^1_{D_\mathfrak{p}} \cong N^1_{D_{\overline{\mathfrak{p}}}}$.

Finally, if $p$ is inert in $E$, then $E_p/\mathbb{Q}_p$ is an unramified quadratic field extension. This is the case if and only if $-3$ is not a square modulo $p$. Then $\mathbb{G}_p \cong \mathbb{G}(\mathbb{Q}_p)$. Therefore, only these primes are "good" primes for us, i.e., leading to Ramanujan bigraphs. By quadratic reciprocity, for a prime $p > 3$,

$$\left(\frac{-3}{p}\right) = \begin{cases} 1, & p \equiv 1, 7 \pmod{12} \\ -1 & p \equiv 5, 11 \pmod{12}. \end{cases}$$

Thus the "good" primes are the primes $p$ such that $p \equiv 5, 11 \pmod{12}$.

Fix a prime $p \equiv 5, 11 \pmod{12}$ and let $q$ be a prime not equal to $p$. We follow the notation in Ballantine and Ciubotaru (2011, 4.3). Let $\mathbb{Z}[p^{-1}]$ be the subring of $\mathbb{Q}$ consisting of rational numbers with powers of $p$ in the denominator. Notice that $\mathbb{G}_\infty$ and $\mathbb{G}_p$ are matrix groups with coefficients in $\mathbb{R}$ and $\mathbb{Q}_p$, respectively. By abuse of notation, we denote by $\mathbb{G}_\infty(\mathbb{Z}[p^{-1}])$ and $\mathbb{G}_p(\mathbb{Z}[p^{-1}])$ the obvious subgroups in $\mathbb{G}_\infty$ and $\mathbb{G}_p$, respectively. It is clear that $\mathbb{G}_\infty(\mathbb{Z}[p^{-1}])$ and $\mathbb{G}_p(\mathbb{Z}[p^{-1}])$ are isomorphic. Define $\mathbb{G}(\mathbb{Z}[p^{-1}]) := \mathbb{G}_\infty(\mathbb{Z}[p^{-1}]) \times \mathbb{G}_p(\mathbb{Z}[p^{-1}])$ to be their product in $\mathbb{G}_\infty \times \mathbb{G}_p$. It follows from Borel (1963) that $\mathbb{G}(\mathbb{Z}[p^{-1}])$ is a lattice in $\mathbb{G}_\infty \times \mathbb{G}_p$. For each positive integer $n$, we define the kernel modulo $q^n$,

$$\Gamma(q^n) := \ker(\mathbb{G}(\mathbb{Z}[p^{-1}]) \to \mathbb{G}(\mathbb{Z}[p^{-1}]/q^n\mathbb{Z}[p^{-1}]),$$

and

$$\Gamma_p(q^n) := \Gamma(q^n) \cap \mathbb{G}_p.$$

Then, as shown in Ballantine and Ciubotaru (2011), each $\Gamma_p(q^n)$ is a discrete co-compact subgroup of $\mathbb{G}_p$. It has finite index and no non-trivial elements of finite order. Thus, each subgroup $\Gamma_p(q^n)$ acts on the Bruhat–Tits tree of $\mathbb{G}_p$ without fixed points and the quotient building is a finite biregular graph of bidegree $(p^3+1, p+1)$.

# 5 An Infinite Family of Ramanujan Bigraphs

Let $\mathbb{G}$ be the inner form of $SU_3$ constructed using the division algebra and involution of Example 3.3. Let $p$ be a prime congruent to 5 or 11 modulo 12 and $q$ a prime not equal to $p$. We denote by $\tilde{X}$ the Bruhat–Tits tree associated with $\mathbb{G}_p$. For each positive integer $n$, let $\Gamma_p(q^n)$ be the subgroup of $\mathbb{G}_p$ constructed in the previous section and let $X_n$ be the quotient of $\tilde{X}$ by the action of $\Gamma_p(q^n)$. By Ballantine and Ciubotaru (2011, Corollary 4.6), $X_n$ is a Ramanujan bigraph. Thus, we have constructed an infinite family of Ramanujan bigraphs. As $\Gamma_p(q^{n+1}) \subsetneq \Gamma_p(q^n)$, the number of vertices of $X_n$ tends to infinity as $n \to \infty$. Moreover, for each $n$, the graph $X_n$ is a subgraph of $X_{n+1}$.

**Acknowledgements** The authors would like to thank Dan Ciubotaru, Judith Ludwig, and Rachel Newton for helpful discussions. This work was initiated at the CIRM workshop, "Femmes en nombre" in October 2013 and the authors would like to thank CIRM, Microsoft Research, the National Science Foundation DMS-1303457, the Clay Mathematics Institute, and the Number Theory Foundation for supporting the workshop.
This work was partially supported by a grant from the Simons Foundation (#245997 to Cristina Ballantine).
This work was partially supported by grants from the National Science Foundation (DMS-1201446) and from PSC-CUNY (Brooke Feigon).
This work was partially supported by the European Social Fund (Kathrin Maurischat).

## References

Alon, N.: Eigenvalues and expanders. Combinatorica **6**(2), 83–96 (1986). Theory of computing, Singer Island (1984). MR 875835 (88e:05077)

Ballantine, C., Ciubotaru, D.: Ramanujan bigraphs associated with SU(3) over a $p$-adic field. Proc. Am. Math. Soc. **139**(6), 1939–1953 (2011). MR 2775370 (2012b:22020)

Borel, A.: Some finiteness properties of adele groups over number fields. Inst. Hautes Études Sci. Publ. Math. **16**, 5–30 (1963). MR 0202718 (34 #2578)

Clozel, L., Harris, M., Taylor, R.: Automorphy for some $l$-adic lifts of automorphic mod $l$ Galois representations. Publ. Math. Inst. Hautes Études Sci. **108**, 1–181 (2008). With Appendix A, summarizing unpublished work of Russ Mann, and Appendix B by Marie-France Vignéras. MR 2470687 (2010j:11082)

Davidoff, G., Sarnak, P., Valette, A.: Elementary Number Theory, Group Theory, and Ramanujan Graphs. London Mathematical Society Student Texts, vol. 55. Cambridge University Press, Cambridge (2003). MR 1989434 (2004f:11001)

Feng, K., Winnie Li, W.-C.: Spectra of hypergraphs and applications. J. Number Theory **60**(1), 1–22 (1996). MR 1405722 (97f:05128)

Garrett, P.: Buildings and Classical Groups. Chapman & Hall, London (1997). MR 1449872 (98k:20081)

Hashimoto, K.-i.: Zeta functions of finite graphs and representations of $p$-adic groups. In: Automorphic Forms and Geometry of Arithmetic varieties. Advanced Studies in Pure Mathematics, vol. 15, pp. 211–280. Academic, Boston (1989). MR 1040609 (91i:11057)

Hashimoto, K.-i., Hori, A.: Selberg-Ihara's zeta function for $p$-adic discrete groups. In: Automorphic Forms and Geometry of Arithmetic Varieties. Advanced Studies in Pure Mathematics, vol. 15, pp. 171–210. Academic, Boston (1989). MR 1040608 (91g:11053)

Harris, M., Labesse, J.-P.: Conditional base change for unitary groups. Asian J. Math. **8**(4), 653–684 (2004)

Hoory, S., Linial, N., Wigderson, A.: Expander graphs and their applications. Bull. Am. Math. Soc. (N.S.) **43**(4), 439–561 (electronic) (2006). MR 2247919 (2007h:68055)

Knus, M.-A., Merkurjev, A., Rost, M., Tignol, J.-P.: The Book of Involutions. American Mathematical Society Colloquium Publications, vol. 44. American Mathematical Society, Providence (1998). With a preface in French by J. Tits. MR 1632779 (2000a:16031)

Lubotzky, A.: Expander graphs in pure and applied mathematics. Bull. Am. Math. Soc. (N.S.) **49**(1), 113–162 (2012). MR 2869010 (2012m:05003)

Lubotzky, A., Phillips, R.L., Sarnak, P.: Explicit expanders and the ramanujan conjectures. In: STOC '86 Proceedings of the eighteenth annual ACM symposium on Theory of computing, pp. 240–246 (1986)

Lubotzky, A., Phillips, R.L., Sarnak, P.: Ramanujan graphs. Combinatorica **8**(3), 261–277 (1988). MR 963118 (89m:05099)

Marcus, A., Spielman, D.A., Srivastava, N.: Interlacing families I: Bipartite Ramanujan graphs of all degrees. Ann. Math. **182**(1), 307–325 (2015)

Margulis, G.A.: Explicit group-theoretic constructions of combinatorial schemes and their applications in the construction of expanders and concentrators. Problemy Peredachi Informatsii **24**(1), 51–60 (1988). MR 939574 (89f:68054)

Pierce, R.S.: Associative Algebras. Graduate Texts in Mathematics, vol. 88. Springer, New York, 1982. Studies in the History of Modern Science, 9. MR 674652 (84c:16001)

Platonov, V., Rapinchuk, A.: Algebraic Groups and Number Theory. Pure and Applied Mathematics, vol. 139. Academic, Boston (1994). Translated from the 1991 Russian original by Rachel Rowen. MR 1278263 (95b:11039)

Rogawski, J.D.: Automorphic representations of unitary groups in three variables. Annals of Mathematics Studies, vol. 123. Princeton University Press, Princeton (1990). MR 1081540 (91k:22037)

Solé, P.: Ramanujan hypergraphs and Ramanujan geometries. In: Emerging applications of number theory (Minneapolis, MN, 1996), The IMA Volumes in Mathematics and its Applications, vol. 109, pp. 583–590. Springer, New York (1999). MR 1691550 (2000c:05108)

Tits, J.: Reductive groups over local fields. In: Automorphic Forms, Representations and *L*-Functions (Proc. Sympos. Pure Math., Oregon State Univ., Corvallis, Ore., 1977), Part 1, Proc. Sympos. Pure Math., XXXIII, pp. 29–69. American Mathematical Society, Providence (1979) MR 546588 (80h:20064)

# Classifications of Elliptic Fibrations of a Singular K3 Surface

**Marie José Bertin, Alice Garbagnati, Ruthi Hortsch, Odile Lecacheux, Makiko Mase, Cecília Salgado, and Ursula Whitcher**

**Abstract** We classify, up to automorphisms, the elliptic fibrations on the singular K3 surface $X$ whose transcendental lattice is isometric to $\langle 6 \rangle \oplus \langle 2 \rangle$.

**MSC 2010:** Primary 14J28, 14J27; Secondary 11G05, 11G42, 14J33

## 1 Introduction

We classify elliptic fibrations on the singular K3 surface $X$ associated with the Laurent polynomial

M.J. Bertin (✉)
Jussieu Institute of Mathematics, Pierre and Marie Curie University, Paris, France
e-mail: marie-jose.bertin@imj-prg.fr

A. Garbagnati
Department of Mathematics, University of Milan, Milan, Italy
e-mail: alice.garbagnati@unimi.it

R. Hortsch
Department of Mathematics, Massachusetts Institute of Technology, Cambridge, MA, USA
e-mail: rhortsch@math.mit.edu

O. Lecacheux
Jussieu Institute of Mathematics, Pierre and Marie Curie University, Paris, France
e-mail: odile.lecacheux@imj-prg.fr

M. Mase
Tokyo Metropolitan University, Tokyo, Japan
e-mail: mtmase@arion.ocn.ne.jp

C. Salgado
Institute for Mathematics, Federal University of Rio de Janeiro, Rio de Janeiro, RJ, Brazil
e-mail: salgado@im.ufrj.br

U. Whitcher
Department of Mathematics, University of Wisconsin-Eau Claire, Eau Claire, WI, USA
e-mail: whitchua@uwec.edu



M.J. Bertin et al. (eds.), *Women in Numbers Europe*, Association for Women in Mathematics Series 2, DOI 10.1007/978-3-319-17987-2_2

$$x+\frac{1}{x}+y+\frac{1}{y}+z+\frac{1}{z}+\frac{x}{y}+\frac{y}{x}+\frac{y}{z}+\frac{z}{y}+\frac{z}{x}+\frac{x}{z}.$$

In order to compute the Néron–Severi lattice, the Picard number, and other basic properties of an algebraic surface, it is useful to identify an elliptic fibration on the surface. Moreover, in view of different applications, one may be interested in finding all the elliptic fibrations of a certain type. The fibrations of rank 0 and maximal torsion lead more easily to the determination of the $L$-series of the variety (Bertin 2010). Those of positive rank lead to symplectic automorphisms of infinite order of the variety. Lenstra's Elliptic Curve Method (ECM) for finding small factors of large numbers originally used elliptic curves on $\mathbb{Q}$ with a torsion-group of order 12 or 16 and rank $\geq 1$ on $\mathbb{Q}$ (Montgomery 1987; Atkin and Morain 1993). One way to obtain infinite families of such curves is to use fibrations of modular surfaces, as explained by Elkies (2007).

If the Picard number of a K3 surface is large, there may be an infinite number of elliptic fibrations, but there is only a finite number of fibrations up to automorphisms, as proved by Sterk (1985). Oguiso used a geometric method to classify elliptic fibrations in Oguiso (1989). Some years later, Nishiyama (1996) proposed a lattice-theoretic technique to produce such classifications, recovering Oguiso's results and classifying other Kummer and K3 surfaces. Since then, results of the same type have been obtained by various authors (Kumar 2014; Elkies and Schütt 2014; Bertin and Lecacheux 2013).

Recently, the work of Braun et al. (2013) described three possible classifications of elliptic fibrations on a K3 surface, shining a new light on the meaning of what is a class of equivalence of elliptic fibrations. In particular, they proposed a $\mathcal{J}_1$-classification of elliptic fibrations up to automorphisms of the surface and a $\mathcal{J}_2$-classification of the frame lattices of the fibrations. For our K3 surface, the two classifications coincide. Thus, it is particularly interesting to exhibit here an $\mathcal{J}_2$-classification by the Kneser–Nishiyama method, since in general it is not easy to obtain the $\mathcal{J}_1$-classification. This topic will be explained in detail in Section 2.

Section 3 is devoted to a toric presentation of the surface $X$, following ideas of Karp et al. (2013), based on the classification of reflexive polytopes in dimension 3. More precisely, the Newton polytope of $X$ is in the same class as the reflexive polytope of index 1529. Since, according to Karp et al. (2013), there is an $S_4$ action on the vertices of polytope 1529 and its polar dual, there is a symplectic action of $S_4$ on $X$. This action will be described on specific fibrations. One of them gives the transcendental lattice $T_X = \langle 6 \rangle \oplus \langle 2 \rangle$. We may use these fibrations to relate $X$ to a modular elliptic surface analyzed by Beauville in Beauville (1982). We also describe a presentation of $X$ found in Garbagnati and Sarti (2009), which represents $X$ as a K3 surface with a prescribed abelian symplectic automorphism group.

The main results of the paper are obtained by Nishiyama's method and are summarized in Section 4, Theorem 4.2.

**Theorem 1.1.** *The classification up to automorphisms of the elliptic fibrations on $X$ is given in Table 1. Each elliptic fibration is given with the Dynkin diagrams* *characterizing its reducible fibers and the rank and torsion of its Mordell–Weil group. More precisely, we obtained* 52 *elliptic fibrations on X, including 17 fibrations of rank* 2 *and one of rank* 3.

Due to the high number of different elliptic fibrations, we give only a few cases of computing the torsion. These cases have been selected to give an idea of the various methods involved. Notice the case of fibrations #22 and #22b exhibiting two elliptic fibrations with the same singular fibers and torsion but not isomorphic. Corresponding to these different fibrations we give some particularly interesting Weierstrass models; it is possible to make an exhaustive list.

## 2 Classification of Elliptic Fibrations on K3 Surfaces

Let $S$ be a smooth complex compact projective surface.

**Definition 2.1.** A surface $S$ is a K3 surface if its canonical bundle and its irregularity are trivial, that is, if $\mathcal{K}_S \simeq \mathcal{O}_S$ and $h^{1,0}(S) = 0$.

**Definition 2.2.** A flat surjective map $\mathcal{E} : S \to \mathbb{P}^1$ is called an *elliptic fibration* if:

1) the generic fiber of $\mathcal{E}$ is a smooth curve of genus 1;
2) there exists at least one section $s : \mathbb{P}^1 \to S$ for $\mathcal{E}$.

In particular, we choose one section of $\mathcal{E}$, which we refer to as the zero section. We always denote by $F$ the class of the fiber of an elliptic fibration and by $O$ the curve (and the class of this curve) which is the image of $s$ in $S$.

The group of the sections of an elliptic fibration $\mathcal{E}$ is called the Mordell–Weil group and is denoted by $\mathrm{MW}(\mathcal{E})$.

A generic K3 surface does not admit elliptic fibrations, but if the Picard number of the K3 is sufficiently large, it is known that the surface must admit at least one elliptic fibration (see Proposition 2.3). On the other hand, it is known that a K3 surface admits a finite number of elliptic fibrations up to automorphisms (see Proposition 2.5). Thus, a very natural problem is to classify the elliptic fibrations on a given K3 surface. This problem has been discussed in several papers, starting in the Eighties. There are essentially two different ways to classify elliptic fibrations on K3 surfaces described in Oguiso (1989) and Nishiyama (1996). In some particular cases, a third method can be applied; see Kumar (2014). First, however, we must introduce a different problem: "What does it mean to 'classify' elliptic fibrations?" A deep and interesting discussion of this problem is given in Braun et al. (2013), where the authors introduce three different types of classifications of elliptic fibrations and prove that under certain (strong) conditions these three different classifications collapse to a unique one. We observe that it was already known by Oguiso (1989) that in general these three different classifications do not collapse to a unique one. We now summarize the results by Braun et al. (2013) and the types of classifications.

## 2.1 Types of Classifications of Elliptic Fibrations on K3 Surfaces

In this section we recall some of the main results on elliptic fibrations on K3 surfaces (for example, compare Schütt and Shioda 2010), and we introduce the different classifications of elliptic fibrations discussed in Braun et al. (2013).

### 2.1.1 The Sublattice $U$ and the $\mathcal{J}_0$-Classification

Let $S$ be a K3 surface and $\mathcal{E} : S \to \mathbb{P}^1$ be an elliptic fibration on $S$. Let $F \in NS(S)$ be the class of the fiber of $\mathcal{E}$. Then $F$ is a nef divisor which defines the map $\phi_{|F|} : S \to \mathbb{P}(H^0(S,F)^*)$ which sends every point $p \in S$ to $(s_0(p) : s_1(p) : \ldots : s_r(p))$, where $\{s_i\}_{i=1,\ldots,r}$ is a basis of $H^0(S,F)$, i.e. a basis of sections of the line bundle associated to the divisor $F$. The map $\phi_{|F|}$ is the elliptic fibration $\mathcal{E}$. Hence, every elliptic fibration on a K3 surface is uniquely associated with an irreducible nef divisor (with trivial self-intersection). Since $\mathcal{E} : S \to \mathbb{P}^1$ admits a section, there exists a rational curve which intersects every fiber in one point. Its class in $NS(S)$ is denoted by $O$ and has the following intersection properties $O^2 = -2$ (since $O$ is a rational curve) and $FO = 1$ (since $O$ is a section). Thus, the elliptic fibration $\mathcal{E} : S \to \mathbb{P}^1$ (with a chosen section, as in Definition 2.2) is uniquely associated with a pair of divisors $(F, O)$. This pair of divisors spans a lattice which is isometric to $U$, represented by the matrix $\begin{bmatrix} 0 & 1 \\ 1 & 0 \end{bmatrix}$, (considering the basis $F, F + O$). Hence each elliptic fibration is associated to a chosen embedding of $U$ in $NS(S)$.

On the other hand, the following result holds:

**Proposition 2.3 (Kondo (1992, Lemma 2.1) and Nikulin (1980a, Corollary 1.13.15)).** *Let $S$ be a K3 surface, such that there exists a primitive embedding $\varphi : U \hookrightarrow NS(S)$. Then $S$ admits an elliptic fibration.*

*Let $S$ be a K3 surface with Picard number $\rho(S) \geq 13$. Then, there is a primitive embedding of $U$ in $NS(S)$ and hence $S$ admits at least one elliptic fibration.*

A canonical embedding of $U$ in $NS(S)$ is defined as follows: Let us denote by $b_1$ and $b_2$ the unique two primitive vectors of $U$ with trivial self-intersection. An embedding of $U$ in $NS(S)$ is called *canonical* if the image of $b_1$ in $NS(S)$ is a nef divisor and the image of $b_2 - b_1$ in $NS(S)$ is an effective irreducible divisor.

The first naive classification of the elliptic fibrations that one can consider is the classification described above, roughly speaking: two fibrations are different if they correspond to different irreducible nef divisors with trivial self-intersections. This essentially coincides with the classification of the canonical embeddings of $U$ in $NS(S)$.

Following Braun et al. (2013) we call this classification the $\mathcal{J}_0$-classification of the elliptic fibrations on $S$.

Clearly, it is possible (and indeed likely, if the Picard number is sufficiently large) that there is an infinite number of irreducible nef divisors with trivial self-intersection and also infinitely many copies of $U$ canonically embedded in $NS(S)$. Thus, it is possible that there is an infinite number of fibrations in curves of genus 1 on $S$ and moreover an infinite number of elliptic fibrations on $S$.

### 2.1.2 Automorphisms and the $\mathcal{J}_1$-Classification

The automorphism group of a variety transforms the variety to itself preserving its structure, but moves points and subvarieties on the variety. Thus, if one is considering a variety with a nontrivial automorphism group, one usually classifies objects on the variety up to automorphisms.

Let $S$ be a K3 surface with a sufficiently large Picard number (at least 2). Then the automorphism group of $S$ is in general nontrivial, and it is often of infinite order. More precisely, if $\rho(S) = 2$, then the automorphism group of $S$ is finite if and only if there is a vector with self-intersection either 0 or $-2$ in the Néron–Severi group. If $\rho(S) \geq 3$, then the automorphism group of $S$ is finite if and only if the Néron–Severi group is isometric to a lattice contained in a known finite list of lattices, cf. Kondo (1989). Let us assume that $S$ admits more than one elliptic fibration (up to the $\mathcal{J}_0$-classification defined above). This means that there exist at least two elliptic fibrations $\mathcal{E} : S \to \mathbb{P}^1$ and $\mathcal{E}' : S \to \mathbb{P}^1$ such that $F \neq F' \in NS(S)$, where $F$ (resp. $F'$) is the class of the fiber of the fibration $\mathcal{E}$ (resp. $\mathcal{E}'$). By the previous observation, it seems very natural to consider $\mathcal{E}$ and $\mathcal{E}'$ equivalent if there exists an automorphism of $S$ which sends $\mathcal{E}$ to $\mathcal{E}'$. This is the idea behind the $\mathcal{J}_1$-classification of the elliptic fibrations introduced in Braun et al. (2013).

**Definition 2.4.** The $\mathcal{J}_1$-classification of the elliptic fibrations on a K3 surface is the classification of elliptic fibrations up to automorphisms of the surface. To be more precise: $\mathcal{E}$ is $\mathcal{J}_1$-equivalent to $\mathcal{E}'$ if and only if there exists $g \in \mathrm{Aut}(S)$ such that $\mathcal{E} = \mathcal{E}' \circ g$.

We observe that if two elliptic fibrations on a K3 surface are equivalent up to automorphism, then all their geometric properties (the type and the number of singular fibers, the properties of the Mordell–Weil group and the intersection properties of the sections) coincide. This is true essentially by definition, since an automorphism preserves all the "geometric" properties of subvarieties on $S$.

The advantages of the $\mathcal{J}_1$-classification with respect to the $\mathcal{J}_0$-classification are essentially two. The first is more philosophical: in several contexts, to classify an object on varieties means to classify the object up to automorphisms of the variety. The second is more practical and is based on an important result by Sterk: the $\mathcal{J}_1$-classification must have a finite number of classes:

**Proposition 2.5 (Sterk (1985)).** *Up to automorphisms, there exists a finite number of elliptic fibrations on a K3 surface.*

### 2.1.3 The Frame Lattice and the $\mathcal{J}_2$-Classification

The main problem of the $\mathcal{J}_1$-classification is that it is difficult to obtain a $\mathcal{J}_1$-classification of elliptic fibrations on K3 surfaces, since it is in general difficult to give a complete description of the automorphism group of a K3 surface and the orbit of divisors under this group. An intermediate classification can be introduced, the $\mathcal{J}_2$-classification. The $\mathcal{J}_2$-classification is not as fine as the $\mathcal{J}_1$-classification, and its geometric meaning is not as clear as the meanings of the classifications introduced above. However, the $\mathcal{J}_2$-classification can be described in a very natural way in the context of lattice theory, and there is a standard method to produce it.

Since the $\mathcal{J}_2$-classification is essentially the classification of certain lattices strictly related to the elliptic fibrations, we recall here some definitions and properties of lattices related to an elliptic fibration.

We have already observed that every elliptic fibration on $S$ is associated with an embedding $\eta : U \hookrightarrow NS(S)$.

**Definition 2.6.** The orthogonal complement of $\eta(U)$ in $NS(S)$, $\eta(U)^{\perp_{NS(S)}}$, is denoted by $W_{\mathcal{E}}$ and called the frame lattice of $\mathcal{E}$.

The frame lattice of $\mathcal{E}$ encodes essentially all the geometric properties of $\mathcal{E}$, as we explain now. We recall that the irreducible components of the reducible fibers which do not meet the zero section generate a root lattice, which is the direct sum of certain Dynkin diagrams. Let us consider the root lattice $(W_{\mathcal{E}})_{\text{root}}$ of $W_{\mathcal{E}}$. Then the lattice $(W_{\mathcal{E}})_{\text{root}}$ is exactly the direct sum of the Dynkin diagram corresponding to the reducible fibers. To be more precise if the lattice $E_8$ (resp. $E_7$, $E_6$, $D_n$, $n \geq 4$, $A_m$, $m \geq 3$) is a summand of the lattice $(W_{\mathcal{E}})_{\text{root}}$, then the fibration $\mathcal{E}$ admits a fiber of type $II^*$ (resp. $IV^*$, $III^*$, $I^*_{n-4}$, $I_{m+1}$). However, the lattices $A_1$ and $A_2$ can be associated with two different types of reducible fibers, i.e. with $I_2$ and $III$ and with $I_3$ and $IV$, respectively. We cannot distinguish between these two different cases using lattice theory. Moreover, the singular non-reducible fibers of an elliptic fibration can be either of type $I_1$ or of type $II$.

Given an elliptic fibration $\mathcal{E}$ on a K3 surface $S$, the lattice $Tr(\mathcal{E}) := U \oplus (W_{\mathcal{E}})_{\text{root}}$ is often called the *trivial lattice* (see Schütt and Shioda 2010, Lemma 8.3 for a more detailed discussion).

Let us now consider the Mordell–Weil group of an elliptic fibration $\mathcal{E}$ on a K3 surface $S$: its properties are also encoded in the frame $W_{\mathcal{E}}$, indeed $\text{MW}(\mathcal{E}) = W_{\mathcal{E}}/(W_{\mathcal{E}})_{\text{root}}$. In particular,

$$\text{rank}(\text{MW}(\mathcal{E})) = \text{rank}(W_{\mathcal{E}}) - \text{rank}((W_{\mathcal{E}})_{\text{root}}) \text{ and}$$
$$(\text{MW}(\mathcal{E}))_{\text{tors}} = \overline{(W_{\mathcal{E}})_{\text{root}}}/(W_{\mathcal{E}})_{\text{root}},$$

where, for every sublattice $L \subset NS(S)$, $\overline{L}$ denotes the primitive closure of $L$ in $NS(S)$, i.e. $\overline{L} := (L \otimes \mathbb{Q}) \cap NS(S)$.

**Definition 2.7.** The $\mathcal{J}_2$-classification of elliptic fibrations on a K3 surface is the classification of their frame lattices.

It appears now clear that if two elliptic fibrations are identified by the $\mathcal{J}_2$-classification, they have the same trivial lattice and the same Mordell–Weil group (since these objects are uniquely determined by the frame of the elliptic fibration).

We observe that if $\mathcal{E}$ and $\mathcal{E}'$ are identified by the $\mathcal{J}_1$-classification, then there exists an automorphism $g \in Aut(S)$, such that $\mathcal{E} = \mathcal{E}' \circ g$. The automorphism $g$ induces an isometry $g^*$ on $NS(S)$ and it is clear that $g^* : W_{\mathcal{E}} \rightarrow W_{\mathcal{E}'}$ is an isometry. Thus the elliptic fibrations $\mathcal{E}$ and $\mathcal{E}'$ have isometric frame lattices and so are $\mathcal{J}_2$-equivalent.

The $\mathcal{J}_2$-classification is not as fine as the $\mathcal{J}_1$-classification; indeed, if $h : W_{\mathcal{E}} \rightarrow W_{\mathcal{E}'}$ is an isometry, a priori there is no reason to conclude that there exists an automorphism $g \in Aut(S)$ such that $g^*_{|W_{\mathcal{E}}} = h$; indeed, comparing the $\mathcal{J}_1$-classification given in Oguiso (1989) and the $\mathcal{J}_2$-classification given in Nishiyama (1996) for the Kummer surface of the product of two non-isogenous elliptic curves, one can check that the first one is more fine than the second one.

The advantage of the $\mathcal{J}_2$-classification sits in its strong relation with the lattice theory; indeed, there is a method which allows one to obtain the $\mathcal{J}_2$-classification of elliptic fibration on several K3 surfaces. This method is presented in Nishiyama (1996) and will be described in this paper in Section 4.1.

#### 2.1.4 Results on the Different Classification Types

One of the main results of Braun et al. (2013) is about the relations among the various types of classifications of elliptic fibrations on K3 surfaces. First we observe that there exists two surjective maps $\mathcal{J}_0 \rightarrow \mathcal{J}_1$ and $\mathcal{J}_0 \rightarrow \mathcal{J}_2$, which are in fact quotient maps (cf. Braun et al. 2013, Formulae (54) and (57)). This induces a map $\mathcal{J}_1 \rightarrow \mathcal{J}_2$ which is not necessarily a quotient map.

The Braun et al. (2013, Proposition C') gives a bound for the number of different elliptic fibrations up to the $\mathcal{J}_1$-classification, which are identified by the $\mathcal{J}_2$-classification. As a Corollary the following is proved:

**Corollary 2.8 (Braun et al. (2013, Corollary D)).** *Let $S_{(a,b,c)}$ be a K3 surface such that the transcendental lattice of $S$ is isometric to $\begin{bmatrix} 2a & b \\ b & 2c \end{bmatrix}$. If $(a, b, c)$ is one of the following $(1, 0, 1)$, $(1, 1, 1)$, $(2, 0, 1)$, $(2, 1, 1)$, $(3, 0, 1)$, $(3, 1, 1)$, $(4, 0, 1)$, $(5, 1, 1)$, $(6, 1, 1)$, $(3, 2, 1)$, then $\mathcal{J}_1 \simeq \mathcal{J}_2$.*

### 2.2 *A Classification Method for Elliptic Fibrations on K3 Surfaces*

The first paper about the classification of elliptic fibrations on K3 surfaces is due to Oguiso (1989). He gives a $\mathcal{J}_1$-classification of the elliptic fibrations on the Kummer surface of the product of two non-isogenous elliptic curves. The method proposed in Oguiso (1989) is very geometric: it is strictly related to the presence of a certain automorphism (a non-symplectic involution) on the K3 surface. Since one has to require that the K3 surface admits this special automorphism, the method suggested in Oguiso (1989) can be generalized only to certain special K3 surfaces (see Kloosterman 2006; Comparin and Garbagnati 2014).

Seven years after the paper (Oguiso 1989), a different method was proposed by Nishiyama in Nishiyama (1996). This method is less geometric and more related to the lattice structure of the K3 surfaces and of the elliptic fibrations. Nishiyama applied this method in order to obtain a $\mathcal{J}_2$-classification of the elliptic fibrations, both on the K3 surface already considered in Oguiso (1989) and on other K3 surfaces (cyclic quotients of the product of two special elliptic curves) to which the method by Oguiso cannot be applied. Later, in Bertin and Lecacheux (2013), the method is used to give a $\mathcal{J}_2$-classification of elliptic fibrations on a K3 surface whose transcendental lattice is $\langle 4\rangle \oplus \langle 2\rangle$.

The main idea of Nishiyama's method is the following: we consider a K3 surface $S$ and its transcendental lattice $T_S$. Then we consider a lattice $T$ such that: $T$ is negative definite; $\text{rank}(T) = \text{rank}(T_S) + 4$; the discriminant group and form of $T$ are the same as the ones of $T_S$. We consider primitive embeddings of $\phi : T \hookrightarrow L$, where $L$ is a Niemeier lattice. The orthogonal complement of $\phi(T)$ in $L$ is in fact the frame of an elliptic fibration on $S$.

The classification of the primitive embeddings of $T$ in $L$ for every Niemeier lattice $L$ coincides with the $\mathcal{J}_2$-classification of the elliptic fibrations on $S$. We will give more details on Nishiyama's method in Section 4.1.

Since this method is related only to the lattice properties of the surface, a priori one cannot expect to find a $\mathcal{J}_1$-classification by using only this method.

Thanks to Corollary 2.8, (see Braun et al. 2013) the results obtained by Nishiyama's method are sometimes stronger than expected. In particular, we will see that in our case (as in the case described in Bertin and Lecacheux 2013) the classification that we obtain for the elliptic fibrations on a certain K3 surface using the Nishiyama's method is in fact a $\mathcal{J}_1$-classification (and not only a $\mathcal{J}_2$-classification).

## 2.3 Torsion Part of the Mordell–Weil Group of an Elliptic Fibration

In Section 4.2, we will classify elliptic fibrations on a certain K3 surface, determining both the trivial lattice and the Mordell–Weil group. A priori, steps (8) and (9) of the algorithm presented in Section 4.1 completely determine the Mordell–Weil group. In any case, we can deduce some information on the torsion part of the Mordell–Weil group by considering only the properties of the reducible fibers of the elliptic fibration. This makes the computation easier, so here we collect some results on the relations between the reducible fibers of a fibration and the torsion part of the Mordell–Weil group.

First, we recall that a section meets every fiber in exactly one smooth point, so a section meets every reducible fiber in one point of a component with multiplicity 1 (we recall that the fibers of type $I_n^*$, $II^*$, $III^*$, $IV^*$ have reducible components with multiplicity greater than 1). We will call the component of a reducible fiber which meets the zero section the *zero component* or *trivial component*.

Every section (being a rational point of an elliptic curve defined over $k(\mathbb{P}^1)$) induces an automorphism of every fiber, in particular of every reducible fiber. Thus, the presence of an $n$-torsion section implies that all the reducible fibers of the fibration admit $\mathbb{Z}/n\mathbb{Z}$ as subgroup of the automorphism group. In particular, this implies the following (well-known) result:

**Proposition 2.9 (cf. Schütt and Shioda (2010, Section 7.2)).** *Let $\mathcal{E} : S \to \mathbb{P}^1$ be an elliptic fibration and let $\mathrm{MW}(\mathcal{E})_{\mathrm{tors}}$ the torsion part of the Mordell–Weil group.*

*If there is a fiber of type $II^*$, then $\mathrm{MW}(\mathcal{E})_{\mathrm{tors}} = 0$.*

*If there is a fiber of type $III^*$, then $\mathrm{MW}(\mathcal{E})_{\mathrm{tors}} \leq (\mathbb{Z}/2\mathbb{Z})$.*

*If there is a fiber of type $IV^*$, then $\mathrm{MW}(\mathcal{E})_{\mathrm{tors}} \leq (\mathbb{Z}/3\mathbb{Z})$.*

*If there is a fiber of type $I_n^*$ and n is an even number, then $\mathrm{MW}(\mathcal{E})_{\mathrm{tors}} \leq (\mathbb{Z}/2\mathbb{Z})^2$.*

*If there is a fiber of type $I_n^*$ and n is an odd number, then $\mathrm{MW}(\mathcal{E})_{\mathrm{tors}} \leq (\mathbb{Z}/4\mathbb{Z})$.*

### 2.3.1 Covers of Universal Modular Elliptic Surfaces

The theory of universal elliptic surfaces parametrizing elliptic curves with prescribed torsion can also be useful when finding the torsion subgroup of a few elliptic fibrations on the list. It relies on the following definition/proposition.

**Proposition 2.10 (See Couveignes and Edixhoven (2011, 2.1.4) or Shioda (1972)).** *Let $\pi : X \to B$ be an elliptic fibration on a surface X. Assume $\pi$ has a section of order N, for some $N \in \mathbb{N}$, with $N \geq 4$. Then X is a cover of the universal modular elliptic surface, $\mathcal{E}_N$, of level N.*

After studying the possible singular fibers of the universal surfaces above, one gets the following.

**Proposition 2.11.** *Let $\mathcal{E}_N$ be the universal modular elliptic surface of level N. The following hold:*

*i) If $N \geq 5$, then $\mathcal{E}_N$ admits only semi-stable singular fibers. They are all of type $I_m$ with $m|N$.*

*ii) The surface $\mathcal{E}_4$ is a rational elliptic surface with singular fibers $I_1^*, I_4, I_1$.*

### 2.3.2 Height Formula for Elliptic Fibrations

The group structure of the Mordell–Weil group is the group structure of the rational points of the elliptic curve defined over the function field of the basis of the fibration. It is also possible to equip the Mordell–Weil group of a pairing taking values in

$\mathbb{Q}$, which transforms the Mordell–Weil group to a $\mathbb{Q}$-lattice. Here we recall the definitions and the main properties of this pairing. For a more detailed description, we refer to Schütt and Shioda (2010) and to the original paper (Shioda 1990).

**Definition 2.12.** Let $\mathcal{E} : S \to C$ be an elliptic fibration and let $O$ be the zero section. The height pairing is the $\mathbb{Q}$-valued pairing, $< -,- >: \mathrm{MW}(\mathcal{E}) \times \mathrm{MW}(\mathcal{E}) \to \mathbb{Q}$ defined on the sections of an elliptic fibration as follows:

$$< P, Q >= \chi(S) + P \cdot O + Q \cdot O - \sum_{c \in \mathcal{C}} contr_c(P, Q),$$

where $\chi(S)$ is the holomorphic characteristic of the surface $S$, $\cdot$ is the intersection form on $NS(S)$, $\mathcal{C} = \{c \in C$ such that the fiber $\mathcal{E}^{-1}(c)$ is reducible$\}$ and $contr_c(P, Q)$ is a contribution which depends on the type of the reducible fiber and on the intersection of $P$ and $Q$ with such a fiber as described in Schütt and Shioda (2010, Table 4).

The value $h(P) :=< P, P >= 2\chi(S) + 2P \cdot O - \sum_{c \in \mathcal{C}} contr_c(P, P)$, is called the *height* of the section $P$.

We observe that the height formula is induced by the projection of the intersection form on $NS(S) \otimes \mathbb{Q}$ to the orthogonal complement of the trivial lattice $Tr(\mathcal{E})$ (cf. Schütt and Shioda 2010, Section 11).

**Proposition 2.13 (Schütt and Shioda (2010, Section 11.6)).** *Let $P \in \mathrm{MW}(\mathcal{E})$ be a section of the elliptic fibration $\mathcal{E} : S \to C$. The section $P$ is a torsion section if and only if $h(P) = 0$.*

# 3 The K3 Surface $X$

The goal of this paper is the classification of the elliptic fibrations on the unique K3 surface $X$ such that $T_X \simeq \langle 6 \rangle \oplus \langle 2 \rangle$. This surface is interesting for several reasons, and we will present it from different points of view.

## *3.1 A Toric Hypersurface and the Symmetric Group $\mathcal{S}_4$*

Let $N$ be a lattice isomorphic to $\mathbb{Z}^n$. The dual lattice $M$ of $N$ is given by $\mathrm{Hom}(N, \mathbb{Z})$; it is also isomorphic to $\mathbb{Z}^n$. We write the pairing of $v \in N$ and $w \in M$ as $\langle v, w \rangle$.

Given a lattice polytope $\diamond$ in $N$, we define its *polar polytope* $\diamond^\circ$ to be $\diamond^\circ = \{w \in \mathrm{M} \mid \langle v, w \rangle \geq -1 \;\forall\, v \in K\}$. If $\diamond^\circ$ is also a lattice polytope, we say that $\diamond$ is a reflexive polytope and that $\diamond$ and $\diamond^\circ$ are a mirror pair. A reflexive polytope must contain $\mathbf{0}$; furthermore, $\mathbf{0}$ is the only interior lattice point of the polytope.

Reflexive polytopes have been classified in one, two, three, and four dimensions. In three dimensions, there are 4319 reflexive polytopes, up to an overall isomorphism preserving lattice structure (Kreuzer and Skarke 1998, 2000). The database of reflexive polytopes is incorporated in the open-source computer algebra software (Stein et al. 2014).

Now, consider the one-parameter family of K3 surfaces given by

$$x + \frac{1}{x} + y + \frac{1}{y} + z + \frac{1}{z} + \frac{x}{y} + \frac{y}{x} + \frac{y}{z} + \frac{z}{y} + \frac{x}{z} + \frac{z}{x} + \lambda. \tag{1}$$

This family of K3 surfaces was first studied in Verrill (1996), where its Picard–Fuchs equation was computed. A general member of the family has Picard lattice given by $U \oplus \langle 6 \rangle$.

The Newton polytope $\diamond^\circ$ determined by the family of polynomials in Equation 1 is a reflexive polytope with 12 vertices and 14 facets. This polytope has the greatest number of facets of any three-dimensional reflexive polytope; furthermore, there is a unique three-dimensional reflexive polytope with this property, up to isomorphism. In the database of reflexive polytopes found in Stein et al. (2014), this polytope has index 1529.

We illustrate its polar polytope $\diamond$ next to $\diamond^\circ$ in Figures 1 and 2.

Let us recall some standard constructions and notations involving toric varieties. A *cone* in $N$ is a subset of the real vector space $N_{\mathbb{R}} = N \otimes \mathbb{R}$ generated by nonnegative $\mathbb{R}$-linear combinations of a set of vectors $\{v_1, \ldots, v_m\} \subset N$. We assume that cones are strongly convex, that is, they contain no line through the origin. Note that each face of a cone is a cone. *fan* $\Sigma$ consists of a finite collection of cones such that each face of a cone in the fan is also in the fan, and any pair of cones in the fan intersects in a common face. We say $\Sigma$ is *simplicial* if the generators of

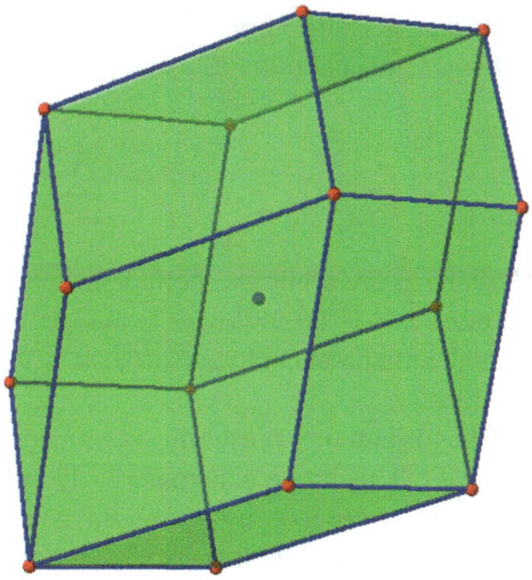

**Fig. 1** $\diamond$ (reflexive polytope 2355)

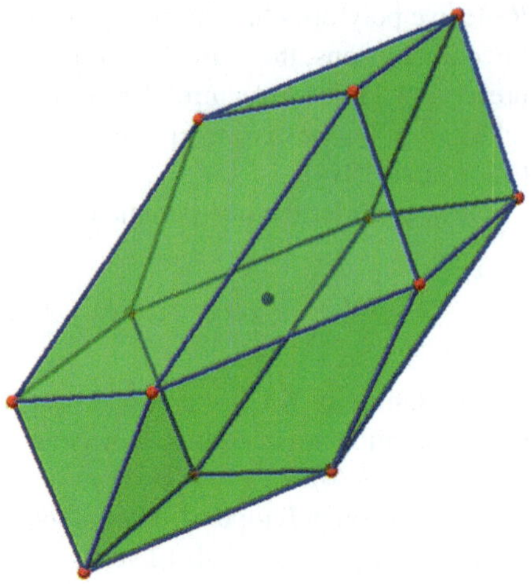

**Fig. 2** $\diamond^\circ$ (reflexive polytope 1529)

each cone in $\Sigma$ are linearly independent over $\mathbb{R}$. If every element of $N_{\mathbb{R}}$ belongs to some cone in $\Sigma$, we say $\Sigma$ is *complete*. A fan $\Sigma$ defines a toric variety $V_\Sigma$. If the fan is complete, we may describe $V_\Sigma$ using homogeneous coordinates, in a process analogous to the construction of $\mathbb{P}^n$ as a quotient space of $(\mathbb{C}^*)^n$. The homogeneous coordinates have one coordinate $z_j$ for each generator of a one-dimensional cone of $\Sigma$.

We may obtain a fan $R$ from a mirror pair of reflexive polytopes in two equivalent ways. We may take cones over the faces of $\diamond \subset N_{\mathbb{R}}$, or we may take the *normal fan* to the polytope $\diamond^\circ \subset M_{\mathbb{R}}$. Let $\Sigma$ be a simplicial refinement of $R$ such that the one-dimensional cones of $\Sigma$ are generated by the nonzero lattice points $v_k$, $k = 1 \ldots q$, of $\diamond$; we call such a refinement a *maximal projective subdivision*. Then the variety $V_\Sigma$ is an orbifold. Then in homogeneous coordinates, we have one coordinate $z_k$ for each nonzero lattice point in $\diamond$. We may describe the anticanonical hypersurfaces in homogeneous coordinates using polynomials of the form:

$$p = \sum_{x \in \diamond^\circ \cap M} c_x \prod_{k=1}^{q} z_k^{\langle v_k, x \rangle + 1}. \tag{2}$$

Here the $c_x$ are arbitrary coefficients. Note that $p$ has one monomial for each lattice point of $\diamond^\circ$. If the reflexive polytope $\diamond$ is three-dimensional, $V_\Sigma$ is smooth and smooth anticanonical hypersurfaces in $V_\Sigma$ are K3 surfaces (see Cox and Katz 1999 for details).

The orientation-preserving symmetry group of $\diamond$ and $\diamond^\circ$ is the symmetric group $\mathcal{S}_4$. This group acts transitively on the vertices of $\diamond^\circ$. As the authors of Karp et al. (2013) observe, by setting the coefficients $c_x$ corresponding to the vertices of $\diamond^\circ$ to 1 and the coefficient corresponding to the origin to a parameter $\lambda$, we obtain a naturally one-parameter family of K3 hypersurfaces with generic Picard rank 19:

$$p = \left( \sum_{x \in \mathrm{vertices}(\diamond^\circ)} \prod_{k=1}^{q} z_k^{\langle v_k, x\rangle + 1} \right) + \lambda z_1 \dots z_q. \tag{3}$$

Equation 3 is simply Equation 1 in homogeneous coordinates.

If we view $\mathcal{S}_4$ as acting on the vertices of $\diamond$ rather than the vertices of $\diamond^\circ$, we obtain a permutation of the homogeneous coordinates $z_k$. The authors of Karp et al. (2013) show that this action of $\mathcal{S}_4$ restricts to a symplectic action on each K3 surface in the pencil given by Equation 3; in particular, we have a symplectic action of $\mathcal{S}_4$ on $X$. In the affine coordinates of Equation 1, the group action is generated by an element $s_2$ of order 2 which acts by $(x, y, z) \mapsto (1/x, 1/z, 1/y)$ and an element $s_4$ of order 4 which acts by $(x, y, z) \mapsto (x/y, x/z, x)$.

## 3.2 *The K3 Surface X*

**Definition 3.1.** Let $X$ be the K3 surface defined by $F = 0$, where $F$ is the numerator of

$$x + \frac{1}{x} + y + \frac{1}{y} + z + \frac{1}{z} + \frac{x}{y} + \frac{y}{x} + \frac{y}{z} + \frac{z}{y} + \frac{x}{z} + \frac{z}{x}.$$

The K3 surface $X$ is the special member of the family of K3 surfaces described in (1) which is obtained by setting $\lambda = 0$.

We will use three elements of the symplectic group $\mathcal{S}_4$: the three-cycle $s_3$ given by $(x, y, z) \mapsto (y, z, x)$, the four-cycle $s_4$ and the two-cycle $s_2$.

We describe explicitly a first elliptic fibration, which gives the main properties of $X$.

## 3.3 *A Fibration Invariant by $s_3$*

We use the following factorizations

$$F = (x + y + z + 1)(xy + yz + zx) + (x + y + z - 3)\,xyz \tag{4}$$

$$F\,(x + y + z) = (x + y + z - 1)^2 xyz + (x + y + z + 1)(x + y)(y + z)(z + x). \tag{5}$$

If $w = x + y + z$, we see that $w$ is invariant under the action of $s_3$. If we substitute $w - x - y$ for $z$, we obtain the equation of an elliptic curve, so the morphism

$$\mathcal{E} : X \to \mathbb{P}^1_w \tag{6}$$

$$(x, y, z) \mapsto w = x + y + z$$

is an elliptic fibration of $X$.

We use the birational transformation

$$x = -\frac{v + (w+1)u}{u(w-3)}, \quad y = -\frac{v(w+1) - u^2}{v(w-3)}$$

with inverse

$$u = -((w-3)x + (w+1))((w-3)y + (w+1)),$$
$$v = ((w-3)x + (w+1))^2((w-3)y + (w+1))$$

to obtain the Weierstrass equation

$$v^2 + \left(w^2 + 3\right) uv + \left(w^2 - 1\right)^2 v = u^3. \tag{7}$$

Notice the torsion points $(u = 0, v = 0)$ and $\left(u = 0, v = -(w^2 - 1)^2\right)$ of order 3 and the 3 points of order 2 with $u$-coordinate $\frac{-1}{4}\left(w^2 - 1\right)^2$,$-\left(w - 1\right)^2$, and $-\left(w + 1\right)^2$.

We use also the Weierstrass form

$$\xi^2 = \eta\left(\eta - (w-3)(w+1)^3\right)\left(\eta - (w+3)(w-1)^3\right) \tag{8}$$

with

$$u = \frac{1}{4}\left(\eta - \left(w^2 - 1\right)^2\right), \quad v = \frac{1}{8}\left(\xi - \left(w^2 + 3\right)\eta + \left(w^2 - 1\right)^3\right).$$

The singular fibers are of type $I_6$ for $w = -1, 1, \infty$ and $I_2$ for $w = 3, -3, 0$. So the trivial lattice of this fibration is $\mathrm{Tr}(\mathcal{E}) = U \oplus A_5^{\oplus 3} \oplus A_1^{\oplus 3}$. Hence the Picard number of $X$ is 20 and $\mathrm{rank}(MW(\mathcal{E})) = 0$. So $X$ is a singular K3 surface. This elliptic fibration is contained in the Shimada and Zhang (2001, Table 2 line 4) and thus its transcendental lattice is $\langle 6 \rangle \oplus \langle 2 \rangle$.

Moreover, all the fibers have split multiplicative reduction and thus the Néron Severi group is generated by curves defined on $\mathbb{Q}$.

*Remark 3.2.* We have already observed that $X$ is a special member of the one-dimensional family of K3 surfaces defined by Equation 1. Indeed, the transcendental lattice of $X$ is primitively embedded in $U \oplus \langle 6 \rangle$ by the vectors $(1, 1, 0), (0, 0, 1)$.

This gives the following proposition:

**Proposition 3.3.** *The Néron–Severi group of the K3 surface $X$ has rank 20 and is generated by divisors which are defined over $\mathbb{Q}$. The transcendental lattice of $X$ is* $T_X \simeq \begin{pmatrix} 2 & 0 \\ 0 & 6 \end{pmatrix}$.

*Remark 3.4.* The equation (8) is the universal elliptic curve with torsion group $\mathbb{Z}/2\mathbb{Z} \times \mathbb{Z}/6\mathbb{Z}$ and is in fact equivalent to the equation given in Kubert (1976).Thus this fibration can be called modular: we can view the base curve $(\mathbb{P}^1_w)$ as the modular curve $X_\Gamma$ with $\Gamma = \{ \begin{pmatrix} a & b \\ c & d \end{pmatrix} \in Sl_2(\mathbb{Z}), a \equiv 1 \bmod 6, c \equiv 0 \bmod 6, b \equiv 0 \bmod 2\}$.
By (5) we can easily obtain the equation

$$(w-1)^2 xyz + (w+1)(x+y)(y+z)(z+x) = 0$$

and realize $X$ by a base change of the modular rational elliptic surface $\mathcal{E}_6$ described by Beauville in Beauville (1982). We can prove that on the fiber, the automorphism $s_3$ corresponds to adding a 3-torsion point.

## 3.4 A Fibration Invariant by $s_4$

If $t = \frac{y}{zx}$, we see that $t$ is invariant under the action of $s_4$. Substituting $tzx$ for $y$ in $F$, we obtain the equation of an elliptic curve. Using standard transformations (as in Kumar 2014(39.2); Atkin and Morain 1993 or Cassels 1991) we obtain the Weierstrass model

$$v^2 = u\left(u^2 - 2t\left(t^2+1\right)u + t^2(t+1)^4\right).$$

The point $Q_t = \left(u = t(t+1)^2, v = 2t^2(t+1)^2\right)$ is of order 4.
The point $P_t = \left(u = (t+1)^2, v = (t^2+1)(t+1)^2\right)$ is of infinite order.
The singular fibers are $2I_1^* (t = 0, \infty) + I_8 (t = -1) + 2I_1 \left(t^2 + t + 1 = 0\right)$.
One can prove that on the fiber, $s_4$ corresponds to the translation by a 4-torsion point. Moreover, the translation by the point $P_t$ defines an automorphism of infinite order on $X$.

*Remark 3.5.* If we compute the height of $P_t$ we can show, using Shioda formula, Shioda (1990) that $P_t$ and $Q_t$ generate the Mordell–Weil group.

## 3.5 A Fibration Invariant by $s_2$

If $r = \frac{y}{z}$, we see that $r$ is invariant under the action of $s_2$. Substituting $rz$ for $y$ we obtain the equation of an elliptic curve and the following Weierstrass model

$$v^2 - \left(r^2-1\right)vu = u\left(u - 2r(r+1)\right)\left(u - 2r^2(r+1)\right).$$

The point $(0,0)$ is a two-torsion point. The point $(2r(r+1), 0)$ is of infinite order.
The singular fibers are
$2I_6 (r = 0, \infty) + I_0^* (r = -1) + I_4 (r = 1) + 2I_1 \left(r^2 - 14r + 1 = 0\right)$.

*Remark 3.6.* From Elkies results (Elkies 2010; Schütt 2010) there is a unique K3 surface $X/\mathbb{Q}$ with Néron–Severi group of rank 20 and discriminant $-12$ that consists entirely of classes of divisors defined over $\mathbb{Q}$. Indeed it is $X$. Moreover, in Elkies (2010) a Weierstrass equation for an elliptic fibration on $X$ defined over $\mathbb{Q}$, is given:

$$y^2 = x^3 - 75x + (4t - 242 + \frac{4}{t}).$$

*Remark 3.7.* The surface $X$ is considered also in a slightly different context in Garbagnati and Sarti (2009) because of its relation with the study of the moduli space of K3 surfaces with a symplectic action of a finite abelian group. Indeed, the aim of the paper (Garbagnati and Sarti 2009) is to study elliptic fibrations $\mathcal{E}_G : S_G \to \mathbb{P}^1$ on K3 surfaces $S_G$ such that $\mathrm{MW}(\mathcal{E}_G) = G$ is a torsion group. Since the translation by a section is a symplectic automorphism of $S_G$, if $\mathrm{MW}(\mathcal{E}_G) = G$, then $G$ is a group which acts symplectically on $S_G$. In Garbagnati and Sarti (2009) it is shown how one can describe both a basis for the Néron–Severi group of $S_G$ and the action induced by the symplectic action of $G$ on this basis. In particular, one can directly compute the lattices $NS(S_G)^G$ and $\Omega_G := NS(S_G)^{\perp}$. The latter does not depend on $S_G$ but only on $G$ and its computation plays a central role in the description of the moduli space of the K3 surfaces admitting a symplectic action of $G$ (see Nikulin 1979; Garbagnati and Sarti 2009). In particular, the case $G = \mathbb{Z}/6\mathbb{Z} \times Z/2\mathbb{Z}$ is considered: in this case, the K3 surface $S_G$ is $X$, and the elliptic fibration $\mathcal{E}_G$ is (6). Comparing the symplectic action of $\mathbb{Z}/6\mathbb{Z} \times \mathbb{Z}/2\mathbb{Z}$ on $X$ given in Garbagnati and Sarti (2009) with the symplectic group action of $\mathcal{S}_4$ described in Section 3.1, we find that the two groups intersect in the subgroup of order 3 generated by the map $s_3$ given by $(x, y, z) \mapsto (z, y, x)$.

## 4 Main Result

This section is devoted to the proof of our main result:

**Theorem 4.2.** *The classification up to automorphisms of the elliptic fibrations on $X$ is given in Table 1. Each elliptic fibration is given with the Dynkin diagrams characterizing its reducible fibers and the rank and torsion of its Mordell–Weil group. More precisely, we obtained* 52 *elliptic fibrations on $X$, including* 17 *fibrations of rank* 2 *and one of rank* 3.

We denote by $r$ the rank($\mathrm{MW}(\mathcal{E})$), and we use Bourbaki notations for $A_n, D_n, E_k$ as in Bertin and Lecacheux (2013).

*Outline of the Proof* The proof consists of an application of Nishiyama's method: the details of this method will be described in Section 4.1. Its application to our case is given in Section 4.2. The application of Nishiyama's method gives us a $\mathcal{J}_2$-classification, which coincides in our cases with a classification up to automorphisms of the surface by Corollary 2.8.

**Table 1** The elliptic fibrations of $X$

| $L_{\text{root}}$ | No. | | | $N_{\text{root}}$ | $r$ | $MW(\mathcal{E})_{\text{tors}}$ |
|---|---|---|---|---|---|---|
| $E_8^3$ | 1 | $A_5 \oplus A_1 \subset E_8$ | | $A_1E_8E_8$ | 1 | (0) |
| | 2 | $A_1 \subset E_8$ | $A_5 \subset E_8$ | $A_1A_2E_7E_8$ | 0 | (0) |
| $E_8D_{16}$ | 3 | $A_5 \oplus A_1 \subset E_8$ | | $A_1D_{16}$ | 1 | $\mathbb{Z}/2\mathbb{Z}$ |
| | 4 | $A_5 \oplus A_1 \subset D_{16}$ | | $A_1D_8E_8$ | 1 | (0) |
| | 5 | $A_5 \subset E_8$ | $A_1 \subset D_{16}$ | $A_1A_1A_2D_{14}$ | 0 | $\mathbb{Z}/2\mathbb{Z}$ |
| | 6 | $A_1 \subset E_8$ | $A_5 \subset D_{16}$ | $E_7D_{10}$ | 1 | (0) |
| $E_7^2D_{10}$ | 7 | $A_5 \oplus A_1 \subset E_7$ | | $E_7D_{10}$ | 1 | $\mathbb{Z}/2\mathbb{Z}$ |
| | 8 | $A_5 \oplus A_1 \subset D_{10}$ | | $A_1A_1A_1E_7E_7$ | 1 | $\mathbb{Z}/2\mathbb{Z}$ |
| | 9 | $A_1 \subset E_7$ | $A_5 \subset E_7$ | $D_6A_1D_{10}$ | 1 | $\mathbb{Z}/2\mathbb{Z}$ |
| | 10 | $A_1 \subset E_7$ | $A_5 \subset E_7$ | $D_6A_2D_{10}$ | 0 | $\mathbb{Z}/2\mathbb{Z}$ |
| | 11 | $A_5 \subset E_7$ | $A_1 \subset D_{10}$ | $A_1A_1D_8E_7$ | 1 | $\mathbb{Z}/2\mathbb{Z}$ |
| | 12 | $A_5 \subset E_7$ | $A_1 \subset D_{10}$ | $A_1A_2D_8E_7$ | 0 | $\mathbb{Z}/2\mathbb{Z}$ |
| | 13 | $A_1 \subset E_7$ | $A_5 \subset D_{10}$ | $E_7D_6D_4$ | 1 | $\mathbb{Z}/2\mathbb{Z}$ |
| $E_7A_{17}$ | 14 | $A_5 \oplus A_1 \subset E_7$ | | $A_{17}$ | 1 | $\mathbb{Z}/3\mathbb{Z}$ |
| | 15 | $A_5 \oplus A_1 \subset A_{17}$ | | $A_9E_7$ | 2 | (0) |
| | 16 | $A_5 \subset E_7$ | $A_1 \subset A_{17}$ | $A_1A_{15}$ | 2 | (0) |
| | 17 | $A_5 \subset E_7$ | $A_1 \subset A_{17}$ | $A_2A_{15}$ | 1 | (0) |
| | 18 | $A_1 \subset E_7$ | $A_5 \subset A_{17}$ | $D_6A_{11}$ | 1 | (0) |
| $D_{24}$ | 19 | $A_5 \oplus A_1 \subset D_{24}$ | | $A_1D_{16}$ | 1 | (0) |
| $D_{12}^2$ | 20 | $A_5 \oplus A_1 \subset D_{12}$ | | $A_1D_4D_{12}$ | 1 | $\mathbb{Z}/2\mathbb{Z}$ |
| | 21 | $A_1 \subset D_{12}$ | $A_5 \subset D_{12}$ | $A_1D_{10}D_6$ | 1 | $\mathbb{Z}/2\mathbb{Z}$ |
| $D_8^3$ | 22 | $A_5 \oplus A_1 \subset D_8$ | | $A_1D_8D_8$ | 1 | $\mathbb{Z}/2\mathbb{Z}$ |
| | 22($b$) | $A_5 \oplus A_1 \subset D_8$ | | $A_1D_8D_8$ | 1 | $\mathbb{Z}/2\mathbb{Z}$ |
| | 23 | $A_1 \subset D_8$ | $A_5 \subset D_8$ | $A_1^3D_6D_8$ | 1 | $(\mathbb{Z}/2\mathbb{Z})^2$ |
| $D_9A_{15}$ | 24 | $A_5 \oplus A_1 \subset D_9$ | | $A_1A_{15}$ | 2 | $\mathbb{Z}/2\mathbb{Z}$ |
| | 25 | $A_5 \oplus A_1 \subset A_{15}$ | | $D_9A_7$ | 2 | (0) |
| | 26 | $A_5 \subset D_9$ | $A_1 \subset A_{15}$ | $A_3A_{13}$ | 2 | (0) |
| | 27 | $A_1 \subset D_9$ | $A_5 \subset A_{15}$ | $A_1A_9D_7$ | 1 | (0) |
| $E_6^4$ | 28 | $A_1 \subset E_6$ | $A_5 \subset E_6$ | $A_1A_5E_6^2$ | 0 | $\mathbb{Z}/3\mathbb{Z}$ |
| $A_{11}E_6D_7$ | 29 | $A_5 \oplus A_1 \subset A_{11}$ | | $A_3D_7E_6$ | 2 | (0) |
| | 30 | $A_5 \subset E_6$ | $A_1 \subset D_7$ | $A_1^2A_{11}D_5$ | 0 | $\mathbb{Z}/4\mathbb{Z}$ |
| | 31 | $A_5 \subset E_6$ | $A_1 \subset A_{11}$ | $A_1A_9D_7$ | 1 | (0) |
| | 32 | $A_1 \subset E_6$ | $A_5 \subset D_7$ | $A_5A_{11}$ | 2 | $\mathbb{Z}/3\mathbb{Z}$ |
| | 33 | $A_5 \subset D_7$ | $A_1 \subset A_{11}$ | $A_9E_6$ | 3 | (0) |
| | 34 | $A_1 \subset D_7$ | $A_5 \subset A_{11}$ | $A_1A_5D_5E_6$ | 1 | (0) |
| | 35 | $A_1 \subset E_6$ | $A_5 \subset A_{11}$ | $A_5^2D_7$ | 1 | (0) |
| $D_6^4$ | 36 | $A_1 \subset D_6$ | $A_5 \subset D_6$ | $A_1D_4D_6^2$ | 1 | $(\mathbb{Z}/2\mathbb{Z})^2$ |

(continued)

**Table 1** (continued)

| $L_{\text{root}}$ | No. | | | $N_{\text{root}}$ | $r$ | $\text{MW}(\mathcal{E})_{\text{tors}}$ |
|---|---|---|---|---|---|---|
| $D_6A_9^2$ | 37 | $A_5 \oplus A_1 \subset A_9$ | | $A_1A_9D_6$ | 2 | $\mathbb{Z}/2\mathbb{Z}$ |
| | 38 | $A_5 \subset A_9$ | $A_1 \subset A_9$ | $A_3A_7D_6$ | 2 | (0) |
| | 39 | $A_5 \subset A_9$ | $A_1 \subset D_6$ | $A_1A_3A_9D_4$ | 1 | $\mathbb{Z}/2\mathbb{Z}$ |
| | 40 | $A_1 \subset A_9$ | $A_5 \subset D_6$ | $A_7A_9$ | 2 | (0) |
| $D_5^2A_7^2$ | 41 | $A_5 \oplus A_1 \subset A_7$ | | $A_7D_5^2$ | 1 | $\mathbb{Z}/4\mathbb{Z}$ |
| | 42 | $A_5 \subset A_7$ | $A_1 \subset A_7$ | $A_1A_5D_5^2$ | 2 | (0) |
| | 43 | $A_5 \subset A_7$ | $A_1 \subset D_5$ | $A_1^2A_3A_7D_5$ | 1 | $\mathbb{Z}/4\mathbb{Z}$ |
| $A_8^3$ | 44 | $A_5 \oplus A_1 \subset A_8$ | | $A_8^2$ | 2 | $\mathbb{Z}/3\mathbb{Z}$ |
| | 45 | $A_1 \subset A_8$ | $A_5 \subset A_8$ | $A_2A_6A_8$ | 2 | (0) |
| $A_{24}$ | 46 | $A_5 \oplus A_1 \subset A_{24}$ | | $A_{16}$ | 2 | (0) |
| $A_{12}^2$ | 47 | $A_5 \oplus A_1 \subset A_{12}$ | | $A_4A_{12}$ | 2 | (0) |
| | 48 | $A_5 \subset A_{12}$ | $A_1 \subset A_{12}$ | $A_6A_{10}$ | 2 | (0) |
| $D_4A_5^4$ | 49 | $A_5 = A_5$ | $A_1 \subset A_5$ | $A_3A_5^2D_4$ | 1 | $\mathbb{Z}/2\mathbb{Z}$ |
| | 50 | $A_5 = A_5$ | $A_1 \subset D_4$ | $A_1^3A_5^3$ | 0 | $\mathbb{Z}/2\mathbb{Z} \times \mathbb{Z}/6\mathbb{Z}$ |
| $A_6^4$ | 51 | $A_5 \subset A_6$ | $A_1 \subset A_6$ | $A_4A_6^2$ | 2 | (0) |

*Remark 4.2.* The fibration given in Section 3.3 is # 50 in Table 1, the one given in Section 3.4 is # 41, the one given in Section 3.5 is # 49, the one given in Remark 3.6 is #1. The fibrations of rank 0 may be found also in Shimada and Zhang (2001).

*Remark 4.3.* We observe that there exists a primitive embedding of $T_X \simeq \langle 6 \rangle \oplus \langle 2 \rangle$ in $U(2) \oplus \langle 2 \rangle$ given by the vectors $\langle (1,1,1), (0,-1,1) \rangle$. Thus, $X$ is a special member of the one-dimensional family of K3 surfaces whose transcendental lattice is isometric to $U(2) \oplus \langle 2 \rangle$. The elliptic fibrations on the generic member $Y$ of this family have already been classified (cf. Comparin and Garbagnati 2014), and indeed the elliptic fibrations in Table 1 specialize the ones in Comparin and Garbagnati (2014, Table 4.5 and Section 8.1 case $r = 19$), either because the rank of the Mordell–Weil group increases by 1 or because two singular fibers glue together producing a different type of reducible fiber.

## *4.1 Nishiyama's Method in Detail: An Algorithm*

This section is devoted to a precise description of Nishiyama's method. Since the method is very well described both in the original paper (Nishiyama 1996) and in some other papers where it is applied, e.g. Bertin and Lecacheux (2013) and Braun et al. (2013), we summarize it in an algorithm which allows us to compute all the results given in Table 1. In the next section we will describe in detail some peculiar cases, in order to show how this algorithm can be applied.

**Definition 4.4.** A Niemeier lattice is an even unimodular negative definite lattice of rank 24.

There are 24 Niemeier lattices. We will denote by $L$ an arbitrary Niemeier lattice. Each of them corresponds uniquely to its root lattice $L_{\text{root}}$.

In Table 2 we list the Niemeier lattices, giving both the root lattices of each one and a set of generators for $L/L_{\text{root}}$. To do this we recall the following notation, introduced in Bertin and Lecacheux (2013):

| | |
|---|---|
| $\alpha_n = \frac{1}{n+1}\sum_{j=1}^{n}(n-j+1)a_j$ | $\delta_l = \frac{1}{2}\left(\sum_{i=1}^{l-2} id_i + \frac{1}{2}(l-2)d_{l-1} + \frac{1}{2}ld_l\right)$ |
| $\overline{\delta}_l = \sum_{i=1}^{l-2} d_i + \frac{1}{2}(d_{l-1}+d_l)$ | $\tilde{\delta}_l = \frac{1}{2}\left(\sum_{i=1}^{l-2} id_i + \frac{1}{2}ld_{l-1} + \frac{1}{2}(l-2)d_l\right)$ |
| $\eta_6 = -\frac{2e_1+3e_2+4e_3+6e_4+5e_5+4e_6}{3}$ | $\eta_7 = -\frac{(2e_1+3e_2+4e_3+6e_4+5e_5+4e_6+3e_7)}{2}$ |

Now let us consider a K3 surface $S$ such that $\rho(S) \geq 12$. Let us denote by $T_S$ its transcendental lattice. We describe an algorithm which gives a $\mathcal{J}_2$-classification of the elliptic fibration on $S$.

(1) **The lattice $T$:** We define the lattice $T$ to be a negative definite lattice such that $\text{rank}(T) = \text{rank}(T_S) + 4$ and the discriminant group and form of $T$ are the same as the ones of $T_S$. The lattice $T$ is not necessarily unique. If it is not, we choose one lattice with this property (the results obtained do not depend on this choice).
(2) **Assumption:** We assume that one can choose $T$ to be a root lattice.
(3) **The embeddings $\phi$:** Given a Niemeier lattice $L$ we choose a set of primitive embeddings $\phi : T \hookrightarrow L_{\text{root}}$ not isomorphic by an element of the Weyl group.
(4) **The lattices $N$ and $N_{\text{root}}$:** Given a primitive embedding $\phi$ we compute the orthogonal complement $N$ of $\phi(T)$ in $L_{\text{root}}$, i.e. $N := \phi(T)^{\perp_{L_{\text{root}}}}$ and $N_{\text{root}}$ its root lattice.
(5) **The lattices $W$ and $W_{\text{root}}$:** We denote by $W$ the orthogonal complement of $\phi(T)$ in $L$, i.e. $W := \phi(T)^{\perp_L}$ and by $W_{\text{root}}$ its root lattice. We observe that $N_{\text{root}} = W_{\text{root}}$ and $N \hookrightarrow W$ with finite index.
(6) **The elliptic fibration $\mathcal{E}_\phi$:** The frame of any elliptic fibration on $S$ is a lattice $W$ obtained as in step 5. Moreover, the trivial lattice of any elliptic fibration on $S$ is of the form $U \oplus N_{\text{root}} = U \oplus W_{\text{root}}$ where $W_{\text{root}}$ and $N_{\text{root}}$ are obtained as above. Hence, we find a $\mathcal{J}_2$-classification of the elliptic fibration on $S$. In particular every elliptic fibration on $S$ is uniquely associated with a primitive embedding $\phi : T \hookrightarrow L$. Let us denote by $\mathcal{E}_\phi$ the elliptic fibration associated with $\phi$.
(7) **The singular fibers:** We already observed (cf. Section 2.1) that almost all the properties of the singular fibers are encoded in the trivial lattice, so it is clear that every $N_{\text{root}}(:= (\phi(T)^{\perp_{L_{\text{root}}}})_{\text{root}})$ determines almost all the properties of the reducible fibers of $\mathcal{E}_\phi$.
(8) **The rank of the Mordell–Weil group:** Let $\phi$ be a given embedding. Let $r := \text{rank}(MW(\mathcal{E}_\phi))$. Then $r = \text{rank}(NS(S)) - 2 - \text{rank}(N_{\text{root}}) = 20 - \text{rank}(T_S) - \text{rank}(N_{\text{root}})$.
(9) **The torsion of the Mordell–Weil group:** The torsion part of the Mordell–Weil group is $\overline{W_{\text{root}}}/W_{\text{root}}(\subset W/N)$ and can be computed in the following way: let

**Table 2** The Niemeier lattices $L$: $L_{\text{root}}$ and $L/L_{\text{root}}$

| $L_{\text{root}}$ | $L/L\text{root}$ |
|---|---|
| $E_8^3$ | $\langle(0)\rangle$ |
| $E_8D_{16}$ | $\mathbb{Z}/2\mathbb{Z} = \langle\delta_{16}\rangle$ |
| $E_7^2D_{10}$ | $(\mathbb{Z}/2\mathbb{Z})^2 = \langle\eta_7^{(1)} + \delta_{10}, \eta_7^{(1)} + \eta_7^{(2)} + \overline{\delta_{10}}\rangle$ |
| $E_7A_{17}$ | $\mathbb{Z}/6\mathbb{Z} = \langle\eta_7 + 3\alpha_{17}\rangle$ |
| $D_{24}$ | $\mathbb{Z}/2\mathbb{Z} = \langle\delta_{24}\rangle$ |
| $D_{12}^2$ | $(\mathbb{Z}/2\mathbb{Z})^2 = \langle\delta_{12}^{(1)} + \overline{\delta_{12}^{(2)}}, \overline{\delta_{12}^{(1)}} + \delta_{12}^{(2)}\rangle$ |
| $D_8^3$ | $(\mathbb{Z}/2\mathbb{Z})^3 = \langle\delta_8^{(1)} + \overline{\delta}_8^{(2)} + \overline{\delta}_8^{(3)}, \overline{\delta}_8^{(1)} + \delta_8^{(2)} + \overline{\delta}_8^{(3)}, \overline{\delta}_8^{(1)} + \overline{\delta}_8^{(2)} + \delta_8^{(3)}\rangle$ |
| $D_9A_{15}$ | $\mathbb{Z}/8\mathbb{Z} = \langle\delta_9 + 2\alpha_{15}\rangle$ |
| $E_6^4$ | $(\mathbb{Z}/3\mathbb{Z})^2 = \langle\eta_6^{(1)} + \eta_6^{(2)} + \eta_6^{(3)}, -\eta_6^{(1)} + \eta_6^{(3)} + \eta_6^{(4)}\rangle$ |
| $A_{11}E_6D_7$ | $\mathbb{Z}/12\mathbb{Z} = \langle\alpha_{11} + \eta_6 + \delta_7\rangle$ |
| $D_6^4$ | $(\mathbb{Z}/2\mathbb{Z})^4 = \langle\delta_6^{(2)} + \overline{\delta}_6^{(3)} + \tilde{\delta}_6^{(4)}, \overline{\delta_6^{(1)}} + \tilde{\delta}_6^{(2)} + \delta_6^{(4)},$<br>$\delta_6^{(1)} + \overline{\delta}_6^{(2)} + \tilde{\delta}_6^{(4)}, \delta_6^{(1)} + \tilde{\delta}_6^{(3)} + \overline{\delta}_6^{(4)}\rangle$ |
| $D_6A_9^2$ | $\mathbb{Z}/2\mathbb{Z} \times \mathbb{Z}/10\mathbb{Z} = \langle\widetilde{\delta_6} + 5\alpha_9^{(2)}, \delta_6 + \alpha_9^{(1)} + 2\alpha_9^{(2)}\rangle$ |
| $D_5^2A_7^2$ | $\mathbb{Z}/4\mathbb{Z} \times \mathbb{Z}/8\mathbb{Z} = \langle\delta_5^{(1)} + \delta_5^{(2)} + 2\alpha_7^{(1)}, \delta_5^{(1)} + 2\delta_5^{(2)} + \alpha_7^{(1)} + \alpha_7^{(2)}\rangle$ |
| $A_8^3$ | $\mathbb{Z}/3\mathbb{Z} \times \mathbb{Z}/9\mathbb{Z} = \langle 3\alpha_8^{(1)} + 3\alpha_8^{(2)}, \alpha_8^{(1)} + 2\alpha_8^{(2)} + 2\alpha_8^{(3)}\rangle$ |
| $A_5^4D_4$ | $(\mathbb{Z}/6\mathbb{Z})^2 \times \mathbb{Z}/2\mathbb{Z} = \langle 5\alpha_5^{(1)} + 2\alpha_5^{(2)} + \alpha_5^{(3)} + \overline{\delta}_4,$<br>$5\alpha_5^{(1)} + 3\alpha_5^{(2)} + 2\alpha_5^{(3)} + 4\alpha_5^{(4)} + \delta_4,$<br>$3\alpha_5^{(1)} + 3\alpha_5^{(4)} + \tilde{\delta}_4\rangle$ |
| $A_6^4$ | $(\mathbb{Z}/7\mathbb{Z})^2 = \langle\alpha_6^{(1)} + 2\alpha_6^{(2)} + \alpha_6^{(3)} + 6\alpha_6^{(4)}, \alpha_6^{(1)} + 6\alpha_6^{(2)} + 2\alpha_6^{(3)} + \alpha_6^{(4)}\rangle$ |
| $D_4^6$ | $(\mathbb{Z}/2\mathbb{Z})^6 = \langle\delta_4^{(1)} + \delta_4^{(i)},\ i = 1, \ldots, 5,\ \sum_{i=1}^{6}\delta_4^{(i)}\rangle$ |
| $A_{24}$ | $\mathbb{Z}/5\mathbb{Z} = \langle 5\alpha_{24}\rangle$ |
| $A_{12}^2$ | $\mathbb{Z}/13\mathbb{Z} = \langle 2\alpha_{13}^{(1)} + 3\alpha_{13}^{(2)}\rangle$ |
| $A_4^6$ | $(\mathbb{Z}/5\mathbb{Z})^3 = \langle\alpha_4^{(1)} + \alpha_4^{(2)} + \alpha_4^{(3)} + 4\alpha_4^{(4)} + 4\alpha_4^{(5)},$<br>$\alpha_4^{(1)} + \alpha_4^{(2)} + 4\alpha_4^{(3)} + \alpha_4^{(5)} + 4\alpha_4^{(6)},$<br>$\alpha_4^{(1)} + 4\alpha_4^{(3)} + \alpha_4^{(4)} + 4\alpha_4^{(5)} + \alpha_4^{(6)}\rangle$ |
| $A_3^8$ | $(\mathbb{Z}/4\mathbb{Z})^4 = \langle 3\alpha_3^{(1)} + \sum_{i=2}^{8} c_i\alpha_3^{(i)}$ such that $(c_2, \ldots, c_8)$<br>is a cyclic permutation of (2001011)$\rangle$ |
| $A_2^{12}$ | $(\mathbb{Z}/3\mathbb{Z})^6 = \langle 2\alpha_2^{(1)} + \sum_{i=2}^{12} c_i\alpha_2^{(i)}$ such that $(c_2, \ldots, c_{12})$<br>is a cyclic permutation of (11211122212)$\rangle$ |
| $A_1^{24}$ | $(\mathbb{Z}/2\mathbb{Z})^{12} = \langle\alpha_1^{(1)} + \sum_{i=2}^{24} c_i\alpha_2^{(i)}$ such that $(c_2, \ldots, c_{24})$<br>is a cyclic permutation of (00000101001100110101111)$\rangle$ |
| — | $\Lambda_{24} \simeq L$ |

$l + L_{\text{root}}$ be a non-trivial element of $L/L_{\text{root}}$. If there exist $k \neq 0$ and $u \in L_{\text{root}}$ such that $k(l + u) \in N_{\text{root}}$, then $l + u \in W$ and the class of $l$ is a torsion element.

*Remark 4.5.* It is not always true that the lattice $T$ can be chosen to be a root lattice, and the method can be applied with some modifications without this assumption, see Braun et al. (2013). Since everything is easier under this assumption and in our case we can require that $T$ is a root lattice, we described the method with the assumption (2). In particular, if $T$ is not a root lattice, then one has to consider the primitive embeddings of $T$ in $L$, but one cannot use the results in Nishiyama (1996, Sections 4 and 5), so the points (3), (4), and (5) are significantly more complicated.

## 4.2 Explicit Computations

Here we apply the algorithm described in Section 4.1 to the K3 surface $X$.

### 4.2.1 Step 1

From Proposition 3.3, we find that the transcendental lattice of $X$ is

$$T_X = \begin{pmatrix} 6 & 0 \\ 0 & 2 \end{pmatrix}.$$

According to Nishiyama (1996), Schütt and Shioda (2010) and Bertin and Lecacheux (2013), $T_X(-1)$ admits a primitive embedding in $E_8$ and we can take $T$ as its orthogonal complement in $E_8$, that is

$$T = A_5 \oplus A_1.$$

### 4.2.2 Step 2

We observe that $T$ is a root lattice.

### 4.2.3 Step 3

We must find all the primitive embeddings $\phi : T \hookrightarrow L_{\text{root}}$ not Weyl isomorphic. This has been done by Nishiyama (1996) for the primitive embeddings of $A_k$ in $A_m$, $D_n$, $E_l$ and for the primitive embeddings of $A_5 \oplus A_1$ into $E_7$ and $E_8$. So we have to determine the primitive embeddings not isomorphic of $A_5 \oplus A_1$ in $A_m$ and $D_n$. This will be achieved using Corollary 4.7 and Lemma 4.10. First we recall some notions used in order to prove these results.

Let $B$ be a negative-definite even lattice, let $a \in B_{\text{root}}$ a root of $B$. The reflection $R_a$ is the isometry $R_a(x) = x + (a \cdot x)\, a$ and the Weyl group of $B$, $W(B)$, is the group generated by $R_a$ for $a \in B_{\text{root}}$.

**Proposition 4.6.** *Let $A$ be a sublattice of $B$. Suppose there exists a sequence of roots $x_1, x_2, \ldots, x_n$ of $A^{\perp_B}$ with $x_i \cdot x_{i+1} = \varepsilon_i$ and $\varepsilon_i^2 = 1$ then the two lattices $A \oplus x_1$ and $A \oplus x_n$ are isometric by an element of the Weyl group of $B$.*

*Proof.* First we prove the statement for $n = 2$. Since the two sublattices $A \oplus \langle x_1 \rangle$ and $A \oplus \langle -x_1 \rangle$ are isometric by $R_{x_1}$ we can suppose that $x_1 \cdot x_2 = 1$ (i.e. $\varepsilon_1 = 1$).

Then $x_1 + x_2$ is also a root and is in $A^{\perp_B}$. So the reflection $R_{x_1+x_2}$ is equal to $I_d$ on $A$. Let $g := R_{x_1} \circ R_{x_1+x_2}$ then $g \in W(B)$ is equal to $I_d$ on $A$. Moreover $g(x_2) = R_{x_1}(x_2 + ((x_1 + x_2) \cdot x_2)(x_1 + x_2)) = x_1$, and so $g$ fits. The case $n > 2$ follows by induction. □

**Corollary 4.7.** *Suppose $n \geq 9$, $p \geq 6$, up to an element of the Weyl group $W(D_n)$ or $W(A_p)$ there is a unique primitive embedding of $A_5 \oplus A_1$ in $D_n$ or $A_p$.*

*Proof.* From Nishiyama (1996) up to an element of the Weyl group there exists one primitive embedding of $A_5$ in $D_n$ or $A_p$. Fix this embedding. If $M$ is the orthogonal of this embedding then $M_{\text{root}}$ is $D_{n-6}$ or $A_{p-6}$. So for two primitive embeddings of $A_1$ in $M_{\text{root}}$ we can apply the previous proposition. □

We study now the primitive embeddings of $A_5 \oplus A_1$ in $D_8$ (which are not considered in the previous corollary, since the orthogonal complement of the unique primitive embedding of $A_5$ in $D_8$ is $\langle -6 \rangle \oplus \langle -2 \rangle^2$).

We denote by $\{\varepsilon_i, 1 \leq i \leq n\}$ the canonical basis of $\mathbb{R}^n$.

We can identify $D_n(-1)$ with $\mathbb{D}_n$, the set of vectors of $\mathbb{Z}^n$ whose coordinates have an even sum.

First we recall the two following propositions, see, for example, Martinet (2002).

**Proposition 4.8.** *The group $Aut(\mathbb{Z}^n)$ is isomorphic to the semi-direct product $\{\pm 1\}^n \rtimes S_n$, where the group $S_n$ acts on $\{\pm 1\}^n$ by permuting the n components.*

**Proposition 4.9.** *If $n \neq 4$, the restriction to $\mathbb{D}_n$ of the automorphisms of $\mathbb{Z}^n$ induces an isomorphism of $Aut(\mathbb{Z}^n)$ onto the group $Aut(\mathbb{D}_n)$. The Weyl group $W(\mathbb{D}_n)$ of index two in $Aut(\mathbb{D}_n)$ corresponds to those elements which induce an even number of changes of signs of the $\varepsilon_i$.*

**Lemma 4.10.** *There are two embeddings of $A_5 \oplus A_1$ in $D_8$ non-isomorphic up to $W(D_8)$.*

*Proof.* Let $d_8 = \varepsilon_1 + \varepsilon_2$ and $d_{8-i+1} = -\varepsilon_{i-1} + \varepsilon_i$ with $2 \leq i \leq 8$ a basis of $\mathbb{D}_8$. We consider the embedding

$$A_5 \hookrightarrow \langle d_7, d_6, d_5, d_4, d_3 \rangle.$$

By Nishiyama's results (Nishiyama 1996), this embedding is unique up to an element of $W(D_8)$ and we have $(A_5)^{\perp_{D_8}} = \langle \sum_{i=1}^{6} \varepsilon_i \rangle \oplus \langle x_7 \rangle \oplus \langle d_1 \rangle$ with $x_7 = \varepsilon_7 + \varepsilon_8$. We see that $\pm x_7$ and $\pm d_1$ are the only roots of $(A_5)^{\perp_{D_8}}$.

We consider the two embeddings

$$A_5 \oplus A_1 \hookrightarrow \langle d_7, d_6, d_5, d_4, d_3 \rangle \oplus \langle x_7 \rangle$$
$$A_5 \oplus A_1 \hookrightarrow \langle d_7, d_6, d_5, d_4, d_3 \rangle \oplus \langle d_1 \rangle.$$

Suppose there exists an element $R'$ of $W(D_8)$ such that $R'(x_7) = d_1$ and $R'(A_5) = A_5$, we shall show that $R'(d_1) = \pm x_7$. If $z := R'(d_1)$, then as $R'$ is an isometry $z \cdot d_1 = R'(d_1) \cdot R'(x_7) = d_1 \cdot x_7 = 0$. Moreover, $z \in (A_5)^{\perp_{D_8}}$ and so $z = \pm x_7$.

Since $R'(A_5) = A_5$, we see that $R'|_{A_5}$ is an element of $O(A_5)$, the group of isometries of $A_5$. We know that $O(A_5)/W(A_5) \sim \mathbb{Z}/2\mathbb{Z}$, generated by the class of $\mu : d_7 \leftrightarrow d_3, d_6 \leftrightarrow d_4, d_5 \leftrightarrow d_5$.

Thus, we have $R'|_{A_5} = \rho \in W(A_5)$ or $R'|_{A_5} = \rho\mu$ with $\rho \in W(A_5)$. We can also consider $\rho$ as an element of the group generated by reflections $R_u$ of $D_8$ with $u \in A_5$. So, for $v$ in $(A_5)^{\perp_{D_8}}$ we have $R_u(v) = v$ if $u \in A_5$ and then $\rho(v) = v$.

Let $R = \rho^{-1}R'$ then $R = R'$ on $(A_5)^{\perp_{D_8}}$. Since $R'(d_1) = \pm x_7$ and $R'(x_7) = d_1$ we have $R'|_{\langle \varepsilon_7, \varepsilon_8 \rangle} = (\varepsilon_7 \to \varepsilon_8, \varepsilon_8 \to -\varepsilon_7)$ or $(\varepsilon_7 \to \varepsilon_7, \varepsilon_8 \to -\varepsilon_8)$. Also we have $R|_{\langle \varepsilon_1, \varepsilon_2, \ldots, \varepsilon_6 \rangle} = I_d$ or $\varepsilon_i \leftrightarrow \varepsilon_{7-i}$.

In the second case $R$ corresponds to a permutation of $\varepsilon_i$ with only one sign minus; thus, $R$ is not an element of $W(D_8)$.

### 4.2.4 Step 4

For each primitive embedding of $A_5 \oplus A_1$ in $L_{\text{root}}$, the computations of $N$ and $N_{\text{root}}$ are obtained in almost all the cases by Nishiyama (1996, Section 5). In the few cases not considered by Nishiyama, one can make the computation directly. The results are collected in Table 3, where we use the following notation. The vectors $x_3$, $x_7$, $z_6$ in $D_n$ are defined by

$$x_3 := d_{n-3} + 2d_{n-2} + d_{n-1} + d_n,$$

$$x_7 := d_{n-7} + 2(d_{n-6} + d_{n-5} + d_{n-4} + d_{n-3} + d_{n-2}) + d_{n-1} + d_n,$$

$$x'_7 := 2(d_{n-6} + d_{n-5} + d_{n-4} + d_{n-3} + d_{n-2}) + d_{n-1} + d_n,$$

$$z_6 := d_{n-5} + 2d_{n-4} + 3d_{n-3} + 4d_{n-2} + 3d_{n-1} + 2d_n,$$

$$\widetilde{z_6} := d_{n-5} + 2d_{n-4} + 3d_{n-3} + 4d_{n-2} + 2d_{n-1} + 3d_n,$$

and the vectors $x$, $y$ in $E_p$ are

$$x := e_1 + e_2 + 2e_3 + 2e_4 + e_5, \; y := e_1 + 2e_2 + 2e_3 + 3e_4 + 2e_5 + e_6.$$

### 4.2.5 Step 5 (An Example: Fibrations 22 and 22(b))

In order to compute $W$ we recall that $W$ is an overlattice of finite index of $N$; in fact, it contains the non-trivial elements of $L/L_{\text{root}}$ which are orthogonal to $\phi(A_5 \oplus A_1)$. Moreover, the index of the inclusion $N \hookrightarrow W$ depends on the discriminant of $N$. Indeed $|d(W)| = |d(NS(X))| = 12$, so the index of the inclusion $N \hookrightarrow W$ is $\sqrt{|d(N)|/12}$.

As example we compute here the lattices $W$ for the two different embeddings of $A_5 \oplus A_1$ in $D_8$ (i.e., for the fibrations 22 and 22(b)). Thus, we consider the Niemeier

**Table 3** The orthogonal complement of the primitive embeddings $A_5 \oplus A_1$ in $L_{\text{root}}$

| No. | Primitive Embedding | Orthogonal Complement |
|---|---|---|
| 1 | $\langle e_1^{(1)}, e_3^{(1)}, \dots, e_6^{(1)} \rangle \oplus \langle e_8^{(1)} \rangle$ | $\langle 4e_1^{(1)} + 6e_2^{(1)} + 8e_3^{(1)} + 12e_4^{(1)} + 10e_5^{(1)} + 8e_6^{(1)} + 6e_7^{(1)} + 3e_8^{(1)} \rangle \oplus \langle y^{(1)} \rangle \oplus \langle e_1^{(2)}, \dots, e_8^{(2)} \rangle \oplus \langle e_1^{(3)}, \dots, e_8^{(3)} \rangle$ |
| 2 | $\langle e_1^{(1)}, e_3^{(1)}, \dots, e_6^{(1)} \rangle \oplus \langle e_1^{(2)} \rangle$ | $\langle e_8^{(1)}, 2e_1^{(1)} + 3e_2^{(1)} + 4e_3^{(1)} + 6e_4^{(1)} + 5e_5^{(1)} + 4e_6^{(1)} + 3e_7^{(1)} + 2e_8^{(1)} \rangle \oplus \langle y^{(1)} \rangle \oplus \langle x^{(2)}, e_2^{(2)}, e_4^{(2)}, \dots, e_8^{(2)} \rangle \oplus \langle e_1^{(3)}, \dots, e_8^{(3)} \rangle$ |
| 3 | $\langle e_1, e_3, \dots, e_6 \rangle \oplus \langle e_8 \rangle$ | $\langle y \rangle \oplus \langle 4e_1 + 6e_2 + 8e_3 + 12e_4 + 10e_5 + 8e_6 + 6e_7 + 3e_8 \rangle \oplus \langle d_1, \dots, d_{16} \rangle$ |
| 4 | $\langle d_{16}, d_{14}, \dots, d_{11} \rangle \oplus \langle d_1 \rangle$ | $\langle z_6 \rangle \oplus \langle x_7, d_9, \dots, d_3, 2d_2 + d_1 \rangle \oplus \langle e_1, \dots, e_8 \rangle$ |
| 5 | $\langle e_1, e_3, \dots, e_6 \rangle \oplus \langle d_{16} \rangle$ | $\langle y \rangle \oplus \langle e_8, 2e_1 + 3e_2 + 4e_3 + 6e_4 + 5e_5 + 4e_6 + 3e_7 + 2e_8 \rangle \oplus \langle d_{15} \rangle \oplus \langle x_3, d_{13}, \dots, d_1 \rangle$ |
| 6 | $\langle d_{16}, d_{14}, \dots, d_{11} \rangle \oplus \langle e_1 \rangle$ | $\langle z_6 \rangle \oplus \langle x_7, d_9, \dots, d_1 \rangle \oplus \langle x, e_2, e_4, \dots, e_8 \rangle$ |
| 7 | $\langle e_2^{(1)}, e_4^{(1)}, \dots, e_7^{(1)} \rangle \oplus \langle e_1^{(1)} \rangle$ | $\langle 3e_1^{(1)} + 4e_2^{(1)} + 6e_3^{(1)} + 8e_4^{(1)} + 6e_5^{(1)} + 4e_6^{(1)} + 2e_7^{(1)} \rangle \oplus \langle d_1, \dots, d_{10} \rangle \oplus \langle e_1^{(2)}, \dots, e_7^{(2)} \rangle$ |
| 8 | $\langle d_{10}, d_8, \dots, d_5 \rangle \oplus \langle d_1 \rangle$ | $\langle z_6 \rangle \oplus \langle x_7 \rangle \oplus \langle d_3 \rangle \oplus \langle d_3 + x_7 + 2d_2 + d_1 \rangle \oplus \langle e_1^{(1)}, \dots, e_7^{(1)} \rangle \oplus \langle e_1^{(2)}, \dots, e_7^{(2)} \rangle$ |
| 9 | $\langle e_1^{(1)}, e_3^{(1)}, \dots, e_6^{(1)} \rangle \oplus \langle e_1^{(2)} \rangle$ | $\langle 2e_1^{(1)} + 3e_2^{(1)} + 4e_3^{(1)} + 6e_4^{(1)} + 5e_5^{(1)} + 4e_6^{(1)} + 3e_7^{(1)} \rangle \oplus \langle y^{(1)} \rangle \oplus \langle x^{(2)}, e_2^{(2)}, e_4^{(2)}, \dots, e_7^{(2)} \rangle \oplus \langle d_1, \dots, d_{10} \rangle$ |
| 10 | $\langle e_2^{(1)}, e_4^{(1)}, \dots, e_7^{(1)} \rangle \oplus \langle e_1^{(2)} \rangle$ | $\langle e_1^{(1)}, 2e_1^{(1)} + 2e_2^{(1)} + 3e_3^{(1)} + 4e_4^{(1)} + 3e_5^{(1)} + 2e_6^{(1)} + e_7^{(1)} \rangle \oplus \langle x^{(2)}, e_2^{(2)}, e_4^{(2)}, \dots, e_7^{(2)} \rangle \oplus \langle d_1, \dots, d_{10} \rangle$ |
| 11 | $\langle e_1^{(1)}, e_3^{(1)}, \dots, e_6^{(1)} \rangle \oplus \langle d_{10} \rangle$ | $\langle 2e_1^{(1)} + 3e_2^{(1)} + 4e_3^{(1)} + 6e_4^{(1)} + 5e_5^{(1)} + 4e_6^{(1)} + 3e_7^{(1)} \rangle \oplus \langle d_9 \rangle \oplus \langle e_1^{(2)}, \dots, e_7^{(2)} \rangle \oplus \langle x_3, d_7, \dots, d_1 \rangle \oplus \langle y^{(1)} \rangle$ |
| 12 | $\langle e_2^{(1)}, e_4^{(1)}, \dots, e_7^{(1)} \rangle \oplus \langle d_{10} \rangle$ | $\langle e_1^{(1)}, 2e_1^{(1)} + 2e_2^{(1)} + 3e_3^{(1)} + 4e_4^{(1)} + 3e_5^{(1)} + 2e_6^{(1)} + e_7^{(1)} \rangle \oplus \langle e_1^{(2)}, \dots, e_7^{(2)} \rangle \oplus \langle d_9 \rangle \oplus \langle x_3, d_7, \dots, d_1 \rangle$ |
| 13 | $\langle d_{10}, d_8, \dots, d_5 \rangle \oplus \langle e_1^{(1)} \rangle$ | $\langle x_7, d_3, d_2, d_1 \rangle \oplus \langle x^{(1)}, e_2^{(1)}, e_4^{(1)}, \dots, e_7^{(1)} \rangle \oplus \langle e_1^{(2)}, \dots, e_7^{(2)} \rangle \oplus \langle z_6 \rangle$ |

(continued)

**Table 3** (continued)

| No. | Primitive Embedding | Orthogonal Complement |
|---|---|---|
| 14 | $\langle e_2, e_4, \ldots, e_6 \rangle \oplus \langle e_1 \rangle$ | $\left\langle \begin{matrix} 2e_1 + 3e_2 + 4e_3 + \\ 6e_4 + 5e_5 + 4e_6 + 3e_7 \end{matrix} \right\rangle \oplus \langle a_1, \ldots, a_{17} \rangle$ |
| 15 | $\langle a_1, \ldots, a_5 \rangle \oplus \langle a_7 \rangle$ | $\langle e_1, \ldots, e_7 \rangle \oplus \left\langle \begin{matrix} a_9, \ldots, a_{17}, \\ \sum_{j=1}^{6} ja_j - 6a_8, \\ a_7 + 2a_8 \end{matrix} \right\rangle$ |
| 16 | $\langle e_1, e_3, \ldots, e_6 \rangle \oplus \langle a_1 \rangle$ | $\langle y \rangle \oplus \left\langle \begin{matrix} 2e_1 + 3e_2 + 4e_3 + \\ 6e_4 + 5e_5 + 4e_6 + 3e_7 \end{matrix} \right\rangle \oplus$ $\langle a_1 + 2a_2, a_3, \ldots, a_{17} \rangle$ |
| 17 | $\langle e_2, e_4, \ldots, e_7 \rangle \oplus \langle a_1 \rangle$ | $\left\langle \begin{matrix} 2e_1 + 2e_2 + 3e_3 + \\ 4e_4 + 3e_5 + 2e_6 + e_7 \end{matrix}, e_1 \right\rangle \oplus$ $\langle a_1 + 2a_2, a_3, \ldots, a_{17} \rangle$ |
| 18 | $\langle a_1, \ldots, a_5 \rangle \oplus \langle e_1 \rangle$ | $\langle x, e_2, e_4, \ldots, e_7 \rangle \oplus \left\langle \begin{matrix} \sum_{j=1}^{6} ja_j, \\ a_7, \ldots a_{17} \end{matrix} \right\rangle$ |
| 19 | $\langle d_{24}, d_{22}, \ldots, d_{19} \rangle \oplus \langle d_1 \rangle$ | $\langle z_6 \rangle \oplus \langle x_7 \rangle \oplus \langle x_7 + d_1 + 2d_2, d_3, \ldots d_{17} \rangle$ |
| 20 | $\left\langle d_{12}^{(1)}, d_{10}^{(1)}, \ldots, d_7^{(1)} \right\rangle \oplus \left\langle d_1^{(1)} \right\rangle$ | $\left\langle \begin{matrix} d_1^{(1)} + 2d_2^{(1)} + d_3^{(1)} + x_7^{(1)}, \\ d_3^{(1)}, d_4^{(1)}, d_5^{(1)} \end{matrix} \right\rangle \oplus$ $\left\langle d_1^{(2)}, \ldots, d_{12}^{(2)} \right\rangle \oplus \left\langle z_6^{(1)} \right\rangle \oplus \left\langle x_7^{(1)} \right\rangle$ |
| 21 | $\left\langle d_{12}^{(1)}, d_{10}^{(1)}, \ldots, d_7^{(1)} \right\rangle \oplus \left\langle d_{12}^{(2)} \right\rangle$ | $\left\langle z_6^{(1)} \right\rangle \oplus \left\langle x_7^{(1)}, d_5^{(1)}, \ldots, d_1^{(1)} \right\rangle \oplus$ $\left\langle d_{11}^{(2)} \right\rangle \oplus \left\langle x_3^{(2)}, d_9^{(2)}, \ldots, d_1^{(2)} \right\rangle$ |
| 22 | $\left\langle d_7^{(1)}, d_6^{(1)}, \ldots, d_3^{(1)} \right\rangle \oplus \left\langle d_1^{(1)} \right\rangle$ | $\langle \widetilde{z_6}^{(1)} \rangle \oplus \left\langle x_7^{(1)} \right\rangle \oplus$ $\left\langle d_1^{(2)}, \ldots, d_8^{(2)} \right\rangle \oplus \left\langle d_1^{(3)}, \ldots, d_8^{(3)} \right\rangle$ |
| 22 $(b)$ | $\left\langle d_7^{(1)}, d_6^{(1)}, \ldots, d_3^{(1)} \right\rangle \oplus \left\langle x_7^{(1)} \right\rangle$ | $\langle \widetilde{z_6}^{(1)} \rangle \oplus \left\langle d_1^{(1)} \right\rangle \oplus$ $\left\langle d_1^{(2)}, \ldots, d_8^{(2)} \right\rangle \oplus \left\langle d_1^{(3)}, \ldots, d_8^{(3)} \right\rangle$ |
| 23 | $\left\langle d_8^{(1)}, d_6^{(1)}, \ldots, d_3^{(1)} \right\rangle \oplus \left\langle d_8^{(2)} \right\rangle$ | $\left\langle z_6^{(1)} \right\rangle \oplus \left\langle x_7^{(1)} \right\rangle \oplus \left\langle d_1^{(1)} \right\rangle \oplus \left\langle d_7^{(2)} \right\rangle \oplus$ $\left\langle x_3^{(2)}, d_5^{(2)}, \ldots, d_1^{(2)} \right\rangle \oplus \left\langle d_1^{(3)}, \ldots, d_8^{(3)} \right\rangle$ |
| 24 | $\langle d_9, d_7, \ldots, d_4 \rangle \oplus \langle d_1 \rangle$ | $\langle z_6 \rangle \oplus \langle x_7, d_1 + 2d_2 \rangle \oplus \langle a_1, \ldots, a_{15} \rangle$ |
| 25 | $\langle a_1, \ldots, a_5 \rangle \oplus \langle a_7 \rangle$ | $\langle d_1, \ldots, d_9 \rangle \oplus \left\langle \begin{matrix} \sum_{j=1}^{6} ja_j - 6a_8, \\ a_7 + 2a_8, \\ a_9, \ldots, a_{15} \end{matrix} \right\rangle$ |
| 26 | $\langle d_9, d_7, \ldots, d_4 \rangle \oplus \langle a_1 \rangle$ | $\langle z_6 \rangle \oplus \langle d_1, d_2, x_7 \rangle \oplus \left\langle \begin{matrix} a_1 + 2a_2, \\ a_3, \ldots, a_{15} \end{matrix} \right\rangle$ |
| 27 | $\langle a_1, \ldots, a_5 \rangle \oplus \langle d_9 \rangle$ | $\langle d_8 \rangle \oplus \langle x_3, d_6, \ldots, d_1 \rangle \oplus \left\langle \begin{matrix} \sum_{j=1}^{6} ja_j, \\ a_7, \ldots, a_{15} \end{matrix} \right\rangle$ |
| 28 | $\left\langle e_1^{(1)}, e_3^{(1)}, \ldots, e_6^{(1)} \right\rangle \oplus \left\langle e_2^{(2)} \right\rangle$ | $\left\langle \begin{matrix} e_2^{(2)} + e_3^{(2)} + 2e_4^{(2)} + e_5^{(2)}, \\ e_1^{(2)}, e_3^{(2)}, e_5^{(2)}, e_6^{(2)} \end{matrix} \right\rangle \oplus$ $\left\langle e_1^{(3)}, \ldots, e_6^{(3)} \right\rangle \oplus \left\langle e_1^{(4)}, \ldots, e_6^{(4)} \right\rangle \oplus \langle y^{(1)} \rangle$ |

(continued)

**Table 3** (continued)

| No. | Primitive Embedding | Orthogonal Complement |
|---|---|---|
| 29 | $\langle a_1, \dots, a_5\rangle \oplus \langle a_7\rangle$ | $\left\langle \begin{array}{c} \sum_{j=1}^6 ja_j - 6a_8, \\ a_7 + 2a_8,\, a_9,\, a_{10},\, a_{11} \end{array}\right\rangle \oplus$ $\langle d_1, \dots, d_7\rangle \oplus \langle e_1, \dots, e_6\rangle$ |
| 30 | $\langle e_1, e_3, \dots, e_6\rangle \oplus \langle d_7\rangle$ | $\langle y\rangle \oplus \langle d_6\rangle \oplus \langle x_3, d_4, \dots, d_1\rangle \oplus \langle a_1, \dots, a_{11}\rangle$ |
| 31 | $\langle e_1, e_3 \dots, e_6\rangle \oplus \langle a_1\rangle$ | $\langle y\rangle \oplus \langle d_1, \dots, d_7\rangle \oplus \langle a_1 + 2a_2, a_3, \dots, a_{11}\rangle$ |
| 32 | $\langle d_7, d_5, \dots, d_2\rangle \oplus \langle e_1\rangle$ | $\langle z_6\rangle \oplus \langle x_7'\rangle \oplus \langle x, e_2, e_4, e_5, e_6\rangle$ $\oplus \langle a_1, \dots, a_{11}\rangle$ |
| 33 | $\langle d_7, d_5, \dots, d_2\rangle \oplus \langle a_1\rangle$ | $\langle z_6\rangle \oplus \langle x_7'\rangle \oplus \langle e_1, \dots, e_6\rangle \oplus$ $\left\langle \begin{array}{c} a_1 + 2a_2, \\ a_3, \dots, a_{11} \end{array}\right\rangle$ |
| 34 | $\langle a_1, \dots, a_5\rangle \oplus \langle d_7\rangle$ | $\langle d_6\rangle \oplus \langle x_3, d_4, \dots, d_1\rangle$ $\oplus \langle e_1, \dots, e_6\rangle \oplus \left\langle \begin{array}{c} \sum_{j=1}^6 ja_j, \\ a_7, \dots, a_{11} \end{array}\right\rangle$ |
| 35 | $\langle a_1, \dots, a_5\rangle \oplus \langle e_1\rangle$ | $\langle x, e_2, e_4, e_5, e_6\rangle \oplus \langle d_1, \dots, d_7\rangle \oplus$ $\left\langle \sum_{j=1}^6 ja_j, a_7, \dots, a_{11}\right\rangle$ |
| 36 | $\left\langle d_6^{(1)}, d_4^{(1)}, \dots, d_1^{(1)}\right\rangle$ $\oplus \left\langle d_6^{(2)}\right\rangle$ | $\left\langle z_6^{(1)}\right\rangle \oplus \left\langle d_5^{(2)}\right\rangle \oplus \left\langle d_1^{(3)}, \dots, d_6^{(3)}\right\rangle \oplus$ $\left\langle x_3^{(2)}, d_3^{(2)}, d_2^{(2)}, d_1^{(2)}\right\rangle \oplus \left\langle d_1^{(4)}, \dots, d_6^{(4)}\right\rangle$ |
| 37 | $\left\langle a_1^{(1)}, \dots, a_5^{(1)}\right\rangle \oplus \left\langle a_7^{(1)}\right\rangle$ | $\left\langle \begin{array}{c} \sum_{j=1}^6 ja_j^{(1)} - 6a_8^{(1)}, \\ a_7^{(1)} + 2a_8^{(1)},\, a_9^{(1)} \end{array}\right\rangle \oplus$ $\left\langle a_1^{(2)}, \dots, a_9^{(2)}\right\rangle \oplus \langle d_1, \dots, d_6\rangle$ |
| 38 | $\left\langle a_1^{(1)}, \dots, a_5^{(1)}\right\rangle \oplus \left\langle a_1^{(2)}\right\rangle$ | $\langle d_1, \dots, d_6\rangle \oplus \left\langle \begin{array}{c} \sum_{j=1}^6 ja_j^{(1)}, \\ a_7^{(1)},\, a_8^{(1)},\, a_9^{(1)} \end{array}\right\rangle \oplus$ $\left\langle a_1^{(2)} + 2a_2^{(2)}, a_3^{(2)}, \dots, a_9^{(2)}\right\rangle$ |
| 39 | $\left\langle a_1^{(1)}, \dots, a_5^{(1)}\right\rangle \oplus \langle d_6\rangle$ | $\left\langle \sum_{j=1}^6 ja_j^{(1)}, a_7^{(1)}, a_8^{(1)}, a_9^{(1)}\right\rangle \oplus \langle d_5\rangle$ $\oplus \langle x_3, d_3, d_2, d_1\rangle \oplus \left\langle a_1^{(2)}, \dots, a_9^{(2)}\right\rangle$ |
| 40 | $\langle d_5, \dots, d_1\rangle \oplus \left\langle a_1^{(1)}\right\rangle$ | $\langle z_6\rangle \oplus \left\langle \begin{array}{c} a_1^{(1)} + 2a_2^{(1)}, \\ a_3^{(1)}, \dots, a_9^{(1)} \end{array}\right\rangle \oplus \left\langle a_1^{(2)}, \dots, a_9^{(2)}\right\rangle$ |
| 41 | $\left\langle a_1^{(1)}, \dots, a_5^{(1)}\right\rangle \oplus \left\langle a_7^{(1)}\right\rangle$ | $\left\langle \sum_{j=1}^6 ja_j^{(1)} + 3a_7^{(1)}\right\rangle \oplus \left\langle a_1^{(2)}, \dots, a_7^{(2)}\right\rangle$ $\oplus \left\langle d_1^{(1)}, \dots, d_7^{(1)}\right\rangle \oplus \left\langle d_1^{(2)}, \dots, d_7^{(2)}\right\rangle$ |
| 42 | $\left\langle a_1^{(1)}, \dots, a_5^{(1)}\right\rangle \oplus \left\langle a_1^{(2)}\right\rangle$ | $\left\langle \sum_{j=1}^6 ja_j^{(1)}, a_7^{(1)}\right\rangle \oplus \left\langle \begin{array}{c} a_1^{(2)} + 2a_2^{(2)}, \\ a_3^{(2)}, \dots, a_7^{(2)} \end{array}\right\rangle$ $\oplus \left\langle d_1^{(1)}, \dots, d_5^{(1)}\right\rangle \oplus \left\langle d_1^{(2)}, \dots, d_5^{(2)}\right\rangle$ |
| 43 | $\left\langle a_1^{(1)}, \dots, a_5^{(1)}\right\rangle \oplus \left\langle d_5^{(1)}\right\rangle$ | $\left\langle \sum_{j=1}^6 ja_j^{(1)}, a_7^{(1)}\right\rangle \oplus \left\langle d_4^{(1)}\right\rangle \oplus$ $\left\langle x_3^{(1)}, d_2^{(1)}, d_1^{(1)}\right\rangle \oplus \left\langle a_1^{(2)}, \dots, a_7^{(2)}\right\rangle$ |
| 44 | $\left\langle a_1^{(1)}, \dots, a_5^{(1)}\right\rangle \oplus \left\langle a_7^{(1)}\right\rangle$ | $\left\langle \sum_{j=1}^6 ja_j^{(1)}, a_7^{(1)} + 2a_8^{(1)}\right\rangle \oplus$ $\left\langle a_1^{(2)}, \dots, a_8^{(2)}\right\rangle \oplus \left\langle a_1^{(3)}, \dots, a_8^{(3)}\right\rangle$ |

(continued)

**Table 3** (continued)

| No. | Primitive Embedding | Orthogonal Complement |
|---|---|---|
| 45 | $\langle a_1^{(1)},\dots,a_5^{(1)}\rangle \oplus \langle a_1^{(2)}\rangle$ | $\left\langle \begin{array}{c}\sum_{j=1}^{6} ja_j^{(1)},\\ a_7^{(1)},\, a_8^{(1)}\end{array}\right\rangle \oplus \left\langle \begin{array}{c} a_1^{(2)}+2a_2^{(2)},\\ a_3^{(2)},\dots,a_8^{(2)}\end{array}\right\rangle \oplus \langle a_1^{(3)},\dots,a_8^{(3)}\rangle$ |
| 46 | $\langle a_1,\dots,a_5\rangle \oplus \langle a_7\rangle$ | $\langle \sum_{j=1}^{6} ja_j - 6a_8, a_7+2a_8, a_9,\dots a_{24}\rangle$ |
| 47 | $\langle a_1^{(1)},\dots,a_5^{(1)}\rangle \oplus \langle a_7^{(1)}\rangle$ | $\left\langle \begin{array}{c}\sum_{j=1}^{6} ja_j^{(1)}-6a_8^{(1)},\\ a_7^{(1)}+2a_8^{(1)},\\ a_9^{(1)},\dots,a_{12}^{(1)}\end{array}\right\rangle \oplus \langle a_1^{(2)},\dots,a_{12}^{(2)}\rangle$ |
| 48 | $\langle a_1^{(1)},\dots,a_5^{(1)}\rangle \oplus \langle a_1^{(2)}\rangle$ | $\left\langle \begin{array}{c}\sum_{j=1}^{6} ja_j^{(1)},\\ a_7^{(1)},\dots,a_{12}^{(1)}\end{array}\right\rangle \oplus \left\langle \begin{array}{c} a_1^{(2)}+2a_2^{(2)},\\ a_3^{(2)},\dots,a_{12}^{(2)}\end{array}\right\rangle$ |
| 49 | $\langle a_1^{(1)},\dots,a_5^{(1)}\rangle \oplus \langle a_1^{(2)}\rangle$ | $\left\langle \begin{array}{c} a_1^{(2)}+2a_2^{(2)},\\ a_3^{(2)},\, a_4^{(2)},\, a_5^{(2)}\end{array}\right\rangle \oplus \langle a_1^{(3)},\dots,a_5^{(3)}\rangle \oplus \langle a_1^{(4)},\dots,a_5^{(4)}\rangle \oplus \langle d_1,\dots,d_4\rangle$ |
| 50 | $\langle a_1^{(1)},\dots,a_5^{(1)}\rangle \oplus \langle d_4\rangle$ | $\langle d_3\rangle \oplus \langle x_3\rangle \oplus \langle d_1\rangle \oplus \langle a_1^{(2)},\dots,a_5^{(2)}\rangle \oplus \langle a_1^{(3)},\dots,a_5^{(3)}\rangle \oplus \langle a_1^{(4)},\dots,a_5^{(4)}\rangle$ |
| 51 | $\langle a_1^{(1)},\dots,a_5^{(1)}\rangle \oplus \langle a_1^{(2)}\rangle$ | $\langle \sum_{j=1}^{6} ja_j^{(1)}\rangle \oplus \left\langle \begin{array}{c} a_1^{(2)}+2a_2^{(2)},\\ a_3^{(2)},\dots,a_6^{(2)}\end{array}\right\rangle \oplus \langle a_1^{(3)},\dots,a_6^{(3)}\rangle \oplus \langle a_1^{(4)},\dots,a_6^{(4)}\rangle$ |

lattice $L$ such that $L_{root} \simeq D_8^3$ and we denote the generators of $L/L_{root}$ as follows: $v_1 := \delta_8^{(1)} + \overline{\delta_8}^{(2)} + \overline{\delta_8}^{(3)}$, $v_2 := \overline{\delta_8}^{(1)} + \delta_8^{(2)} + \overline{\delta_8}^{(3)}$, $v_3 := \overline{\delta_8}^{(1)} + \overline{\delta_8}^{(2)} + \delta_8^{(3)}$.

**Fibration #22:** We consider the embedding $\varphi_1 : A_5 \oplus A_1 \hookrightarrow L$ such that $\varphi_1(A_5 \oplus A_1) = \langle d_7^{(1)}, d_6^{(1)}, d_5^{(1)}, d_4^{(1)}, d_3^{(1)}\rangle \oplus \langle d_1^{(1)}\rangle$. The generators of the lattice $N$ are described in Table 3 and one can directly check that $N \simeq \langle -6\rangle \oplus A_1 \oplus D_8 \oplus D_8$. So, $|d(N)| = 6\cdot 2^5$ and the index of the inclusion $N \hookrightarrow W$ is $2^2 = \sqrt{6\cdot 2^5/12}$. This implies that there is a copy of $(\mathbb{Z}/2\mathbb{Z})^2 \subset (\mathbb{Z}/2\mathbb{Z})^3$ which is also contained in $W$ and so in particular is orthogonal to $\varphi_1(A_5 \oplus A_1)$.

We observe that $v_1$ is orthogonal to the embedded copy of $A_5 \oplus A_1$, $v_2$ and $v_3$ are not. Moreover $v_2 - v_3$ is orthogonal to the embedded copy of $A_5 \oplus A_1$. Hence $v_1$ and $v_2 - v_3$ generates $W/N \simeq (\mathbb{Z}/2\mathbb{Z})^2$. We just observe that $v_2 - v_3 \in W$ is equivalent mod $W_{root}$ to the vector $w_2 := \widetilde{\delta_8^{(2)}} + \widetilde{\delta_8^{(3)}} \in W$, so $W/N \simeq (\mathbb{Z}/2\mathbb{Z})^2 \simeq \langle v_1, w_2\rangle$. We will reconsider this fibration in Section 4.3 comparing it with the fibration #22*b*.

**Fibration #22(*b*):** We consider the other embedding of $A_5 \oplus A_1$ in $L_{root}$, i.e. $\varphi_2 : A_5 \oplus A_1 \hookrightarrow L$ such that $\varphi_2(A_5 \oplus A_1) = \langle d_7^{(1)}, d_6^{(1)}, d_5^{(1)}, d_4^{(1)}, d_3^{(1)}\rangle \oplus \langle x_7^{(1)}\rangle$.

The generators of the lattice $N$ is described in Table 3 and one can directly check that $N \simeq \langle -6\rangle \oplus A_1 \oplus D_8 \oplus D_8$. As above this implies that $W/N \simeq (\mathbb{Z}/2\mathbb{Z})^2$ which is generated by elements in $L/L_{root}$ which are orthogonal to $\varphi_2(A_5 \oplus A_1)$. In particular, $v_1 - v_2$ and $v_2 - v_3$ are orthogonal to $\varphi_2(A_5 \oplus A_1)$ so $v_1 - v_2 \in W$ and $v_2 - v_3 \in W$. Moreover, $v_2 - v_3 = \widetilde{\delta_8^{(2)}} + \widetilde{\delta_8^{(3)}} \mod W_{root}$. So, denoted by $w_2 := \widetilde{\delta_8^{(2)}} + \widetilde{\delta_8^{(3)}}$, we have that $W/N \simeq \langle v_1 - v_2, w_2 \rangle$. We will reconsider this fibration in Section 4.3 comparing it with the fibration #22.

#### 4.2.6 Step 6

We recalled in Section 2.1 that each elliptic fibration is associated with a certain decomposition of the Néron–Severi group as a direct sum of $U$ and a lattice, called $W$. In step 5 we computed all the admissible lattices $W$, so we classify the elliptic fibrations on $X$. We denote all the elliptic fibrations according to their associated embeddings; this gives the first five columns of the Table 1.

#### 4.2.7 Step 7

Moreover, again in Section 2.1, we recalled that each reducible fiber of an elliptic fibration is uniquely associated with a Dynkin diagram and that a Dynkin diagram is associated with at most two reducible fibers of the fibration. This completes step 7.

#### 4.2.8 Step 8

In order to compute the rank of the Mordell–Weil group it suffices to perform the suggested computation, so $r = 18 - \text{rank}(N_{\text{root}})$. This gives the sixth column of Table 1.

For example, in cases 22 and 22(b), the lattice $N_{\text{root}}$ coincides and has rank 17, thus $r = 1$ in both the cases.

#### 4.2.9 Step 9

In order to compute the torsion part of the Mordell–Weil group one has to identify the vectors $v \in W/N$ such that $kv \in N_{\text{root}}$ for a certain nontrivial integer number $k \in \mathbb{Z}$; this gives the last column of Table 1. We will demonstrate this procedure in some examples below (on fibrations #22 and #22$(b)$), but first we remark that in several cases it is possible to use an alternative method either in order to completely determine $MW(\mathcal{E})_{\text{tors}}$ or at least to bound it. We already presented the theoretical aspect of these techniques in Section 2.3.

Probably the easiest case is the one where $r = 0$. In this case $MW(\mathcal{E}) = MW(\mathcal{E})_{tors}$. Since $r = 0$, this implies that $\text{rank}(N_{\text{root}}) = 18 = \text{rank}N$, so $N = N_{\text{root}}$. Hence $W/N = W/N_{\text{root}}$, thus every element $w \in W/N$ is such that a multiple is contained in $N_{\text{root}}$, i.e. every element of $W/N$ contributes to the torsion. Thus, $MW(\mathcal{E}) = W/N = W/N_{\text{root}}$. This immediately allows to compute the torsion for the 7 extremal fibrations #$2, 5, 10, 12, 28, 30, 50$.

**Fibration** #50 $N_{\text{root}} \simeq A_1^{\oplus 3} \oplus A_5^{\oplus 3}$ $(r = 0)$ The lattice $N = N_{\text{root}}$ is $A_1^{\oplus 3} \oplus A_5^{\oplus 3}$, then $|d(N)| = 2^3 6^3$ and $|W/N| = 2 \times 6$. Moreover $W/N \subset L/L_{\text{root}} \simeq (\mathbb{Z}/6\mathbb{Z})^2 \times \mathbb{Z}/2\mathbb{Z}$. This immediately implies that $W/N = \mathbb{Z}/6\mathbb{Z} \times \mathbb{Z}/2\mathbb{Z}$.
**Fibration** #1 $N_{\text{root}} \simeq A_1 \oplus E_8^{\oplus 2}$ **(fibers of special type, Proposition 2.9)** The presence of the lattice $E_8$ as summand of $N_{\text{root}}$ implies that the fibration has a fiber of type $II^*$ (two in this specific case). Hence $\text{MW}(\mathcal{E})_{\text{tors}}$ is trivial.
**Fibration** #29 $N_{\text{root}} \simeq A_3 \oplus D_7 \oplus E_6$ **(fibers of special type, Proposition 2.9)** By Proposition 2.9 if a fibration has a fiber of type $IV^*$, then the Mordell–Weil group is a subgroup of $\mathbb{Z}/3\mathbb{Z}$. On the other hand, a fiber of type $D_7$, i.e., $I_3^*$ can only occur in fibrations with 4 or 2-torsion or trivial torsion group. Therefore $\text{MW}(\mathcal{E})_{\text{tors}}$ is trivial.
**Fibration** #25 $N_{\text{root}} \simeq A_7 \oplus D_9$ **(the height formula, Section 2.3.2)** Suppose there is a non-trivial torsion section $P$. Then, taking into account the possible contributions of the reducible fibers to the height pairing, there is $0 \leq i \leq 7$ such that one of the following holds:

$$4 = \frac{i(8-i)}{8} + 1 \text{ or } 4 = \frac{i(8-i)}{8} + 1 + 5/4.$$

After a simple calculation, one sees that neither of the above can happen and therefore the torsion group $\text{MW}(\mathcal{E})_{\text{tors}}$ is trivial.
**Fibration** #22 $N_{\text{root}} \simeq A_1 \oplus D_8 \oplus D_8$ We already computed the generators of $W/N$ in Section 4.2.5, $W/N \simeq (\mathbb{Z}/2\mathbb{Z})^2 \simeq \langle v_1, w_2 \rangle$. A basis of $N_{\text{root}}$ is $\langle x_7^{(1)} \rangle \oplus \langle d_i^{(j)} \rangle_{i=1,\dots 8, j=2,3}$. So

$$\begin{aligned} 2v_1 &= d_1^{(1)} + 2d_2^{(1)} + 3d_3^{(1)} + 4d_4^{(1)} + 5d_5^{(1)} + 6d_6^{(1)} + 2d_7^{(1)} + 3d_8^{(1)} \\ &\quad + \sum_{i=2}^{3} \left( d_7^{(i)} + d_8^{(i)} + 2 \left( \sum_{j=1}^{7} d_j^{(i)} \right) \right) \end{aligned}$$

and $2v_1 \notin N_{\text{root}}$ since $d_1^{(1)} + 2d_2^{(1)} + 3d_3^{(1)} + 4d_4^{(1)} + 5d_5^{(1)} + 6d_6^{(1)} + 2d_7^{(1)} + 3d_8^{(1)}$ is not a multiple of $x_7^{(1)}$. Vice-versa

$$2w_2 \subset D_8^{(2)} \oplus D_8^{(3)} \in N_{\text{root}}.$$

Thus $\text{MW}(\mathcal{E}) = \mathbb{Z} \times \mathbb{Z}/2\mathbb{Z}$.
**Fibration** #22$(b)$ $N_{\text{root}} \simeq A_1 \oplus D_8 \oplus D_8$ Similarly, we consider the generators of $W/N \simeq (\mathbb{Z}/2\mathbb{Z})^2 \simeq \langle v_1 - v_2, w_2 \rangle$ computed in Section 4.2.5. A basis of $N_{\text{root}}$ is $\langle d_1^{(1)} \rangle \oplus \langle d_i^{(j)} \rangle_{i=1,\dots 8, j=2,3}$. So

$$2w_1 \notin N_{\text{root}} \text{ and } 2w_2 \in N_{\text{root}}.$$

Thus also in this case $\text{MW}(\mathcal{E}) = \mathbb{Z} \times \mathbb{Z}/2\mathbb{Z}$.

### 4.3 Again on Fibrations #22 and #22$(b)$

As we can check in Table 1 and we proved in the previous sections, the fibrations #22 and #22$(b)$ are associated with the same lattice $N$ and to the same Mordell–Weil group. However, we proved in Lemma 4.10 that they are associated to different (up to Weyl group) embeddings in the Niemeier lattices, so they correspond to fibrations which are not identified by the $\mathcal{J}_2$-fibration and in particular they cannot have the same frame. The following question is now natural: what is the difference between these two fibrations? The answer is that the section of infinite order, which generates the free part of the Mordell–Weil group of these two fibrations, has different intersection properties, as we show now in two different ways and contexts.

**Fibration #22:** We use the notation of Section 4.2.5. Moreover we fix the following notation: $\Theta_1^1 := x_7^{(1)}$ and $\Theta_i^{(j)} := d_i^{(j)}$, $i = 1, \ldots 8, j = 2, 3$ are, respectively, the non trivial components of the fibers of type $I_2$, $I_4^*$, and $I_4^*$, respectively.

The class $P := 2F + O - v_1$ is the class of a section of infinite order of the fibration, generating the free part of $\mathrm{MW}(\mathcal{E})$ and the class $Q := 2F + O - w_2$ is the class of the 2-torsion section of the fibration. The section $P$ meets the components $\Theta_1^1, \Theta_1^2, \Theta_1^3$ and $Q$ meets the components $\Theta_0^1, \Theta_7^2, \Theta_7^3$. We observe that $h(P) = 3/2$ and $h(Q) = 0$ which agree with Schütt and Shioda (2010, Formula 22) and the fact that $Q$ is a torsion section, respectively. We also give an explicit equation of this fibration and of its sections, see (10).

**Fibration #22$(b)$:** We use the notation of Section 4.2.5. Moreover we fix the following notation: $\Theta_1^1 := d_1^{(1)}$ and $\Theta_i^{(j)} := d_i^{(j)}$, $i = 1, \ldots 8$, $j = 2, 3$ are respectively the non-trivial components of the fibers of type $I_2$, $I_4^*$, and $I_4^*$, respectively. The class $Q := 2F + O - w_2$ is the class of the 2-torsion section of the fibration. Observe that $Q$ meets the components $\Theta_0^1, \Theta_7^2, \Theta_7^3$. The class

$$P = 2F + O + v_1 - v_2 - \Theta_1^2 - \Theta_2^2 - \Theta_3^2 - \Theta_4^2 - \Theta_5^2 - \Theta_6^2 - \Theta_7^2$$

is the class of a section of infinite order, which intersects the following components of the reducible fibers: $\Theta_1^1, \Theta_7^2, \Theta_0^3$. This agrees with the height formula. We also give an explicit equation of this fibration and of its sections, see (11).

*Remark 4.11.* The generators of the free part of the Mordell–Weil group is clearly defined up to the sum by a torsion section. The section $P \oplus Q$ intersects the reducible fibers in the following components $\Theta_1^1, \Theta_0^2, \Theta_7^3$ (this follows by the group law on the fibers of type $I_2$ (or $III$) and $I_4^*$).

*Remark 4.12.* Comparing the sections of infinite order of the fibration 22 and the one of the fibration 22(b), one immediately checks that their intersection properties are not the same, so the frames of the elliptic fibration 22 and of elliptic fibration 22(b) are not the same and hence these two elliptic fibrations are in fact different under the $\mathcal{J}_2$-classification.

We observe that both the fibrations #22 and #22$(b)$ specialize the same fibration, which is given in Comparin and Garbagnati (2014, Section 8.1, Table $r = 19$, case 11)). Indeed the torsion part of the Mordell–Weil group, which is already present in the more general fibrations analyzed in Comparin and Garbagnati (2014), are the same and the difference between the fibration 22 and the fibration 22(b) is in the free part of the Mordell Weil group, so the difference between these two fibrations involves exactly the classes that correspond to our specialization.

Here we also give an equation for each of the two different fibrations #22 and #22(b). Both these equations are obtained from the equation of the elliptic fibration #8 (9). So first we deduce an equation for #8: Let $c := \frac{v}{(w-1)^2}$. Substituting $v$ by $c\,(w-1)^2$ in (7), we obtain the equation of an elliptic curve depending on $c$, which corresponds to the fibration #8 and with the following Weierstrass equation

$$E_c : \beta^2 = \alpha\left(\alpha^2 + 6c^2\alpha - c^3\,(c-4)\,(4c-1)\right). \tag{9}$$

**Fibration #22** Putting $n' = \frac{\alpha}{c^2(4c-1)}$, $\beta = \frac{yc^2(4c-1)}{4n'^3}$, $\quad c = \frac{x}{4n'^3}$, in (9) we obtain

$$E_{n'} : y^2 = x\left(x^2 - n'\left(n'^2 - 6n^{(\prime)}\right) + 1\right)x + 16n'^4\right) \tag{10}$$

with singular fibers of type $2I_4^*(n = 0, \infty) + I_2 + 2I_1$. We notice the point $P = \left(((n'-1)^2 n', -2n'^2(n'^2-1))\right)$ of height $\frac{3}{2}$, therefore $P$ and $Q = (0,0)$ generate the Mordell–Weil group of $E_{n'}$. To study the singular fiber at $n' = \infty$ we do the transformation $N' = \frac{1}{n}, y = \frac{\beta_1}{N'^6}, x = \frac{\alpha_1}{N'^4}$ and $P = (\alpha_1, \beta_1)$ with $\alpha_1 = (N'-1)^2 N'$ and $\beta_1 = -2N'^2\left(N'^2 - 1\right)$. We deduce that the section $P$ intersects the component of singular fibers at 0 and $\infty$ with the same subscript, so this fibration corresponds to fibration #22.

**Fibration #22(b)** Putting $n = \frac{2\alpha}{c(4c-1)}$, $\beta = \frac{yc(4c-1)}{4n}$, $c = \frac{-x}{2n}$ in (9) we obtain

$$E_n : y^2 = x\left(x^2 + 2n\left(n^2 + 3n + 4\right) + n^4\right) \tag{11}$$

with singular fibers of type $2I_4^*(n = 0, \infty) + I_2 + 2I_1$. We notice the point $P = \left(4, 2(n+2)^2\right)$ of height $\frac{3}{2}$, therefore $P$ and $Q = (0,0)$ generate the Mordell–Weil group of $E_n$. Since $P$ does not meet the node of the Weierstrass model at $n = 0$, the section $P$ intersects the component $\Theta_0$ of the singular fiber for $n = 0$, so this fibration corresponds to #22(b).

*Remark 4.13.* Let us denote by $\mathcal{E}_9$ and $\mathcal{E}_{21}$ the elliptic fibrations #9 and #21, respectively. They satisfy $Tr(\mathcal{E}_9) \simeq Tr(\mathcal{E}_{21})$ and $\mathrm{MW}(\mathcal{E}_9) \simeq \mathrm{MW}(\mathcal{E}_{21})$, but $\mathcal{E}_9$ is not $\mathcal{J}_2$-equivalent to $\mathcal{E}_{21}$ since, as above, the infinite order sections of these two fibrations have different intersection properties with the singular fibers. Indeed these two fibrations correspond to different fibrations (Comparin and Garbagnati 2014, Case 10a) and case 10b), Section 8.1, Table $r = 19$) on the more general family of K3 surfaces considered in Comparin and Garbagnati (2014).

**Acknowledgements** We thank the organizers and all those who supported our project for their efficiency, their tenacity and expertise. The authors of the paper have enjoyed the hospitality of CIRM at Luminy, which helped to initiate a very fruitful collaboration, gathering from all over the world junior and senior women, bringing their skill, experience, and knowledge from geometry and number theory. Our gratitude goes also to the referee for pertinent remarks and helpful comments.

A.G is supported by FIRB 2012 "Moduli Spaces and Their Applications" and by PRIN 2010–2011 "Geometria delle varietà algebriche." C.S is supported by FAPERJ (grant E26/112.422/2012). U.W. thanks the NSF-AWM Travel Grant Program for supporting her visit to CIRM.

## References

Atkin, A.O., Morain, F.: Finding suitable curves for the elliptic curve method of factorization. Math. Comput. **60**, 399–405 (1993)

Beauville, A.: Les familles stables de courbes elliptiques sur $\mathbb{P}^1$ admettant quatre fibres singulières. C. R. Acad. Sci. Paris Sér. I Math. **294**, 657–660 (1982)

Bertin, M.J.: Mesure de Mahler et série $L$ d'une surface K3 singulière. Actes de la Conférence: Fonctions $L$ et Arithmétique. Publ. Math. Besan. Actes de la conférence Algèbre Théorie Nbr., Lab. Math. Besançon, pp. 5–28 (2010)

Bertin, M.J., Lecacheux, O.: Elliptic fibrations on the modular surface associated to $\Gamma_1(8)$. In: Arithmetic and Geometry of K3 Surfaces and Calabi-Yau Threefolds. Fields Institute Communications, vol. 67, pp. 153–199. Springer, New York (2013)

Braun, A.P., Kimura, Y., Watari, T.: On the classification of elliptic fibrations modulo isomorphism on K3 surfaces with large Picard number, Math. AG; High Energy Physics (2013) [arXiv:1312.4421]

Cassels, J.W.S.: Lectures on Elliptic Curves. London Mathematical Society Student Texts, vol. 24. Cambridge University Press, Cambridge (1991)

Comparin, P., Garbagnati, A.: Van Geemen-Sarti involutions and elliptic fibrations on K3 surfaces double cover of $\mathbb{P}^2$. J. Math. Soc. Jpn. **66**, 479–522 (2014)

Couveignes, J.-M., Edixhoven, S.: Computational Aspects of Modular Forms and Galois Representations, Annals of Math Studies 176. Princeton University Press, Princeton

Cox, D., Katz, S.: Mirror Symmetry and Algebraic Geometry. American Mathematical Society, Providence (1999)

Elkies, N.D.: http://www.math.harvard.edu/~elkies/K3_20SI.html#-7[1 (2010)

Elkies, N.D.: Three Lectures on Elliptic Surfaces and Curves of High Rank. Lecture notes, Oberwolfach (2007)

Elkies, N., Schütt, M.: Genus 1 fibrations on the supersingular K3 surface in characteristic 2 with Artin invariant 1. Asian J. Math. (2014) [arXiv:1207.1239]

Garbagnati, A., Sarti, A.: Elliptic fibrations and symplectic automorphisms on K3 surfaces. Commun. Algebra **37**, 3601–3631 (2009)

Karp, D., Lewis, J., Moore, D., Skjorshammer, D., Whitcher, U.: On a family of K3 surfaces with $\mathcal{S}_4$ symmetry. In: Arithmetic and Geometry of K3 Surfaces and Calabi-Yau Threefolds. Fields Institute Communications. Springer, New York (2013)

Kloosterman, R.: Classification of all Jacobian elliptic fibrations on certain K3 surfaces. J. Math. Soc. Jpn. **58**, 665–680 (2006)

Kondo, S.: Algebraic K3 surfaces with finite automorphism group. Nagoya Math. J. **116**, 1–15 (1989)

Kondo, S.: Automorphisms of algebraic K3 surfaces which act trivially on Picard groups. J. Math. Soc. Jpn. **44**, 75–98 (1992)

Kubert, D.S.: Universal bounds on the torsion of elliptic curves. Proc. Lond. Math. Soc. **33**(3), 193–237 (1976)

Kumar, A.: Elliptic fibrations on a generic Jacobian Kummer surface. J. Algebraic Geom. [arXiv:1105.1715] **23**, 599–667 (2014)

Kreuzer, M., Skarke, H.: Classification of reflexive polyhedra in three dimensions. Adv. Theor. Math. Phys. **2**, 853–871 (1998)

Kreuzer, M., Skarke, H.: Complete classification of reflexive polyhedra in four dimensions. Adv. Theor. Math. Phys. **4**, 1209–1230 (2000)

Martinet, J.: Perfect Lattices in Euclidean Spaces, vol. 327. Springer, Berlin/Heidelberg (2002)

Montgomery, P.L.: Speeding the Pollard and Elliptic Curve Methods of Factorization. Math. Comput. **48**. (1987) 243–264.

Nikulin, V.V.: Finite groups of automorphisms of Kählerian K3 surfaces. (Russian) Trudy Moskov. Mat. Obshch. **38**, 75–137 (1979). English translation: Trans. Moscow Math. Soc. **38**, 71–135 (1980)

Nikulin, V.V.: Integral symmetric bilinear forms and some of their applications. Math. USSR Izv. **14**, 103–167 (1980)

Nishiyama, K.: The Jacobian fibrations on some K3 surfaces and their Mordell–Weil groups. Jpn. J. Math. (N.S.) **22**, 293–347 (1996)

Oguiso, K.: On Jacobian fibrations on the Kummer surfaces of the product of nonisogenous elliptic curves. J. Math. Soc. Jpn. **41**, 651–680 (1989)

Schütt, M.: K3 surface with Picard rank 20 over $\mathbb{Q}$. Algebra Number Theory **4**, 335–356 (2010)

Schütt, M., Shioda, T.: Elliptic surfaces. In: Algebraic Geometry in East Asia–Seoul 2008. Advanced Studies in Pure Mathematics, vol. 60, pp. 51–160. The Mathematical Society of Japan, Tokyo (2010)

Shimada, I., Zhang, D.-Q.: Classification of extremal elliptic K3 surfaces and fundamental groups of open K3 surfaces. Nagoya Math. J. **161**, 23–54 (2001)

Shioda, T.: On the Mordell–Weil lattices. Commun. Math. Univ. St. Pauli **39**, 211–240 (1990)

Shioda, T.: On elliptic modular surfaces J. Math. Soc. Jpn. **24**(1), 20–59 (1972)

Stein, W.A., et al.: Sage Mathematics Software (Version 6.1). The Sage Development Team. http://www.sagemath.org (2014)

Sterk, H.: Finiteness results for algebraic K3 surfaces. Math. Z. **180**, 507–513 (1985)

Verrill, H.: Root lattices and pencils of varieties. J. Math. Kyoto Univ. **36**, 423–446 (1996)

# Shalika Germs for $\mathfrak{sl}_n$ and $\mathfrak{sp}_{2n}$ Are Motivic

**Sharon M. Frechette, Julia Gordon, and Lance Robson**

**Abstract** We prove that Shalika germs on the Lie algebras $\mathfrak{sl}_n$ and $\mathfrak{sp}_{2n}$ belong to the class of so-called motivic functions defined by means of a first-order language of logic. It is a well-known theorem of Harish-Chandra that for a Lie algebra $\mathfrak{g}(F)$ over a local field $F$ of characteristic zero, the Shalika germs, normalized by the square root of the absolute value of the discriminant, are bounded on the set of regular semisimple elements $\mathfrak{g}^{\mathrm{rss}}$, however, it is not easy to see how this bound depends on the field $F$. As a consequence of the fact that Shalika germs are motivic functions for $\mathfrak{sl}_n$ and $\mathfrak{sp}_{2n}$, we prove that for these Lie algebras, this bound must be of the form $q^a$, where $q$ is the cardinality of the residue field of $F$, and $a$ is a constant. Our proof that Shalika germs are motivic in these cases relies on the interplay of DeBacker's parametrization of nilpotent orbits with the parametrization using partitions, and the explicit matching between these parametrizations due to Nevins (Algebra Representation Theory **14**, 161–190, 2011). We include two detailed examples of the matching of these parametrizations.

## 1 Introduction

In this paper we prove that Shalika germs for the Lie algebras of type $\mathfrak{sl}_n$ and $\mathfrak{sp}_{2n}$ belong to the class of so-called motivic functions, and explore some of the consequences of this fact.

Shalika germs first appeared in the papers of Shalika (1972) and Harish-Chandra (1973). The survey of their role in harmonic analysis on $p$-adic groups is beyond the scope of this paper; we refer the reader to the beautiful article by Kottwitz (2005), and to Harish-Chandra 1999 for the detailed definitions and main results regarding them. We note that Shalika germs, by definition, are functions on the set of regular semisimple elements in a Lie algebra, yet, except for those defined on a few Lie

S.M. Frechette
Dept. of Mathematics & Computer Science, College of the Holy Cross, 1 College Street, Worcester, MA 01610

J. Gordon (✉) • L. Robson
Mathematics Department, The University of British Columbia, Vancouver, BC, Canada
e-mail: gor@math.ubc.ca

 
M.J. Bertin et al. (eds.), *Women in Numbers Europe*, Association for Women in Mathematics Series 2, DOI 10.1007/978-3-319-17987-2_3

algebras of small rank, their exact values elude computation. Here we use a general theorem about uniform bounds for motivic functions proved in Shin and Templier 2015, Appendix B to estimate the absolute values of the Shalika germs in a uniform way over all local fields of a fixed (sufficiently large) residue characteristic.

First, let us recall the definitions. Let $F$ be a local field, $\mathbf{G}$ be a connected reductive algebraic group over $F$, and $\mathfrak{g}$ its Lie algebra. In our results, $\mathbf{G} = \mathbf{SL}_n$ or $\mathbf{Sp}_{2n}$, although several of the background results hold in greater generality. Let $X \in \mathfrak{g}(F)$, with adjoint orbit $\mathcal{O}_X = \{\mathrm{Ad}(g)X \mid g \in \mathbf{G}(F)\}$ and stabilizer $C_G(X)$. (Since here we are dealing with the classical Lie algebras, the Adjoint action is just conjugation: $\mathrm{Ad}(g)X = gXg^{-1}$.) The space $\mathcal{O}_X$ with the $p$-adic topology is homeomorphic to $\mathbf{G}(F)/C_G(X)$, which carries a $G$-invariant quotient measure. For the fields $F$ of characteristic zero, it was proved by Ranga Rao (1972) that when transported to the orbit of $X$, this measure is a Radon measure on $\mathfrak{g}(F)$, i.e., it is finite on compact subsets of $\mathfrak{g}(F)$. (Strictly speaking, it is the group version of this statement that is proved in Ranga Rao 1972, but in characteristic zero this is equivalent to the Lie algebra version.) Denote this quotient measure on $\mathbf{G}(F)/C_G(X)$ by $d^*g$. The *orbital integral* at $X$ is the distribution $\mu_X$ on $C_c^\infty(\mathfrak{g}(F))$ defined by

$$\mu_X(f) = \int_{\mathbf{G}(F)/C_G(X)} f(\mathrm{Ad}(g)X)d^*g. \tag{1}$$

For the fields of sufficiently large positive characteristic (with an explicit bound on the characteristic), convergence of orbital integrals was proved by McNinch (2004).

There are finitely many nilpotent orbits in $\mathfrak{g}(F)$, provided the field $F$ has characteristic zero or sufficiently large positive characteristic (depending on the root system of $\mathfrak{g}$). The Shalika germ expansion expresses the regular semisimple orbital integrals as linear combinations of nilpotent ones, in a neighbourhood of the origin. More precisely, let $\mathrm{Nil}(F)$ denote the finite set of nilpotent orbits in $\mathfrak{g}(F)$, let $\mathfrak{g}(F)^{\mathrm{rss}}$ denote the set of regular semisimple elements in $\mathfrak{g}(F)$, and for each $\mathcal{O} \in \mathrm{Nil}(F)$ let $\mu_{\mathcal{O}}$ be the orbital integral over $\mathcal{O}$ (it is a linear functional on $C_c^\infty(\mathfrak{g}(F))$). For every $f \in C_c^\infty(\mathfrak{g}(F))$ there exists a neighbourhood $U_f$ of zero in $\mathfrak{g}(F)$, and functions $\Gamma_{\mathcal{O}}(X)$ defined on $\mathfrak{g}(F)^{\mathrm{rss}} \cap U_f$, such that for all $X \in \mathfrak{g}(F)^{\mathrm{rss}} \cap U_f$, we have the expansion

$$\mu_X(f) = \sum_{\mathcal{O}\in\mathrm{Nil}(F)} \Gamma_{\mathcal{O}}(X)\mu_{\mathcal{O}}(f). \tag{2}$$

The functions $\Gamma_{\mathcal{O}}$ are called *provisional Shalika germs*, using the terminology of Kottwitz (2005, §6, §17). These provisional Shalika germs are well defined as germs of functions at the origin; moreover, they possess a natural homogeneity property, and using this they can be extended canonically to the entire set $\mathfrak{g}(F)^{\mathrm{rss}}$ (see Kottwitz 2005, § 17 for details).

The goal of this paper is to prove that provisional Shalika germs belong to the class of the so-called motivic functions. This was proved for $\mathfrak{g} = \mathfrak{sp}_{2n}$ by L. Robson in his M.Sc. essay (Robson 2012); we include this case here since it was not published elsewhere. We also study the case $\mathfrak{g} = \mathfrak{sl}_n$ which is in some ways simpler, but has a technical issue that does not arise in the $\mathfrak{sp}_{2n}$ case (namely, the dependence of the set of nilpotent orbits on the field $F$); this was the content of our WIN project. We present both cases in detail here in preparation for a general proof for all Lie algebras, which will appear elsewhere.

The class of motivic functions was defined by R. Cluckers and F. Loeser (Cluckers and Loeser 2008). Concretely, motivic functions are complex-valued functions on $p$-adic manifolds defined uniformly in $p$ by means of a first-order language of logic, called Denef-Pas language, which we will define below. We include a brief, simplified version of the definition of motivic functions. For the details, as well as a survey of the applications of this class of functions in harmonic analysis on $p$-adic groups, we refer the reader to the survey (Cluckers et al. 2011a) and the original papers (Cluckers and Loeser 2008; Cluckers et al. 2011b). The aim of this paper is to add Shalika germs to the list of functions arising in harmonic analysis that can be studied via motivic integration techniques, in the case of $\mathfrak{g} = \mathfrak{sl}_n$ or $\mathfrak{sp}_{2n}$.

Cluckers, Hales, and Loeser (2011b) prove that regular semisimple orbital integrals are motivic, and in Cluckers et al. (2014a) the same statement is proved for all, and in particular, nilpotent orbital integrals. Thus, once we have shown that the functions $\Gamma_{\mathcal{O}}$ are motivic, we see that both sides of (2) are motivic functions. As an immediate consequence of the Transfer Principle proved by Cluckers and Loeser (2008), this shows the Shalika germ expansion holds for fields of sufficiently large positive characteristic. (This was previously proved by DeBacker (2002a); our results give an alternative proof.) More importantly, the uniform boundedness result from Shin and Templier (2015, Appendix B) then implies the uniform bound on Shalika germs normalized by the square root of the discriminant (see Theorem 17 below).

Our main results are stated and proved in Section 6. The rest of the paper provides a review of all the prerequisites, thus experts may want to turn immediately to the last section. The proof that provisional Shalika germs are motivic functions has two main ingredients: first we must establish a way to describe nilpotent orbits in the motivic context, and second, we find definable test functions which allow us to isolate individual Shalika germs in the linear combination. The first step requires a parametrization of nilpotent orbits that is as field-independent as possible, and this is where partitions are advantageous. For the second step, it is convenient to use DeBacker's parametrization of orbits. The proof of the main theorem essentially works in much greater generality than stated here, except we do not quite have the structure in Denef-Pas language that would capture the set of nilpotent orbits in general. Here, for the special cases of $\mathfrak{sl}_n$ and $\mathfrak{sp}_{2n}$, we use the matching between the two parametrizations of nilpotent orbits that was established by Nevins (2011). Since all three of the authors found this material challenging to absorb, in Section 5 we include detailed examples for $\mathfrak{sl}_3$ and $\mathfrak{sp}_4$; we hope they will be useful for future students.

# 2 Motivic Functions

This section is included in order for the paper to be self-contained. However, this overview of the definitions has appeared in various forms in several papers on the topic; the present version is quoted nearly verbatim from Cluckers et al. (2011a), except for Section 2.3, which is new and specifically adapted for the purposes of this paper.

Informally, motivic functions are built from definable functions in the Denef-Pas language. Thus they are given independently of the field and can be interpreted in any non-Archimedean local field. We first recall the definition of the Denef-Pas language.

## *2.1 Denef-Pas Language*

Denef-Pas language is a first order language of logic designed for working with valued fields. Formulas in this language will allow us to uniformly handle sets and functions for all local fields. We start by defining two sublanguages of the language of Denef-Pas: the language of rings and Presburger language.

### 2.1.1 The Language of Rings

Apart from the symbols for variables $x_1, \ldots, x_n, \ldots$ and the usual logical symbols equality '$=$', parentheses '(', ')', the quantifiers '$\exists$', '$\forall$', and the logical operations conjunction '$\wedge$', negation '$\neg$', disjunction '$\vee$', the language of rings consists of the following symbols:

- constants '0', '1';
- binary functions '$\times$', '$+$'.

A (first-order) formula in the language of rings is any syntactically correct formula built out of these symbols. (One usually omits the words 'first order'.) If a formula in the language of rings has $n$ free variables, then it defines a subset of $R^n$ for any ring $R$. For example, the formula "$\exists x_2\, (x_2 \times x_1 = 1)$" defines the set of units $R^\times$ in any ring $R$. Note that by convention, quantifiers always run over the ring in question. Note also that quantifier-free formulas in the language of rings define constructible sets, as they appear in classical algebraic geometry.

### 2.1.2 Presburger Language

A formula in Presburger language is built out of variables running over $\mathbb{Z}$, the logical symbols (as above) and symbols '$+$', '$\leq$', '0', '1', and for each $d = 2, 3, 4, \ldots$, a symbol '$\equiv_d$' to denote the binary relation $x \equiv y \pmod{d}$. Note the absence of the symbol for multiplication.

### 2.1.3 Denef-Pas Language

The Denef-Pas language is a three-sorted language in the sense that its formulas utilize three different "sorts" of elements: those of the valued field, of the residue field, and of the value group (which will always be $\mathbb{Z}$ in our setting). Each variable in such a formula runs over only the elements of one of the sorts, so there are three disjoint sets of symbols for the variables of the different sorts. To create a syntactically correct formula, one must pay attention to the sorts when composing functions and inserting them into relations.

In addition to the variables and the logical symbols, the formulas use the following symbols:

- In the valued field sort: the language of rings.
- In the residue field sort: the language of rings.
- In the $\mathbb{Z}$-sort: the Presburger language.
- the symbol $\mathrm{ord}(\cdot)$ for the valuation map from the nonzero elements of the valued field sort to the $\mathbb{Z}$-sort, and the symbol $\overline{\mathrm{ac}}(\cdot)$ for the so-called angular component, which is a multiplicative function from the valued field sort to the residue field sort (more about this function below).

A formula in this language can be interpreted in any discretely valued field $F$ which comes with a uniformizing element $\varpi$, by letting the variables range over $F$, over its residue field $k_F$, and over $\mathbb{Z}$, respectively, depending on the sort to which they belong; ord is the valuation map (defined on $F^\times$ and such that $\mathrm{ord}(\varpi) = 1$), and $\overline{\mathrm{ac}}$ is defined as follows: if $x$ is a unit (that is, $\mathrm{ord}(x) = 0$), then $\overline{\mathrm{ac}}(x)$ is the residue of $x$ modulo $\varpi$ (thus, an element of the residue field); for all other nonzero $x$, one puts $\overline{\mathrm{ac}}(x) := \varpi^{-\mathrm{ord}(x)}x \bmod (\varpi)$. Thus, for $x \neq 0$, $\overline{\mathrm{ac}}(x)$ is the residue class of the first non-zero coefficient of the $\varpi$-adic expansion of $x$. Finally, we define $\overline{\mathrm{ac}}(0) = 0$.

Thus, a formula $\varphi$ in this language with $n$ free valued-field variables, $m$ free residue-field variables, and $r$ free $\mathbb{Z}$-variables gives naturally, for each discretely valued field $F$, a subset $\varphi(F)$ of $F^n \times k_F^m \times \mathbb{Z}^r$: namely, $\varphi(F)$ is the set of all the tuples for which the interpretation of $\varphi$ in $F$ is "true".

We will denote this language by $\mathcal{L}_{\mathrm{DP}}$.

## 2.2 *Definable Sets and Motivic Functions*

The $\mathcal{L}_{\mathrm{DP}}$-formulas introduced in the previous section allow us to obtain a field-independent notion of subsets of $F^n \times k_F^m \times \mathbb{Z}^r$ for all local fields $F$ of sufficiently large residue characteristic. The reason behind the restriction on characteristic is explained below in Remark 3.

**Definition 1.** A collection $X = (X_F)_F$ of subsets $X_F \subset F^n \times k_F^m \times \mathbb{Z}^r$ is called a *definable set* if there is an $\mathcal{L}_{\mathrm{DP}}$-formula $\varphi$ and an integer $M$ such that $X_F = \varphi(F)$ for each $F$ with residue characteristic at least $M$ (cf. Remark 3), where $\varphi(F)$ is as described at the end of Section 2.1.3.

By Definition 1, a definable set is actually a collection of sets indexed by non-Archimedean local fields $F$; such practice is not uncommon in model theory and has its analogues in classical algebraic geometry. A particularly simple definable set is $(F^n \times k_F^m \times \mathbb{Z}^r)_F$, for which we introduce the simplified notation $\mathrm{VF}^n \times \mathrm{RF}^m \times \mathbb{Z}^r$, where VF stands for valued field and RF for residue field. We apply the typical set-theoretical notation to definable sets $X, Y$, e.g., $X \subset Y$ (if $X_F \subset Y_F$ for each $F$), $X \times Y$, and so on.

**Definition 2.** For definable sets $X$ and $Y$, a collection $f = (f_F)_F$ of functions $f_F : X_F \to Y_F$ is called a *definable function* and denoted by $f : X \to Y$ if the collection of graphs of the $f_F$ is a definable set.

*Remark 3.* There is a subtle issue here, due to the fact that the same definable set can be defined by different formulas. Technically, it would be more elegant to think of a definable set as an equivalence class of what we have called definable sets in Definition 1, where we call two such definable sets equivalent if they are the same for all $F$ with sufficiently large residue characteristic. To ease notation, we will not emphasize this point, but because of this all results presented in this paper will only be valid *for fields with sufficiently large residue characteristic*. In particular, we assume hereafter that $\mathrm{char}(F) \neq 2$.

We now come to motivic functions, for which definable functions are the building blocks. We note that while definable functions, by definition, must be $\mathrm{VF}^n \times \mathrm{RF}^m \times \mathbb{Z}^r$-valued for some $m, n, r$, the *motivic* functions are built from definable sets and functions, and can be thought of as complex-valued functions (although here they will naturally be $\mathbb{Q}$-valued). This does not require thinking of rational or complex numbers in the context of logic; these are just usual complex-valued functions that happen to be built from definable ingredients as prescribed by the following definition.

**Definition 4.** Let $X = (X_F)_F$ be a definable set. A collection $f = (f_F)_F$ of functions $f_F : X_F \to \mathbb{C}$ is called *a motivic function* on $X$ if and only if there exist integers $N$, $N'$, and $N''$, such that, for all non-Archimedean local fields $F$,

$$f_F(x) = \sum_{i=1}^{N} q_F^{\alpha_{iF}(x)}(\#(Y_{iF})_x)\Big(\prod_{j=1}^{N'} a_{ijF}(x)\Big)\Big(\prod_{\ell=1}^{N''} \frac{1}{1-q_F^{a_{i\ell}}}\Big), \quad \text{for } x \in X_F, \tag{3}$$

for some

- nonzero integers $a_{i\ell}$,
- definable functions $\alpha_i : X \to \mathbb{Z}$ and $\beta_{ij} : X \to \mathbb{Z}$,
- definable sets $Y_i \subset X \times \mathrm{RF}^{r_i}$,

where, for $x \in X_F$, $(Y_{iF})_x$ is the finite set $\{y \in k_F^{r_i} \mid (x, y) \in Y_{iF}\}$, and $q_F$ is the cardinality of the residue field $k_F$.

We call a motivic function on a one-point set a *motivic constant.*

In Theorem 17, we will need to allow the square root of the cardinality of the residue field as a possible value of a motivic function. Hence, we will use the slightly generalized notion of a motivic function introduced in Cluckers et al. (2014a, §B.3.1). Namely, given an integer $r > 0$ and a definable $\mathbb{Z}$-valued function $f$, expressions of the form $q_F^{f/r}h$, where $h$ is a motivic function as above, will also be called motivic functions.

## 2.3 Adding Constants to the Language

We will need to extend Denef-Pas language by adding finitely many constant symbols in the valued field sort, whose role will be to encode units whose angular components form a set of representatives of $k_F^\times/(k_F^\times)^m$, where $m$ is a fixed integer. Such extensions were first used by T.C. Hales, and an extension very similar to the one we define here appears first in J. Diwadkar's thesis (Diwadkar 2006, § 2.2.3).

Specifically, let $m$ be a fixed integer. We add $m$ constant symbols $d_1, \dots, d_m$ to the valued field sort of Denef-Pas language, to obtain the language $\mathcal{L}_{\mathrm{DP}m}$.

Now we need to define their interpretation, given a local field $F$ with a uniformizer $\varpi$ and residue field $k$.

If the set $k^\times/(k^\times)^m$ has $m$ elements, then we want $d_1, \dots, d_m$ to be interpreted as units such that their angular components form a set of representatives of distinct $(k^\times)^m$-cosets. Specifically, we can write a formula

$$\exists y_1, \dots, y_m \in F^\times, \mathrm{ord}(y_i) = 0, \nexists z : y_i = y_j z^m \text{ if } i \neq j.$$

This formula is true for $F$ under our assumption. Then we can set the values of $d_1, \dots, d_m$ in $F$ to be any collection $\{y_1, \dots, y_m\}$ satisfying this formula.

If the cardinality of $k_F^\times/(k_F^\times)^m$ is equal to $\ell < m$, we write a similar formula stating that $\{y_1, \dots, y_\ell\}$ are distinct representatives of $(k_F^\times)^m$-cosets, with the convention that the trivial coset is always represented by the constant symbol 1. More precisely, for every divisor $\ell$ of $m$, let $\phi_{\ell,m}$ be the following formula, with the quantifiers ranging over the residue field sort:

$$\phi_{\ell,m}(y_1, \dots, y_\ell) := \text{`}\nexists z : \ y_i = y_j z^m \text{ for } i \neq j \wedge \forall x \exists z, x = y_i z^m \text{ for some } 1 \leq i \leq \ell.\text{'} \tag{4}$$

(This formula is written slightly informally; in reality, it contains a conjunction of $\ell(\ell-1)/2$ formulas, and a disjunction of $\ell$ formulas.) This formula states that $y_1, \dots, y_\ell$ are distinct representatives of $k_F^\times/(k_F^\times)^m$ in $k_F^\times$.

For a given finite field $k$ and fixed $m$, exactly one of the statements

$$\psi_{\ell,m} := \text{`}\exists y_1,\ldots,y_\ell,\ \phi_{\ell,m}(y_1,\ldots,y_\ell)\text{'}$$

holds, as $\ell$ runs over all divisors of $m$. If $\psi_{\ell,m}$ holds in $k_F$, we interpret the constant symbols $d_1,\ldots,d_\ell$ as units of the valued field such that $\phi_{\ell,m}(\overline{\mathrm{ac}}(d_1),\ldots,\overline{\mathrm{ac}}(d_\ell))$ holds. Set the rest of the $d_i$ equal to 1.

None of the constructions and theorems of motivic integration change if we add finitely many constant symbols. Hereafter, we fix an integer $n$ (coming from a fixed Lie algebra $\mathfrak{sl}_n$ or $\mathfrak{sp}_{2n}$), and say that a set or function is *definable* if it is definable in the language $\mathcal{L}_{\mathrm{DP}m}$ for some $m \le nP(n)$, where $P(n)$ is the number of partitions of $n$. We shall see later that we may need to consider the union of languages $\mathcal{L}_{\mathrm{DP}m}$ as $m$ varies over a set of integers associated with partitions of $n$; however, it does not matter how many constants we add, as long as it is a finite number that is fixed in advance.

In the same way that a non-Archimedean local field $F$ with a choice of the uniformizer is a structure for the language $\mathcal{L}_{\mathrm{DP}}$, we note that a structure for $\mathcal{L}_{\mathrm{DP}m}$ is a non-Archimedean local field $F$ with a choice of the uniformizer of the valuation, and a choice of a collection of units whose angular components form a set of representatives of $k_F^\times/(k_F^\times)^m$.

The theory of motivic integration works as usual in this setting; this is a very special case of the set-up of Cluckers and Loeser (2015), where the new constants can be thought of as part of the theory $\mathcal{T}$ (in the setting of that paper).

With this terminology, we can now state this paper's goal precisely: to show that Shalika germs are motivic functions, up to dividing by a motivic constant, in the sense that they are motivic functions where we use the language $\mathcal{L}_{\mathrm{DP}m}$ with some finite $m$. We note that not all motivic constants are invertible in the ring of motivic functions, which is why we require the "up to motivic constant" provision.

## 3 Classification of Nilpotent Orbits of $\mathfrak{sl}_n$ and $\mathfrak{sp}_{2n}$ Via Partitions

As discussed in the Introduction, we study two parameterizations of nilpotent orbits in $\mathfrak{sl}_n$ and in $\mathfrak{sp}_{2n}$, with a view toward defining these orbits by formulas in Denef-Pas language. In this section we recall a well-known parametrization involving partitions, and in Section 4 we recall a parametrization due to DeBacker (2002b), involving the Bruhat-Tits building for $\mathfrak{g}$. In fact, a proof of definability of the nilpotent orbits for $\mathfrak{sl}_n$ using DeBacker's parametrization is carried out explicitly in Diwadkar's thesis (Diwadkar 2006). Here we recast it in a slightly simpler form, taking advantage of the explicit matching between the two parametrizations, as proved by Nevins (2011), and also of recent developments in the theory of motivic integration that allow us to slightly simplify Diwadkar's terminology.

### 3.1 Notation

Hereafter, $F$ will stand for a non-Archimedean local field with char($F$) $\neq$ 2, and $\overline{F}$ for a separable closure of $F$. The ring of integers of $F$ will be denoted by $\mathfrak{O}$ (or $\mathfrak{O}_F$ if there is a possibility of confusion), the maximal ideal by $\mathfrak{P}$, and the residue field by $k_F$. We will always assume that $F$ comes with a choice of the uniformizer of the valuation $\varpi$, and when talking about the language $\mathcal{L}_{\mathrm{DP}m}$, with a choice of representatives for $\mathfrak{O}^\times/(\mathfrak{O}^\times)^m$ as discussed above in Section 2.3.

### 3.2 Parametrization of Nilpotent Orbits in $\mathfrak{sl}(n)$ Using Partitions

For a positive integer $n$, a *partition* $\lambda = (\lambda_1, \lambda_2, \ldots, \lambda_t)$ of $n$ is a weakly decreasing sequence of positive integers ( i.e. $\lambda_1 \geq \lambda_2 \geq \cdots \geq \lambda_t$) whose sum is $n$. We say the $\lambda_i$ are the *parts* of the partition $\lambda$, and the *length* of $\lambda$ is $t$. For each $1 \leq j \leq n$, the *multiplicity* $m_j(\lambda)$ is the number of parts of $\lambda$ satisfying $\lambda_i = j$. We denote the greatest common divisor of the parts $\lambda_i$ by gcd($\lambda$).

It is well known that when the characteristic of $F$ is greater than $n$, the set of nilpotent orbits of $\mathfrak{sl}_n(\overline{F})$ is in one-to-one correspondence with the set of partitions of $n$ (see Collingwood and McGovern 1993 or Waldspurger 2001, for instance). A nilpotent orbit corresponds to the partition whose parts are determined by the blocks in its Jordan normal form. Specifically, let $\lambda = (\lambda_1, \lambda_2, \ldots, \lambda_t)$ be a partition of $n$, and let $J_{\lambda_i}$ denote the $\lambda_i \times \lambda_i$-matrix whose $(j, j+1)$ entries are equal to 1 for $1 \leq j \leq \lambda_i$, with all remaining entries equal to 0. Let $J_\lambda$ denote the $n \times n$-matrix in Jordan normal form whose Jordan blocks are the $J_{\lambda_i}$, and let $\mathcal{O}_\lambda$ denote the nilpotent orbit in $\mathfrak{sl}_n(\overline{F})$ with representative $J_\lambda$.

The explicit correspondence between partitions and $F$-rational nilpotent orbits is described in the following proposition. The number of $F$-rational nilpotent orbits depends both on the partition $\lambda$ and on the characteristic of $F$, in a controlled way.

**Proposition 5 (Nevins (2011), Prop. 4).** *Let $\lambda$ be a partition of $n$, and $m = \gcd(\lambda)$. For any $d \in F^\times$ define the $n \times n$-matrix $D(d) = \mathrm{diag}(1, 1, \ldots, 1, d)$.*

(1) *For each $d \in F^\times$, the matrix $X_d = J_\lambda D(d)$ represents an $F$-rational orbit in $\mathcal{O}_\lambda(F)$, and conversely every orbit has a representative of this form.*
(2) *The $\mathbf{SL}_n(F)$-orbits represented by $J_\lambda D(d)$ and $J_{\lambda'} D(d')$ coincide if and only if $\lambda = \lambda'$ and $d \equiv d'$ in $F^\times/(F^\times)^m$.*

*Example 6.* In the case of $\mathfrak{sl}_3$, we have three partitions: $\lambda = (3), (2,1)$, and $(1,1,1)$. The corresponding nilpotent orbits $\mathcal{O}_\lambda$ in $\mathfrak{sl}_3(\overline{F})$ have representatives $X_{(3)} = \begin{pmatrix} 0&1&0 \\ 0&0&1 \\ 0&0&0 \end{pmatrix}$, $X_{(2,1)} = \begin{pmatrix} 0&1&0 \\ 0&0&0 \\ 0&0&0 \end{pmatrix}$, and $X_{(1,1,1)} = \begin{pmatrix} 0&0&0 \\ 0&0&0 \\ 0&0&0 \end{pmatrix}$ respectively.

The nilpotent orbits $\mathcal{O}_{(2,1)}$ and $\mathcal{O}_{(1,1,1)}$ do not split further into distinct $F$-rational orbits, since $m = \gcd(\lambda) = 1$ for these partitions. Since $m = 3$ for the first partition, the nilpotent orbit $\mathcal{O}_{(3)}$ splits into $|F^\times/(F^\times)^3|$ distinct $F$-rational orbits, represented by the matrices

$$X_d = J_{(3)}D(d) = \begin{pmatrix} 0 & 1 & 0 \\ 0 & 0 & d \\ 0 & 0 & 0 \end{pmatrix},$$

one for each distinct equivalence class of $d$ in $F^\times/(F^\times)^3$. By our assumptions, $F$ has residue characteristic $\neq 2$ and its residue field $k_F$ has $q = p^k$ elements, where $p$ is prime. By standard results in group theory, the number of cubes in $k_F^\times$ is $\frac{q-1}{\gcd(3,q-1)}$, and so the cardinality of $k_F^\times/(k_F^\times)^3$ is $\gcd(3, q-1)$. Thus the number of distinct $F$-rational orbits in this case is

$$\left|F^\times/(F^\times)^3\right| = 3 \cdot \left|k_F^\times/(k_F^\times)^3\right| = \begin{cases} 9 & \text{if } 3 \mid (q-1), \\ 3 & \text{otherwise.} \end{cases}$$

## 3.3 *Parametrization of Nilpotent Orbits in $\mathfrak{sp}_{2n}$ Using Partitions*

In the case of $\mathfrak{sp}_{2n}$, classes of quadratic forms over $F$ take the place of the cosets $F^\times/(F^\times)^m$ that we have seen in the parametrization of nilpotent orbits in the $\mathfrak{sl}_n$ case. Thus, we begin by recalling the classification of quadratic forms.

### 3.3.1 Quadratic Forms

Let $V$ be a finite-dimensional vector space over $F$, and $Q$ a non-degenerate quadratic form defined on $V$. Recall that the quadratic space $(V, Q)$ over $F$ is *anisotropic* if there is no nonzero $\mathbf{x} \in V$ such that $Q(\mathbf{x}) = 0$, and is *isotropic* otherwise.

Consider the quadratic form $q_0 : F^2 \to F$ that is represented in the standard basis by the matrix $q_0 = \begin{pmatrix} 0 & 1 \\ 1 & 0 \end{pmatrix}$. The quadratic space $(F^2, q_0)$ is the *hyperbolic plane*, a key example in the theory of quadratic forms. Since $\mathrm{char}(F) > 2$, if $(V, Q)$ is a non-degenerate quadratic space over $F$, then by the Witt decomposition (see Lam 2005 for instance), the quadratic form $Q$ can be decomposed into an orthogonal direct sum

$$Q = q_0^m \oplus Q_{\mathrm{aniso}}, \tag{5}$$

for some $m \leq \frac{1}{2}\dim(Q)$, where $(V_{\text{aniso}}, Q_{\text{aniso}})$ is anisotropic and uniquely determined up to isometry. The integer $m$ is called the *Witt index* of $(V, Q)$ and the quadratic form $Q_{\text{aniso}}$ is called the *anisotropic part* of $Q$. Moreover, quadratic forms of a given dimension may be classified by their discriminant and Hasse invariant.

Since $\text{char}(F) \neq 2$, we have $F^\times/(F^\times)^2 \simeq \mathbb{Z}/2\mathbb{Z} \times \mathbb{Z}/2\mathbb{Z}$; thus, there are at most 8 nondegenerate quadratic forms over $F$ of a given dimension. On the other hand, the maximum possible dimension of an anisotropic form over $F$ is four. Thus, by the Witt decomposition, to list all equivalence classes of quadratic forms, it suffices to list the classes of anisotropic forms. Representatives for these classes are given in the following lemma.

**Lemma 7 (Nevins (2011, Lemma 3)).** *Let $F$ be as above. If $-1 \in (F^\times)^2$, let $\alpha = \varepsilon$ be a fixed nonsquare unit in $F$. If $-1 \notin (F^\times)^2$, let $\alpha = 1$ and $\varepsilon = -1$. Given a quadratic form $Q$, its anisotropic part $Q_{aniso}$ is either the zero subspace, or isometric to one of the 15 anisotropic forms in Table 1.*

Given a class of quadratic forms, a matrix representative of the form $Q = q_0^m \oplus Q_{\text{aniso}}$ (or $Q = q_0^m$), where $Q_{\text{aniso}}$ is one of the diagonal matrices given in the above table, will be called a *minimal matrix representative* of the class.

### 3.3.2 Partition Parametrization of the Nilpotent Orbits in $\mathfrak{sp}_{2n}$

Embed $\mathbf{Sp}_{2n}$ into $\text{GL}_{2n}$ as $\mathbf{Sp}_{2n} = \{g \in \text{GL}_{2n} : g^t J g = J\}$, where $J = \left(\begin{smallmatrix} 0 & I \\ -I & 0 \end{smallmatrix}\right)$. Note that this is a different embedding than the one used by Waldspurger (2001), however, the parametrization below follows his methods. Let $V$ denote the vector space of the natural representation of $\mathfrak{sp}_{2n}$, with symplectic form defined by $\langle x, y\rangle = x^t J y$.

**Table 1** Explicit representatives of the 15 nonzero equivalence classes of anisotropic quadratic forms over a local $F$ with residue characteristic not 2

| Dimension | disc($Q$) | Hasse($Q$) | Representative | |
|---|---|---|---|---|
| 1 | 1 | 1 | 1 | |
| 1 | $\varepsilon$ | 1 | $\varepsilon$ | |
| 1 | $\varpi$ | 1 | $\varpi$ | |
| 1 | $\varepsilon\varpi$ | 1 | $\varepsilon\varpi$ | |
| 2 | $\alpha$ | 1 | $\text{diag}(1, \alpha)$ | |
| 2 | $\alpha$ | $-1$ | $\text{diag}(\varpi, \alpha\varpi)$ | |
| 2 | $tt'\varpi$ | $(t, t'\varpi)_F$ | $\text{diag}(t, t'\varpi)$ | $t, t' \in \{1, \varepsilon\}$ |
| 3 | $t$ | $-1$ | $\text{diag}(\alpha t, \varpi, \alpha\varpi)$ | $t \in \{1, \varepsilon\}$ |
| 3 | $\alpha t\varpi$ | $(\alpha, \varpi)_F$ | $\text{diag}(1, \alpha, t\varpi)$ | $t \in \{1, \varepsilon\}$ |
| 4 | 1 | $-1$ | $\text{diag}(1, -\varepsilon, -\varpi, \varepsilon\varpi)$ | |

The nilpotent Adjoint orbits in $\mathfrak{sp}_{2n}(\overline{F})$ are parametrized by partitions $\lambda$ of $2n$ in which the odd parts have even multiplicity (Collingwood and McGovern 1993, Corollary 4.1.8). For such a partition $\lambda$, let $\mathcal{O}_\lambda$ denote the geometric nilpotent orbit corresponding to $\lambda$.

The $F$-points, $\mathcal{O}_\lambda(F)$, of this orbit may fail to be a single $\mathbf{Sp}_{2n}(F)$-orbit, so the set of partitions $\lambda$ is no longer sufficient to parametrize the nilpotent orbits over $F$. Instead, there is a set defined in terms of classes of quadratic forms corresponding to the partition $\lambda$ that parametrizes the $\mathbf{Sp}_{2n}(F)$-orbits in $\mathcal{O}_\lambda(F)$. Let $\overline{\mathcal{Q}} = (\mathcal{Q}_2, \ldots, \mathcal{Q}_{2n})$ be an $n$-tuple of isometry classes of quadratic forms over $F$. We say that $\overline{\mathcal{Q}}$ *corresponds to the partition* $\lambda$ of $2n$ (whose odd parts have even multiplicities) if $\dim(\mathcal{Q}_i) = m_i(\lambda)$ for each $i = 2, \ldots, 2n$.

**Theorem 8 (Nevins (2011, Proposition 5), Due to Waldspurger (2001)).** *Let $\lambda$ be a partition of $2n$, and suppose the odd parts of $\lambda$ have even multiplicity. Then $\mathcal{O}_\lambda(F)$ is a union of $\mathbf{Sp}_{2n}(F)$-orbits parametrized by the $n$-tuples*

$$\overline{\mathcal{Q}} = (\mathcal{Q}_2, \ldots, \mathcal{Q}_{2n})$$

*corresponding to $\lambda$ (as defined above), where $\mathcal{Q}_i$ is an isometry class of a nondegenerate quadratic form over $F$.*

Following Nevins (2011), for each pair $(\lambda, \overline{\mathcal{Q}})$ we give an explicitly defined $X \in \mathfrak{sp}_{2n}(F)$ in the corresponding nilpotent orbit. We first give a decomposition of the vector space $V$ corresponding to the partition $\lambda$, and then define $X$ by its action on each component.

Let $\{p_1, \ldots, p_n, q_1, \ldots, q_n\}$ denote a *symplectic basis* for $V$; that is, a basis such that $\langle p_i, q_j\rangle = \delta_{ij}$, $\langle q_i, p_j\rangle = -\delta_{ij}$, and $\langle p_i, p_j\rangle = \langle q_i, q_j\rangle = 0$. For each $i \in \{1, \ldots, 2n\}$, let $s_i = \sum_{j<i} \frac{1}{2} j m_j$. Then the elements $s_i$ are integers such that $0 = s_1 \leq s_2 \leq \cdots \leq s_{2n} \leq n$. For each $j$ with $m_j \neq 0$, let $V(j)$ be the subspace given by

$$V(j) = \operatorname{span}\{p_{s_j+1}, \ldots, p_{s_j+\frac{1}{2}jm_j}, q_{s_j+1}, \ldots, q_{s_j+\frac{1}{2}jm_j}\}. \tag{6}$$

Then $V = \bigoplus_{j: m_j \neq 0} V(j)$, so we may define $X$ by its action on each subspace $V(j)$.

If $j$ is odd, let $\mu = (j, \ldots, j)$, a partition of $\frac{1}{2} j m_j$, and define the restriction of $X$ to $V(j)$ with respect to the basis given in (6) by

$$X|_{V(j)} = \begin{pmatrix} J_\mu & 0 \\ 0 & -J_\mu^t \end{pmatrix}. \tag{7}$$

If $j = 2N$ is even, define $X|_{V(j)}$ with respect to the basis given in (6) by

$$X|_{V(j)} = \begin{pmatrix} J_{Nm_j}^{m_j} & Z \oplus (-1)^N Q_j \\ 0 & -(J_{Nm_j}^{m_j})^t \end{pmatrix}, \tag{8}$$

where $Z$ is the $m_j(N-1) \times m_j(N-1)$ zero matrix and $Q_j$ is the minimal matrix representative of $\mathcal{Q}_j$. Then we have the following correspondence:

**Theorem 9 (Nevins (2011), Adapted from Proposition 6).** *Let $\lambda$ be as above. The matrix $X \in \mathfrak{sp}_{2n}(F)$ defined by (7) and (8) is a representative of the $\mathbf{Sp}_{2n}(F)$-orbit in $\mathcal{O}_\lambda(F)$ corresponding to the n-tuple $\overline{\mathcal{Q}}$.*

# 4 Parametrization of Nilpotent Orbits Via the Building

## *4.1 Preliminaries Regarding the Building*

Following the notation and terminology of Nevins (2011), we briefly recall the necessary facts about the standard affine apartment of the Bruhat-Tits building $\mathcal{B}(\mathbf{G}) = \mathcal{B}(\mathbf{G}, F)$ for $\mathbf{G}$ a connected reductive algebraic group over $F$. However, since this is the only case we need in this paper, we assume that $\mathbf{G}$ is split over $F$, which simplifies these definitions substantially.

Let $\mathbf{T}$ be a split maximal torus of $\mathbf{G}$. Let $X^*(\mathbf{T})$ be the group of $F$-rational characters of $\mathbf{T}$ and let $X_*(\mathbf{T})$ be the group of $F$-rational cocharacters. Let $\langle\, ,\, \rangle : X^*(\mathbf{T}) \times X_*(\mathbf{T}) \to \mathbb{Z}$ denote the natural pairing. Let $\Phi = \Phi(\mathbf{G}, \mathbf{T})$ denote the set of roots of $\mathbf{T}$ in $\mathbf{G}$; it is a finite subset of $X^*(\mathbf{T})$, and $\mathfrak{g}$ has the root space decomposition

$$\mathfrak{g} = \mathfrak{t} \oplus \bigoplus_{\alpha \in \Phi} \mathfrak{g}_\alpha,$$

where $\mathfrak{t}$ is the Lie algebra of $\mathbf{T}$ and the root subspace $\mathfrak{g}_\alpha$ is defined by

$$\mathfrak{g}_\alpha = \{X \in \mathfrak{g} \mid \mathrm{Ad}(t)X = \alpha(t)X \text{ for all } t \in \mathbf{T}\}.$$

The standard *affine apartment* $\mathcal{A}$ in the building $\mathcal{B}(\mathbf{G})$ is the affine space underlying the vector space $X_*(\mathbf{T}) \otimes_{\mathbb{Z}} \mathbb{R}$, together with a hyperplane structure.

Let $W(\Phi) = W(\mathbf{G}, \mathbf{T})$ denote the Weyl group of $\mathbf{T}$ in $\mathbf{G}$. The Weyl group is generated by reflections through hyperplanes corresponding to each root $\alpha \in \Phi$; its action on $X^*(\mathbf{T})$ preserves $\Phi$.

For each $\alpha \in \Phi$ and $n \in \mathbb{Z}$, we consider the affine functional, or *affine root*, $\alpha + n : \mathcal{A} \to \mathbb{R}$ defined for each $x = \lambda \otimes s \in \mathcal{A}$ by

$$(\alpha + n)(x) = \langle \alpha + n, \lambda \otimes x \rangle = s\langle \alpha, \lambda \rangle + n.$$

Put $\Psi = \{\alpha + n \mid \alpha \in \Phi, n \in \mathbb{Z}\}$, and for each $\psi = \alpha + n \in \Psi$ consider the hyperplane

$$H_\psi = \{x \in \mathcal{A} \mid \psi(x) = 0\}.$$

The set of all such hyperplanes forms a hyperplane structure on $\mathcal{A}$.

#### 4.1.1 The Standard Apartment for $\mathfrak{sl}(n)$

When $\mathbf{G} = \mathbf{SL}_n$, let $\mathbf{T}$ be the diagonal torus and consider the apartment $\mathcal{A}$ corresponding to $\mathbf{T}$. We identify $X_*(\mathbf{T})$ with $\mathbb{Z}^n$, and think of cocharacters explicitly as functions $t \mapsto \mathrm{diag}(t^{x_1}, \dots, t^{x_n})$, for $t \in F^\times$ and $(x_1, \dots, x_n) \in \mathbb{Z}^n$. Identify $\mathcal{A}$ with $X_*(\mathbf{T}) \otimes \mathbb{R}$, and for each $i$ with $1 \leq i \leq n$, define the mapping $e_i : \mathcal{A} \to \mathbb{R}$ by $e_i(f \otimes s) = sx_i$, for $f = (t \mapsto \mathrm{diag}(t^{x_1}, t^{x_2}, \dots, t^{x_n})) \in X_*(\mathbf{T})$ and $s \in \mathbb{R}$. Then the set of roots is given by

$$\Phi = \{e_i - e_j \mid 1 \leq i \neq j \leq n\}. \tag{9}$$

Each of the root spaces $\mathfrak{g}_{e_i - e_j}$ is one-dimensional. We may view $\mathfrak{g}_{e_i - e_j}$ as being spanned by the matrix $E_{ij}$ whose entries are all zero except for the $(i,j)$-entry, which equals 1.

#### 4.1.2 The Standard Apartment for $\mathfrak{sp}_{2n}$

When $\mathbf{G} = \mathbf{Sp}_{2n}$, again let $\mathbf{T}$ be the diagonal torus, whose elements are of the form $\tau = \mathrm{diag}(t_1, t_2, \dots, t_n, t_1^{-1}, t_2^{-1}, \dots, t_n^{-1})$ by our choice of the embedding.

The rank of $\mathfrak{sp}_{2n}$ is $n$, so we have $X^*(\mathbf{T}) \simeq X_*(\mathbf{T}) \simeq \mathbb{Z}^n$, as abelian groups. Similar to the $\mathfrak{sl}_n$ case, for $1 \leq i \leq n$, define the mapping $e_i : \mathcal{A} \to \mathbb{R}$ by $e_i(f \otimes s) = sx_i$, for $f = (t \mapsto \mathrm{diag}(t_1^{x_1}, t_2^{x_2}, \dots, t_n^{x_n}, t_1^{-x_1}, t_2^{-x_2}, \dots, t_n^{-x_n}))$ and $x \in \mathbb{R}$. Then $\Phi$ is the set

$$\Phi = \{e_i - e_j,\ \pm(e_i + e_j),\ \pm 2e_i \mid 1 \leq i \neq j \leq n\}. \tag{10}$$

Finally, let $\mathcal{A}$ be the standard apartment of $\mathfrak{sp}_{2n}$ relative to this root datum.

Below in Sections 5.3 and 5.4, we discuss in detail the examples of $\mathfrak{sl}(3)$ and $\mathfrak{sp}(4)$.

### 4.2 DeBacker's Parametrization Using the Building

Generalizing the work of Barbasch and Moy (1997), DeBacker (2002b) developed a parametrization of nilpotent orbits that relies on facets in the Bruhat-Tits building of $\mathfrak{g}$ and is valid for any reductive Lie algebra over $F$, provided the residue characteristic is sufficiently large. This is actually a family of parametrizations that depends on a real parameter $r$, although for our purposes it suffices to consider the case corresponding to $r = 0$, in the notation of DeBacker (2002b). To ease notation, we omit the $r$-dependence in DeBacker's notation and state suitably modified versions of the relevant theorems with $r = 0$.

Let $\mathcal{A}$ denote the standard affine apartment corresponding to the Lie algebra $\mathfrak{g}$. The set $\mathcal{A}$ has the structure of a simplicial complex (generally, polysimplicial, but the groups we are considering in this paper are simple). Let us define the *facets* (i.e. the simplices) in the apartment $\mathcal{A}$. For $x \in \mathcal{A}$ and $n \in \mathbb{Z}$, define the sets

$$\Phi_x = \{\alpha \in \Phi \mid \alpha(x) \in \mathbb{Z}\} \qquad \text{and} \qquad \mathcal{H}_n = \{x \in \mathcal{A} \mid |\Phi_x| = n\}. \tag{11}$$

For an integer $n$, a *facet* of $\mathcal{A}$ is defined to be any connected component $\mathcal{F}$ of $\mathcal{H}_n$. We denote by $A(\mathcal{F}, \mathcal{A})$ the smallest affine subspace of $\mathcal{A}$ containing $\mathcal{F}$. With this, we define the dimension of a facet to be $\dim(\mathcal{F}) = \dim A(\mathcal{F}, \mathcal{A})$, hence facets of the apartment $\mathcal{A}$ have bounded dimension.

Given a subspace $H$ of $\mathcal{A}$, a facet $\mathcal{F} \subset H$ is said to be *maximal* if the dimension of $\mathcal{F}$ is maximal among the dimensions of facets contained in $H$. An *alcove* is the closure of any facet of maximal dimension in $\mathcal{A}$.

For example, $\mathcal{H}_0$ consists of all $x \in \mathcal{A}$ for which $\alpha(x) \notin \mathbb{Z}$ for all $\alpha \in \Phi$. This is the set of points $x \in \mathcal{A}$ that do not lie on any of the hyperplanes $H_{\alpha-m}$ for any $\alpha \in \Phi$ and any $m \in \mathbb{Z}$. Thus any connected component of $\mathcal{H}_0$ is the interior of some alcove in $\mathcal{A}$. For instance, in the case of $\mathfrak{sl}_3$ or $\mathfrak{sp}_4$, these facets will be two-dimensional. Also for $\mathfrak{sl}_3$ or $\mathfrak{sp}_4$, any connected component of $\mathcal{H}_1$ is the edge of an alcove, and any connected component of $\mathcal{H}_2$ is a vertex of an alcove.

### 4.2.1 Moy-Prasad Filtration Lattices, and Generalized Facets

For each pair $(x, r)$ with $x \in \mathcal{A}$ and $r \in \mathbb{R}$, Moy and Prasad (1994) define certain $\mathfrak{O}$-lattices $\mathfrak{g}_{x,r}$ giving a filtration of $\mathfrak{g}$. The parameter $r$ is referred to as the *depth* of the lattice. Since we consider only the case $r = 0$, we suppress $r$ from the notation throughout.

Associated with each root $\alpha \in \Phi$ we have the root subgroup $\mathbf{U}_\alpha$, a $\mathbf{T}(F)$-invariant, closed one-parameter subgroup of $\mathbf{G}$, and the root subspace $\mathfrak{g}_\alpha$, which coincides with the tangent space of $\mathbf{U}_\alpha$. (For the rest of this section, we reserve boldface letters for algebraic groups, and their non-boldface counterparts for the groups of rational points.) Note that our groups $\mathbf{G}$, $\mathbf{T}$ and $\mathbf{U}_\alpha$ are in fact defined over $\mathbb{Z}$, and thus we can talk about the well-defined subgroups $\mathbf{G}(\mathfrak{O})$, $\mathbf{U}_\alpha(\mathfrak{O})$, etc.

With this notation (see Rabinoff 2005, § 3 for more detail of the notation) for each $x \in \mathcal{A}$, define the *parahoric subgroup* $G_x$ as

$$G_x = \langle \mathbf{T}(\mathfrak{O}), \mathbf{U}_\alpha(\mathfrak{P}^{-\lfloor \alpha(x) \rfloor}) \mid \alpha \in \Phi \rangle;$$

its *pro-unipotent radical* is

$$G_x^+ = \langle \mathbf{T}(\mathfrak{P} + 1), \mathbf{U}_\alpha(\mathfrak{P}^{1-\lceil \alpha(x) \rceil}) \mid \alpha \in \Phi \rangle.$$

(This turns out to be equivalent to the more complicated standard definition, cf. Rabinoff 2005, (3.1).)

Similarly, for each $x \in \mathcal{A}$, we have the corresponding lattices $\mathfrak{g}_x \supset \mathfrak{g}_x^+$ in the Lie algebra:

$$\mathfrak{g}_x = \langle \mathfrak{h}(\mathfrak{O}), \mathfrak{P}^{-\lfloor \alpha(x) \rfloor} X_\alpha \mid \alpha \in \Phi \rangle \tag{12}$$

and

$$\mathfrak{g}_x^+ = \langle \mathfrak{h}(\mathfrak{P}), \mathfrak{P}^{1-\lceil \alpha(x) \rceil} X_\alpha \mid \alpha \in \Phi \rangle, \tag{13}$$

where the root space $\mathfrak{g}_\alpha$ is spanned by the element $X_\alpha$, and $\mathfrak{h} = \mathrm{Lie}(\mathbf{T})$ is the Cartan subalgebra of $\mathfrak{g}$ corresponding to $\mathbf{T}$. (More precisely, by choosing a splitting $(\mathbf{B}, \mathbf{T}, \{x_\alpha\})$ of $\mathbf{G}$, defined over $\mathbb{Z}$, we would then have the corresponding generators $X_\alpha$ of $\mathfrak{g}_\alpha$. For our classical Lie algebras, $X_\alpha$ are the standard generators of the corresponding root spaces; see examples in Section 5 below.)

If $x, y \in \mathcal{A}$ are contained in the same facet $\mathcal{F}$, then we have $G_x = G_y$ and $G_x^+ = G_y^+$, as well as $\mathfrak{g}_x = \mathfrak{g}_y$ and $\mathfrak{g}_x^+ = \mathfrak{g}_y^+$. For a given facet $\mathcal{F}$, we will simply write $\mathfrak{g}_\mathcal{F}$ and $\mathfrak{g}_\mathcal{F}^+$ for the lattices associated with any $x \in \mathcal{F}$. We also have need of the quotient of these lattices, denoted $V_\mathcal{F} = \mathfrak{g}_\mathcal{F}/\mathfrak{g}_\mathcal{F}^+$, which is a Lie algebra over $k_F$.

In order to state DeBacker's parametrization theorem, we need to define an equivalence relation on facets, and in order to do that, we require the notion of a *generalized facet*. For each $x \in \mathcal{B}(\mathbf{G})$, the set

$$\mathfrak{F} = \{y \in \mathcal{B}(\mathbf{G}) \mid \mathfrak{g}_x = \mathfrak{g}_y \text{ and } \mathfrak{g}_x^+ = \mathfrak{g}_y^+\} \tag{14}$$

is called the *generalized facet containing* $x$. We say two generalized facets $\mathfrak{F}_1$ and $\mathfrak{F}_2$ are *strongly associate* if $A(\mathfrak{F}_1 \cap \mathcal{A}, \mathcal{A}) = A(\mathfrak{F}_2 \cap \mathcal{A}, \mathcal{A}) \neq \emptyset$, for some apartment $\mathcal{A}$. If there exists an element $g \in \mathbf{G}$ such that $\mathfrak{F}_1$ and $g\mathfrak{F}_2$ are strongly associate, then we say $\mathfrak{F}_1$ and $\mathfrak{F}_2$ are *associate*. For two facets $\mathcal{F}_1$ and $\mathcal{F}_2$ contained in a given apartment $\mathcal{A}$, we say that $\mathcal{F}_1$ and $\mathcal{F}_2$ are associate if the generalized facets they determine are associate.

*Remark 10.* In this paper, thanks to the explicit parametrization of orbits and Nevins' matching theorem, we need not interpret this notion of associate using Denef-Pas language; the fact that this notion involves the whole building and not just a single apartment is one of the main obstructions we currently perceive to obtaining our main result for general Lie algebras.

### 4.2.2 DeBacker's Parametrization

We say an element $v \in V_{\mathcal{F}}$ is *degenerate* if the coset it parametrizes contains a nilpotent element (i.e. if there exists a nilpotent $X \in \mathfrak{g}_{\mathcal{F}}$ such that $v = X + \mathfrak{g}_{\mathcal{F}}^{+}$). Let $\mathcal{I}(F)$ be the set given by

$$\mathcal{I}(F) = \{(\mathcal{F}, v) \mid \mathcal{F} \subset \mathcal{A} \text{ is a facet, and } v \in V_{\mathcal{F}} \text{ is a degenerate element}\}. \tag{15}$$

DeBacker defines an equivalence relation $\sim$ on $\mathcal{I}(F)$: say that $(\mathcal{F}_1, v_1) \sim (\mathcal{F}_2, v_2)$ if and only if there exists $g \in \mathbf{G}$ such that $A(\mathcal{F}_1, \mathcal{A}) = A(g\mathcal{F}_2, \mathcal{A})$, and such that under the resulting natural identification of $V_{\mathcal{F}_1}$ with $\mathrm{Ad}(g)V_{\mathcal{F}_2}$, the elements $v_1$ and $\mathrm{Ad}(g)v_2$ lie in the same orbit under $G_x$ for any $x \in \mathcal{F}_1$.

Let $\mathrm{Nil}(F)$ denote the set of rational nilpotent orbits in $\mathfrak{g}$. Using the theory of $\mathfrak{sl}_2$-triples, DeBacker proves the following results regarding the relationship between the sets $\mathcal{I}(F)$ and $\mathrm{Nil}(F)$.

**Lemma 11 (DeBacker (2002b)).** *Suppose the residue characteristic of F is sufficiently large, and $(\mathcal{F}, v) \in \mathcal{I}(F)$. Then*

1. *(Lemma 5.3.3, $r = 0$ case) There exists a unique nilpotent orbit of minimal dimension which intersects the coset $v$ nontrivially. We denote this nilpotent orbit by $\mathcal{O}(\mathcal{F}, v)$.*
2. *(Lemma 5.4.1, $r = 0$ case) The map $\gamma : \mathcal{I}(F)/\sim \longrightarrow \mathrm{Nil}(F)$ defined by $(\mathcal{F}, v) \mapsto \mathcal{O}(\mathcal{F}, v)$ is a well-defined, surjective map.*

However, this map is not injective. (A detailed explanation of this phenomenon is given in Nevins 2011.) To obtain a one-to-one correspondence, we must restrict to the subset of *distinguished pairs*. We say a pair $(\mathcal{F}, v) \in \mathcal{I}(F)$ is *distinguished* if $v$ is not an element of any proper Levi subalgebra of the $V_{\mathcal{F}}$. Let

$$\mathcal{I}^d(F) = \{(\mathcal{F}, v) \in \mathcal{I}(F) \mid (\mathcal{F}, v) \text{ is distinguished}\}. \tag{16}$$

**Theorem 12 (DeBacker (2002b), $r = 0$ Case of Theorem 5.6.1).** *Suppose the residue characteristic of F is sufficiently large. Then there is a bijective correspondence between $\mathcal{I}^d(F)/\sim$ and the set of nilpotent orbits in $\mathfrak{g}(F)$ given by the map which sends $(\mathcal{F}, v)$ to $\mathcal{O}(\mathcal{F}, v)$.*

*Proof.* Theorem 5.6.1 from DeBacker (2002b) contains this statement for a slightly different parameter space, allowing the facets $\mathcal{F}$ of $\mathcal{I}^d(F)$ to run over the enlarged Bruhat-Tits building of $\mathfrak{g}$. By virtue of Theorem 5.6 from Nevins (2011), one may substitute the parameter space $\mathcal{I}^d(F)$ defined above.

# 5 Explicit Matching Between the Two Parametrizations

Following Nevins (2011), we give a correspondence between the parametrization involving partitions and DeBacker's parametrization defined in terms of the building, for the cases $\mathfrak{g} = \mathfrak{sl}_n$ and $\mathfrak{sp}_{2n}$. (In order for both parametrizations to be valid, we must again assume that the residue characteristic of $F$ is not 2, and the characteristic of $F$ itself is zero, or sufficiently large.) We also work out the examples of $\mathfrak{sl}_3$ and $\mathfrak{sp}_4$ in complete detail.

In Section 6 we will associate certain definable functions with each nilpotent orbit. For that purpose it would be very convenient to use DeBacker's parametrization, however, the set $\mathrm{Nil}(F)$ itself is more easily understood through Waldspurger's parametrization via partitions. Thus, it is necessary to understand the explicit matching between these two parametrizations.

## 5.1 The Matching for $\mathfrak{sl}_n$

For each partition $\lambda = (\lambda_1, \lambda_2, \ldots, \lambda_t)$ of $n$, and each diagonal matrix $D = \mathrm{diag}(d_1, d_2, \ldots, d_n) \in \mathbf{T}$, define the set

$$I_\lambda = \{1, 2, \ldots, n\} \setminus \{\lambda_1, \lambda_1 + \lambda_2, \ldots, \sum_i \lambda_i = n\}.$$

$I_\lambda$ represents the set of locations of the nonzero entries of the matrices $J_\lambda D(d) \in \mathcal{O}_\lambda(F)$ described in Proposition 5. For each $i \in I_\lambda$, the value $d_{i+1}$ is the $(i, i+1)$-entry of $X$, and all remaining entries are zero.

Recall the hyperplanes $H_{\varphi+n} = \{x \in \mathcal{A} \mid \varphi(x) = -n\}$, defined for roots $\varphi \in \Phi$. Also recall the standard notation $\alpha_i = e_i - e_{i+1}$ for the simple roots of $\mathbf{SL}_n$. With the notation as above, define

$$H_{\lambda,D} = \bigcap_{i \in I_\lambda} H_{\alpha_i + \mathrm{val}(d_{i+1})} \subseteq \mathcal{A}.$$

Note that when $\lambda = (1, 1, \ldots, 1)$, we have $X = 0$ and $I_\lambda = \emptyset$. Nevins (2011) states that the zero orbit corresponds to the associate class of the interior of any alcove in the apartment. The following theorem of Nevins establishes the remainder of the correspondence between the two parametrizations of the nilpotent orbits for $\mathfrak{sl}_n$.

**Theorem 13 (Nevins (2011), Theorem 2 with $r = 0$).** *Let $\lambda, D$, and $H_{\lambda,D} \subset \mathcal{A}$ be as above, and let $\mathcal{F}$ be any facet of maximal dimension in $H_{\lambda,D}$. For any $x \in \mathcal{F}$, we have $X = J_\lambda D \in \mathfrak{g}_\mathcal{F}$; set $v$ to be its image in $V_\mathcal{F}$. Then $(\mathcal{F}, v) \in \mathcal{I}^d(F)$ and $\mathcal{O}(\mathcal{F}, v) = \mathrm{Ad}(\mathfrak{sl}_n(F))X$.*

## 5.2 The Matching for $\mathfrak{sp}_{2n}$

Let $X \in \mathcal{O}_\lambda(F)$ be the nilpotent element corresponding to the $n$-tuple $\overline{\mathcal{Q}}$, as in Theorem 9. For each odd $j$, let

$$I_j = \{1, \ldots, \tfrac{1}{2}jm_j\} \setminus \{j, 2j, \ldots, \tfrac{1}{2}jm_j\}$$

and let $S_j = S_j^1$ denote the set of simple roots

$$S_j^1 = \{e_{s_j+k} - e_{s_j+k+1} \mid k \in I_j\}.$$

For each even $j$, suppose $Q_j = q_0^m \oplus Q_{\text{aniso}}$ is the minimal matrix representative for $\mathcal{Q}_j$, where $m$ is the Witt index of $Q_j$ ($0 \le 2m \le m_j$), and set $M_j = (\frac{1}{2}j - 1)m_j$. Then we take $S_j = S_j^1 \cup S_j^2$, where

$$S_j^1 = \{e_{s_j+k} - e_{s_j+k+m_j} \mid 1 \le k \le M_j\} \cup \{e_{s_j+M_j+2i-1} + e_{s_j+M_j+2i} \mid 1 \le i \le m\}$$

and

$$S_j^2 = \{2e_{s_j+M_j+i} \mid 2m < i \le m_j\}.$$

If $Q_{\text{aniso}} = \text{diag}(a_{2m+1}, \ldots, a_{m_j})$, define for each root $\alpha_i = 2e_{s_j+M_j+i}$ the integer $v_{\alpha_i} = \text{val}(a_i)$ for $2m + 1 \le i \le m_j$. Let $H_{\lambda,\overline{\mathcal{Q}}}$ be the common intersection (over all $j$) of the hyperplanes $H_\alpha$ for $\alpha \in S_j^1$ and $H_{\alpha+v_\alpha}$ for $\alpha \in S_{2j}^2$. Finally, the following theorem of Nevins gives the correspondence between the two parametrizations of nilpotent orbits for $\mathfrak{sp}_{2n}$:

**Theorem 14 (Nevins (2011), Theorem 4 with $r = 0$).** *The affine subspace $H_{\lambda,\overline{\mathcal{Q}}} \subset \mathcal{A}$ is a nonempty union of facets. Let $\mathcal{F}$ be any maximal facet in $H_{\lambda,\overline{\mathcal{Q}}}$, and let $v$ denote the projection of $X$ in $V_\mathcal{F}$. Then $(\mathcal{F}, v) \in \mathcal{I}^d(F)$ and $\mathcal{O}(\mathcal{F}, v) =$* $\text{Ad}(\mathfrak{sp}_{2n}(F))X$.

## 5.3 Example: The Correspondence, in the Case of $\mathfrak{sl}_3$

We examine these parametrizations and their correspondence in the case of the Lie algebra $\mathfrak{g} = \mathfrak{sl}_3$. Representatives $X_\lambda$ for the nilpotent orbits $\mathcal{O}_\lambda$ are given above in Example 6. Following the construction given in Section 5.1, we compute the sets $H_{\lambda,D(d)}$ as in Theorem 13, where $D = D(d) = \text{diag}(1, 1, \ldots, 1, d)$, for $d \in F^\times$. We then determine the maximal facet $\mathcal{F} \subseteq H_{\lambda,D}$ corresponding to each orbit $\mathcal{O}_{(1,1,1)}$ and $\mathcal{O}_{(2,1)}$, and to each of the $F$-rational nilpotent orbits contained in $\mathcal{O}_{(3)}$.

For the partition $\lambda = (1, 1, 1)$, we have $I_{(3)} = \emptyset$ and $X = 0$. In this trivial case, the corresponding maximal facet is the interior of any alcove in the apartment $\mathcal{A}$.

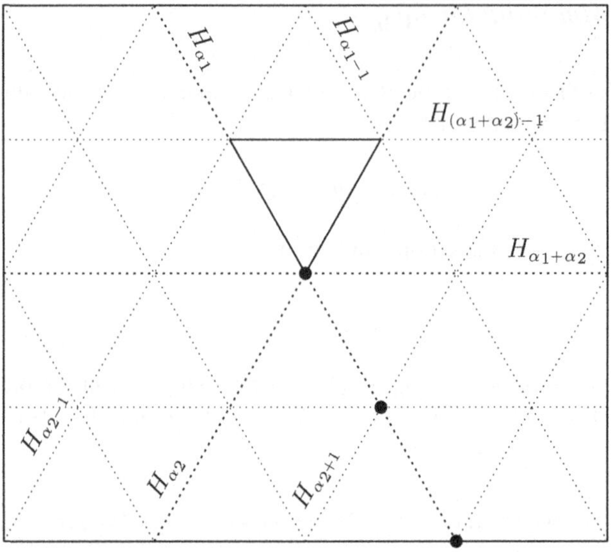


**Fig. 1** Standard affine apartment of $\mathfrak{sl}_3(F)$. *Solid edges* outline an alcove, and *dotted lines* indicate affine hyperplanes. *Dots* indicate the sets $H_{(3),D(d)}$

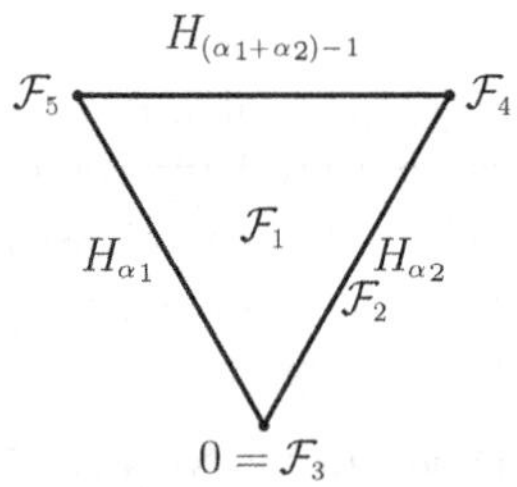


**Fig. 2** Facets $\mathcal{F}_i$ in the given alcove of the standard affine apartment of $\mathfrak{sl}_3(F)$. The facet $\mathcal{F}_2$ is associate with the other edges

The standard apartment for $\mathfrak{sl}_3$ is shown above in Figure 1. We may choose the alcove given by the outlined region, which is bounded by the hyperplanes $H_{\alpha_1}$, $H_{\alpha_2}$, and $H_{(\alpha_1+\alpha_2)-1}$. This is denoted by $\mathcal{F}_1$ above in Figure 2.

When $\lambda = (2,1)$, we have $I_{(2,1)} = \{1\}$ and $H_{(2,1),D(d)} = H_{\alpha_1}$ for any $d \in F^\times$. Any facet of maximal dimension in $H_{(2,1),D}$, is therefore an edge in an alcove of $\mathcal{A}$. Since the three edges of any alcove are associates, it suffices to consider a single edge, denoted $\mathcal{F}_2$.

For $\lambda = (3)$, we have $I_{(3)} = \{1,2\}$ and $H_{(3),D(d)} = H_{\alpha_1} \cap H_{\alpha_2+\mathrm{val}(d)}$. Recall that we have $X_{(3)} = \left(\begin{smallmatrix} 0&1&0 \\ 0&0&1 \\ 0&0&0 \end{smallmatrix}\right)$ and this orbit splits into $3 \cdot \gcd(3, q-1)$ orbits $\mathcal{O}_\lambda(F)$ whose representatives are given by

$$X_d = \begin{pmatrix} 0&1&0 \\ 0&0&d \\ 0&0&0 \end{pmatrix}, \quad \text{one for each distinct equivalence class of } d \text{ in } F^\times/(F^\times)^3. \tag{17}$$

When $3 \nmid (q-1)$ we have $d \in \{1, \varpi, \varpi^2\}$. When $d \mid (q-1)$, we fix a non-cubic unit $\varepsilon \in F^\times$, and have $d \in \{1, \varepsilon, \varepsilon^2, \varpi, \varepsilon\varpi, \varepsilon^2\varpi, \varpi^2, \varepsilon\varpi^2, \varepsilon^2\varpi^2\}$. In any case, $H_{(3),D(d)}$ will be a single point, thus any facet of maximal dimension in $H_{(3),D(d)}$ will consist of a single element.

Specifically, when $\mathrm{val}(d) = 0$, then $H_{(3),D(d)} = \{0\}$, with corresponding facet denoted by $\mathcal{F}_3$ in Figure 2. Taking $\mathrm{val}(d) = 1$ gives $H_{(3),D(d)} = H_{\alpha_1} \cap H_{\alpha_2+1}$ which is not in the chosen alcove. For our purposes in handling definability, it is convenient to fix a single alcove. With this in mind, we note that $\mathbf{G}(F)$ acts on $\mathcal{A}$ via the affine Weyl group; and so, reflecting this point across the hyperplane $H_{\alpha_1+\alpha_2}$ to the upper-right vertex of the alcove, we see that this facet and the facet denoted by $\mathcal{F}_4$ are associates. Similarly, $\mathrm{val}(d) = 2$ gives $H_{(3),D(d)} = H_{\alpha_1} \cap H_{\alpha_2+2}$, which maps to facet $\mathcal{F}_5$ under the affine Weyl group action (e.g. by reflecting across $H_{\alpha_2}$ and then $H_{\alpha_1+1}$).

In order to calculate the lattices associated with each facet $\mathcal{F}_i$, we first determine the root spaces $\mathfrak{g}_\alpha$ for $\alpha \in \Phi$. By (9) (with $\alpha_i = e_i - e_{i+1}$), we have

$$\Phi = \{\alpha_1, \alpha_2, \alpha_1 + \alpha_2, -\alpha_1, -\alpha_2, -(\alpha_1 + \alpha_2)\}.$$

Each root space is one-dimensional, and the generators for the six root spaces are given by the matrices

$$X_{\alpha_1} = \begin{pmatrix} 0 & 1 & 0 \\ 0 & 0 & 0 \\ 0 & 0 & 0 \end{pmatrix} \quad X_{\alpha_2} = \begin{pmatrix} 0 & 0 & 0 \\ 0 & 0 & 1 \\ 0 & 0 & 0 \end{pmatrix} \quad X_{\alpha_1+\alpha_2} = \begin{pmatrix} 0 & 0 & 1 \\ 0 & 0 & 0 \\ 0 & 0 & 0 \end{pmatrix}$$

$$X_{-\alpha_1} = \begin{pmatrix} 0 & 0 & 0 \\ 1 & 0 & 0 \\ 0 & 0 & 0 \end{pmatrix} \quad X_{-\alpha_2} = \begin{pmatrix} 0 & 0 & 0 \\ 0 & 0 & 0 \\ 0 & 1 & 0 \end{pmatrix} \quad X_{-(\alpha_1+\alpha_2)} = \begin{pmatrix} 0 & 0 & 0 \\ 0 & 0 & 0 \\ 1 & 0 & 0 \end{pmatrix}$$

Sample calculations for the lattices associated with the facets $\mathcal{F}_1$ and $\mathcal{F}_4$ are given in detail below, followed by a table with full results for each of the five facets mentioned above. Although the lattices associated with $H_{(3),D(d)}$ when $\mathrm{val}(d) = 1$ will be different from the lattices $\mathfrak{g}_{\mathcal{F}_4}$ and $\mathfrak{g}^+_{\mathcal{F}_4}$, the quotients will be identical. Therefore we may consider $\mathcal{F}_4$ when determining the image $v$ of $X_d$ in the quotient. The case $\mathrm{val}(d) = 2$ is handled similarly. In the same spirit, we compute $\mathfrak{g}_\mathcal{F}$ for the edge that is labelled $\mathcal{F}_2$ in Figure 2; though the lattices $\mathfrak{g}_\mathcal{F}$ and $\mathfrak{g}^+_\mathcal{F}$ are different for the other two associate edges, the quotients $V_\mathcal{F}$ for them are isomorphic.

**Facet $\mathcal{F}_1$:** $x \in \mathcal{F}_1$ if and only if $0 < \alpha_1(x), \alpha_2(x), (\alpha_1 + \alpha_2)(x) < 1$. For any $x \in \mathcal{F}_1$ this gives $\lfloor \alpha_1(x) \rfloor = 0$ and $\lceil \alpha_1(x) \rceil = 1$, while $\lfloor -\alpha_1(x) \rfloor = -1$ and $\lceil -\alpha_1(x) \rceil = 0$. We have $\mathfrak{P}^{-\lfloor \alpha_1(x) \rfloor} = \mathfrak{P}^0 = \mathfrak{O}$ and $\mathfrak{P}^{-\lfloor -\alpha_1(x) \rfloor} = \mathfrak{P}^1 = \mathfrak{P}$, hence

$$\mathfrak{P}^{-\lfloor \alpha_1(x) \rfloor} X_{\alpha_1} = \begin{pmatrix} 0 & \mathfrak{O} & 0 \\ 0 & 0 & 0 \\ 0 & 0 & 0 \end{pmatrix} \text{ and } \mathfrak{P}^{-\lfloor -\alpha_1(x) \rfloor} X_{-\alpha_1} = \begin{pmatrix} 0 & 0 & 0 \\ \mathfrak{P} & 0 & 0 \\ 0 & 0 & 0 \end{pmatrix}.$$

We may similarly calculate $\mathfrak{P}^{-\lfloor \varphi(x) \rfloor} X_\phi$ for each of the roots $\varphi = \pm\alpha_2, \pm(\alpha_1 + \alpha_2)$. Finally, $\mathfrak{h}(\mathfrak{O}) = \begin{pmatrix} \mathfrak{O} & 0 & 0 \\ 0 & \mathfrak{O} & 0 \\ 0 & 0 & \mathfrak{O} \end{pmatrix}$, and so we obtain

$$\mathfrak{g}_{\mathcal{F}_1} = \begin{pmatrix} \mathfrak{O} & \mathfrak{O} & \mathfrak{O} \\ \mathfrak{P} & \mathfrak{O} & \mathfrak{O} \\ \mathfrak{P} & \mathfrak{P} & \mathfrak{O} \end{pmatrix}.$$

Next, $\mathfrak{P}^{1-\lceil \alpha_1(x) \rceil} = \mathfrak{O}$ and $\mathfrak{P}^{1-\lceil -\alpha_1(x) \rceil} = \mathfrak{P}$, and similarly for $\pm\alpha_2$ and $\pm(\alpha_1 + \alpha_2)$. Computing the corresponding lattice representative for each root, and using the fact that $\mathfrak{h}(\mathfrak{P}) = \begin{pmatrix} \mathfrak{P} & 0 & 0 \\ 0 & \mathfrak{P} & 0 \\ 0 & 0 & \mathfrak{P} \end{pmatrix}$, we find that $\mathfrak{g}^+_{\mathcal{F}_1} = \begin{pmatrix} \mathfrak{P} & \mathfrak{O} & \mathfrak{O} \\ \mathfrak{P} & \mathfrak{P} & \mathfrak{O} \\ \mathfrak{P} & \mathfrak{P} & \mathfrak{P} \end{pmatrix}$. Since $k_F = \mathfrak{O}/\mathfrak{P}$, taking the quotient gives

$$V_{\mathcal{F}_1} = \mathfrak{g}_{\mathcal{F}_1}/\mathfrak{g}^+_{\mathcal{F}_1} = \begin{pmatrix} k_F & 0 & 0 \\ 0 & k_F & 0 \\ 0 & 0 & k_F \end{pmatrix}.$$

**Facet $\mathcal{F}_4$:** This vertex lies on the hyperplanes $H_{\alpha_1-1}$, $H_{\alpha_2}$, and $H_{(\alpha_1+\alpha_2)-1}$, so $x \in \mathcal{F}_4$ if and only if $\alpha_1(x) = (\alpha_1 + \alpha_2)(x) = 1$ and $\alpha_2(x) = 0$. Thus $\mathfrak{P}^{-\lfloor \varphi(x) \rfloor} = \mathfrak{P}^{-1}$ and $\mathfrak{P}^{1-\lceil \varphi(x) \rceil} = \mathfrak{O}$ for $\varphi = \alpha_1$ and $\alpha_1 + \alpha_2$, and similarly $\mathfrak{P}^{-\lfloor \varphi(x) \rfloor} = \mathfrak{P}$ and $\mathfrak{P}^{1-\lceil \varphi(x) \rceil} = \mathfrak{P}^2$ for $\varphi = -\alpha_1$ and $-(\alpha_1 + \alpha_2)$. For the remaining roots, we have $\mathfrak{P}^{-\lfloor \pm\alpha_2(x) \rfloor} = \mathfrak{O}$ and $\mathfrak{P}^{1-\lceil \pm\alpha_2(x) \rceil} = \mathfrak{P}$. Adding the corresponding matrix representatives, we obtain

$$\mathfrak{g}_{\mathcal{F}_4} = \begin{pmatrix} \mathfrak{O} & \mathfrak{P}^{-1} & \mathfrak{P}^{-1} \\ \mathfrak{P} & \mathfrak{O} & \mathfrak{O} \\ \mathfrak{P} & \mathfrak{O} & \mathfrak{O} \end{pmatrix} \text{ and } \mathfrak{g}^+_{\mathcal{F}_4} = \begin{pmatrix} \mathfrak{P} & \mathfrak{O} & \mathfrak{O} \\ \mathfrak{P}^2 & \mathfrak{P} & \mathfrak{P} \\ \mathfrak{P}^2 & \mathfrak{P} & \mathfrak{P} \end{pmatrix}.$$

Identifying $\mathfrak{P}^a$ with $\varpi^a\mathfrak{O}$, we get isomorphisms of $\mathfrak{P}^a/\mathfrak{P}^{a+1}$ with $k_F = \mathfrak{O}/\mathfrak{P}$. Thus taking the quotient gives $V_{\mathcal{F}_4} = \begin{pmatrix} k_F & k_F & k_F \\ k_F & k_F & k_F \\ k_F & k_F & k_F \end{pmatrix}$.

| Parahoric $\mathfrak{g}_{\mathcal{F}}$ | Pro-unipotent $\mathfrak{g}_{\mathcal{F}}^{+}$ | $V_{\mathcal{F}}$ |
|---|---|---|
| $\mathfrak{g}_{\mathcal{F}_1} = \begin{pmatrix} \mathfrak{O} & \mathfrak{O} & \mathfrak{O} \\ \mathfrak{P} & \mathfrak{O} & \mathfrak{O} \\ \mathfrak{P} & \mathfrak{P} & \mathfrak{O} \end{pmatrix}$ | $\mathfrak{g}_{\mathcal{F}_1}^{+} = \begin{pmatrix} \mathfrak{P} & \mathfrak{O} & \mathfrak{O} \\ \mathfrak{P} & \mathfrak{P} & \mathfrak{O} \\ \mathfrak{P} & \mathfrak{P} & \mathfrak{P} \end{pmatrix}$ | $V_{\mathcal{F}_1} = \begin{pmatrix} k_F & 0 & 0 \\ 0 & k_F & 0 \\ 0 & 0 & k_F \end{pmatrix}$ |
| $\mathfrak{g}_{\mathcal{F}_2} = \begin{pmatrix} \mathfrak{O} & \mathfrak{O} & \mathfrak{O} \\ \mathfrak{P} & \mathfrak{O} & \mathfrak{O} \\ \mathfrak{P} & \mathfrak{O} & \mathfrak{O} \end{pmatrix}$ | $\mathfrak{g}_{\mathcal{F}_2}^{+} = \begin{pmatrix} \mathfrak{P} & \mathfrak{O} & \mathfrak{O} \\ \mathfrak{P} & \mathfrak{P} & \mathfrak{P} \\ \mathfrak{P} & \mathfrak{P} & \mathfrak{P} \end{pmatrix}$ | $V_{\mathcal{F}_2} = \begin{pmatrix} k_F & 0 & 0 \\ 0 & k_F & k_F \\ 0 & k_F & k_F \end{pmatrix}$ |
| $\mathfrak{g}_{\mathcal{F}_3} = \begin{pmatrix} \mathfrak{O} & \mathfrak{O} & \mathfrak{O} \\ \mathfrak{O} & \mathfrak{O} & \mathfrak{O} \\ \mathfrak{O} & \mathfrak{O} & \mathfrak{O} \end{pmatrix}$ | $\mathfrak{g}_{\mathcal{F}_3}^{+} = \begin{pmatrix} \mathfrak{P} & \mathfrak{P} & \mathfrak{P} \\ \mathfrak{P} & \mathfrak{P} & \mathfrak{P} \\ \mathfrak{P} & \mathfrak{P} & \mathfrak{P} \end{pmatrix}$ | $V_{\mathcal{F}_3} = \begin{pmatrix} k_F & k_F & k_F \\ k_F & k_F & k_F \\ k_F & k_F & k_F \end{pmatrix}$ |
| $\mathfrak{g}_{\mathcal{F}_4} = \begin{pmatrix} \mathfrak{O} & \mathfrak{P}^{-1} & \mathfrak{P}^{-1} \\ \mathfrak{P} & \mathfrak{O} & \mathfrak{O} \\ \mathfrak{P} & \mathfrak{O} & \mathfrak{O} \end{pmatrix}$ | $\mathfrak{g}_{\mathcal{F}_4}^{+} = \begin{pmatrix} \mathfrak{P} & \mathfrak{O} & \mathfrak{O} \\ \mathfrak{P}^2 & \mathfrak{P} & \mathfrak{P} \\ \mathfrak{P}^2 & \mathfrak{P} & \mathfrak{P} \end{pmatrix}$ | $V_{\mathcal{F}_4} = \begin{pmatrix} k_F & k_F & k_F \\ k_F & k_F & k_F \\ k_F & k_F & k_F \end{pmatrix}$ |
| $\mathfrak{g}_{\mathcal{F}_5} = \begin{pmatrix} \mathfrak{O} & \mathfrak{O} & \mathfrak{P}^{-1} \\ \mathfrak{O} & \mathfrak{O} & \mathfrak{P}^{-1} \\ \mathfrak{P} & \mathfrak{P} & \mathfrak{O} \end{pmatrix}$ | $\mathfrak{g}_{\mathcal{F}_5}^{+} = \begin{pmatrix} \mathfrak{P} & \mathfrak{P} & \mathfrak{O} \\ \mathfrak{P} & \mathfrak{P} & \mathfrak{O} \\ \mathfrak{P}^2 & \mathfrak{P}^2 & \mathfrak{P} \end{pmatrix}$ | $V_{\mathcal{F}_5} = \begin{pmatrix} k_F & k_F & k_F \\ k_F & k_F & k_F \\ k_F & k_F & k_F \end{pmatrix}$ |

Finally, we determine the image $v$ of $X$ in $V_{\mathcal{F}}$ for each representative $X_\lambda$ and $X_d$ as above. The nilpotent orbit $\mathcal{O}_{(1,1,1)}$ has representative $X_{(1,1,1)} = \mathbf{0}$, with corresponding facet $\mathcal{F}_1$. Its image $v$ in $V_{\mathcal{F}_1}$ is simply the zero matrix. Similarly, $\mathcal{O}_{(2,1)}$ has representative $X_{(2,1)} = \begin{pmatrix} 0 & 1 & 0 \\ 0 & 0 & 0 \\ 0 & 0 & 0 \end{pmatrix}$, whose corresponding facet is an associate of $\mathcal{F}_2$. Its image in $V_{\mathcal{F}_2}$ is $v_{(2,1)} = \begin{pmatrix} 0 & 1 & 0 \\ 0 & 0 & 0 \\ 0 & 0 & 0 \end{pmatrix}$. For $\lambda = (3)$, with $d = \varepsilon^a \varpi^b$ $(0 \le a, b \le 2)$ and $X_d$ as in Equation (17), the image of $X_d$ in the quotient $V_{F_{b+2}}$ is

$$v_d = \begin{pmatrix} 0 & 1 & 0 \\ 0 & 0 & \overline{\mathrm{ac}}(d) \\ 0 & 0 & 0 \end{pmatrix}.$$

## 5.4 *Example:* $\mathfrak{sp}_4$

We now examine the two parametrizations and their correspondence in the case of the Lie algebra $\mathfrak{g} = \mathfrak{sp}_4$. As in Theorem 8, consider only the partitions of 4 whose odd parts have even multiplicity, i.e. $\lambda = (4), (2,2), (2,1,1)$, and $(1,1,1,1)$. Each orbit $\mathcal{O}_\lambda$ splits into a certain number of $F$-rational orbits, depending on $\lambda$. (For details, see Nevins (2011), Table 1.) Below we give the details for the rational nilpotent orbits contained in the algebraic orbit $\mathcal{O}_{(4)}$.

The partition $\lambda = (4)$ corresponds to the nilpotent matrix

$$X = \begin{pmatrix} 0 & 1 & 0 & 0 \\ 0 & 0 & 0 & 1 \\ 0 & 0 & 0 & 0 \\ 0 & 0 & -1 & 0 \end{pmatrix} = YJ_{(4)}Y^{-1}, \quad \text{where } Y = \begin{pmatrix} 1 & 1 & 1 & 0 \\ 0 & 1 & 1 & 1 \\ 0 & 0 & 0 & -1 \\ 0 & 0 & 1 & 1 \end{pmatrix}.$$

We have $m_4(\lambda) = 1$ and $m_i(\lambda) = 0$ for $i = 1, 2, 3$, and therefore the vector space $V \simeq F^4$ satisfies $V = V(4)$. By Theorem 9, we may use Equation (8) with $X = X\,|_{V(4)}$, $j = 4$, $N = 2$, and $m_j = 1$ to determine representatives of the $F$-rational nilpotent orbits in $\mathcal{O}_\lambda(F)$. By the results of Section 3.3, there are four minimal matrix representatives of quadratic forms of dimension $m_4 = 1$, given by

$$X_a = \begin{pmatrix} 0 & 1 & 0 & 0 \\ 0 & 0 & 0 & a \\ 0 & 0 & 0 & 0 \\ 0 & 0 & -1 & 0 \end{pmatrix},$$

such that $a$ runs over the set $\{1,\ \varepsilon,\ \varpi,\ \varepsilon\varpi\}$, where $\varepsilon$ is a fixed non-square unit in $F$.

We now turn to the correspondence in Section 5.2. Fixing $X = X_a$, we have $S_j = \emptyset$ for $j \neq 4$. For $j = 4$, by the given construction we see that the Witt index $m$ of $Q_4$ is equal to 0, and $M_4 = 1$. Thus

$$S_4^1 = \{e_1 - e_2\}, \quad S_4^2 = \{2e_2\}.$$

We have $Q_{\text{aniso}} = \text{diag}(a)$, so for $\alpha_1 = 2e_{M_4+1} = 2e_2$, we have $v_{\alpha_1} = \text{val}(a)$. Thus

$$H_{\lambda,\overline{Q}} = H_{e_1-e_2} \cap H_{2e_2+\text{val}(a)}.$$

Now $\text{val}(a)$ is either 0 or 1, since $\varepsilon$ is a unit. Figure 3 shows the standard apartment of $\mathfrak{sp}_4$, along with the hyperplanes $H_{e_1-e_2}$, $H_{2e_2}$, and $H_{2e_2+1}$ and the associated intersections $H_{\lambda,\overline{Q}}$ of these hyperplanes. From the diagram, it is clear that there is a unique maximal facet $\mathcal{F}_a$ (vertex) in each set $H_{\lambda,\overline{Q}}$, and $\mathcal{F}_a$ consists of a single element.

In order to calculate the associated lattices, we first determine the root spaces $\mathfrak{g}_\alpha$ for $\alpha \in \Phi$. By (10), we have

$$\Phi = \{\pm(e_1 - e_2),\ \pm(e_1 + e_2),\ \pm 2e_1,\ \pm 2e_2\}.$$

Each root space is one-dimensional, and the generators for the root spaces are given by the matrices

$$X_{e_1-e_2} = \begin{pmatrix} 0 & 1 & 0 & 0 \\ 0 & 0 & 0 & 0 \\ 0 & 0 & 0 & 0 \\ 0 & 0 & -1 & 0 \end{pmatrix} \quad X_{e_2-e_1} = \begin{pmatrix} 0 & 0 & 0 & 0 \\ 1 & 0 & 0 & 0 \\ 0 & 0 & 0 & -1 \\ 0 & 0 & 0 & 0 \end{pmatrix}$$

$$X_{e_1+e_2} = \begin{pmatrix} 0 & 0 & 0 & 1 \\ 0 & 0 & 1 & 0 \\ 0 & 0 & 0 & 0 \\ 0 & 0 & 0 & 0 \end{pmatrix} \quad X_{-(e_1+e_2)} = \begin{pmatrix} 0 & 0 & 0 & 0 \\ 0 & 0 & 0 & 0 \\ 0 & 1 & 0 & 0 \\ 1 & 0 & 0 & 0 \end{pmatrix}$$

$$X_{2e_1} = \begin{pmatrix} 0 & 0 & 1 & 0 \\ 0 & 0 & 0 & 0 \\ 0 & 0 & 0 & 0 \\ 0 & 0 & 0 & 0 \end{pmatrix} \quad X_{-2e_1} = \begin{pmatrix} 0 & 0 & 0 & 0 \\ 0 & 0 & 0 & 0 \\ 1 & 0 & 0 & 0 \\ 0 & 0 & 0 & 0 \end{pmatrix}$$

$$X_{2e_2} = \begin{pmatrix} 0 & 0 & 0 & 0 \\ 0 & 0 & 0 & 1 \\ 0 & 0 & 0 & 0 \\ 0 & 0 & 0 & 0 \end{pmatrix} \quad X_{-2e_2} = \begin{pmatrix} 0 & 0 & 0 & 0 \\ 0 & 0 & 0 & 0 \\ 0 & 0 & 0 & 0 \\ 0 & 1 & 0 & 0 \end{pmatrix}$$

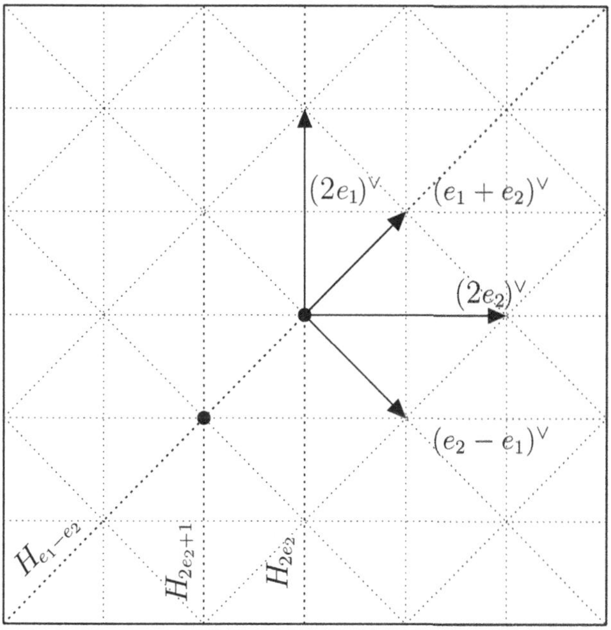


**Fig. 3** The standard affine apartment of $\mathfrak{sp}_4(F)$. *Arrows* indicate positive co-roots, and *dotted lines* indicate affine hyperplanes. *Dots* indicate the sets $H_{\lambda,\overline{\mathcal{Q}}}$

Identify $\mathcal{F}_a$ with the single element that it contains. By referencing Figure 3, we can calculate $\alpha(\mathcal{F}_a)$ for $\alpha \in \Phi$. Clearly $\alpha(\mathcal{F}_a) = 0$ for all $\alpha$ when $\mathcal{F}_a = 0$ (i.e. when $a = 1, \varepsilon$). Suppose $a = \varpi$ or $\varepsilon\varpi$, so that $\mathcal{F}_a = \frac{-1}{2}(e_1 + e_2)^\vee$. This gives the values $\pm 2e_1(\mathcal{F}_a) = \mp 1$, $\pm 2e_2(\mathcal{F}_a) = \mp 1$, $\pm(e_1 + e_2)(\mathcal{F}_a) = \mp 1$, and $\pm(e_1 - e_2)(\mathcal{F}_a) = 0$.

Using the definition of the lattices $\mathfrak{g}_{\mathcal{F}}$ and $\mathfrak{g}_{\mathcal{F}}^+$ given in (12) and (13), respectively, we compute the following:

If $a = 1$ or $\varepsilon$, we have

$$\mathfrak{g}_{\mathcal{F}_a} = \begin{pmatrix} \mathfrak{O} & \mathfrak{O} & \mathfrak{O} & \mathfrak{O} \\ \mathfrak{O} & \mathfrak{O} & \mathfrak{O} & \mathfrak{O} \\ \mathfrak{O} & \mathfrak{O} & \mathfrak{O} & \mathfrak{O} \\ \mathfrak{O} & \mathfrak{O} & \mathfrak{O} & \mathfrak{O} \end{pmatrix} \quad \text{and} \quad \mathfrak{g}_{\mathcal{F}_a}^+ = \begin{pmatrix} \mathfrak{P} & \mathfrak{P} & \mathfrak{P} & \mathfrak{P} \\ \mathfrak{P} & \mathfrak{P} & \mathfrak{P} & \mathfrak{P} \\ \mathfrak{P} & \mathfrak{P} & \mathfrak{P} & \mathfrak{P} \\ \mathfrak{P} & \mathfrak{P} & \mathfrak{P} & \mathfrak{P} \end{pmatrix}.$$

If $a = \varpi$ or $\varepsilon\varpi$, we have

$$\mathfrak{g}_{\mathcal{F}_a} = \begin{pmatrix} \mathfrak{O} & \mathfrak{O} & \mathfrak{P} & \mathfrak{P} \\ \mathfrak{O} & \mathfrak{O} & \mathfrak{P} & \mathfrak{P} \\ \mathfrak{P}^{-1} & \mathfrak{P}^{-1} & \mathfrak{O} & \mathfrak{O} \\ \mathfrak{P}^{-1} & \mathfrak{P}^{-1} & \mathfrak{O} & \mathfrak{O} \end{pmatrix} \quad \text{and} \quad \mathfrak{g}_{\mathcal{F}_a}^+ = \begin{pmatrix} \mathfrak{P} & \mathfrak{P} & \mathfrak{P}^2 & \mathfrak{P}^2 \\ \mathfrak{P} & \mathfrak{P} & \mathfrak{P}^2 & \mathfrak{P}^2 \\ \mathfrak{O} & \mathfrak{O} & \mathfrak{P} & \mathfrak{P} \\ \mathfrak{O} & \mathfrak{O} & \mathfrak{P} & \mathfrak{P} \end{pmatrix}.$$

It is clear that we have $k_F$ in each entry of the quotient in both cases, hence $V_{\mathcal{F}} \simeq \mathfrak{sp}_4(k_F)$. Finally, we determine the image $v_a$ of $X_a$ in $V_{\mathcal{F}}$. Then we have

$$v_a = \begin{pmatrix} 0 & 1 & 0 & 0 \\ 0 & 0 & 0 & \overline{\mathrm{ac}}(a) \\ 0 & 0 & 0 & 0 \\ 0 & 0 & -1 & 0 \end{pmatrix} \quad \text{for each } a \in \{1, \varepsilon, \varpi, \varepsilon\varpi\}.$$

# 6 Shalika Germs

## 6.1 The Main Results

Here we prove that the so-called *provisional* Shalika germs are motivic (in the terminology of Kottwitz 2005, § 6). Harish-Chandra defined Shalika germs on the full Lie algebra, using their homogeneity (see Kottwitz 2005, § 17 for a detailed discussion). Here we will show, roughly, that for every nilpotent orbit there exists a motivic function that coincides (up to a motivic constant) with the Shalika germ corresponding to that orbit on a definable neighbourhood of the origin. However, rescaling any given element of the Lie algebra so that it would fall into this neighbourhood presents a slight problem from the definable point of view, and so we shall address the full question of homogeneity elsewhere. It turns out that the existence of motivic functions that represent the Shalika germs in some small neighbourhood of the origin is sufficient for the application we have in mind, namely, the uniform-in-$p$ bound on the normalized Shalika germs, which appears in Theorem 17 below.

**Theorem 15.** *Let $\mathfrak{g} = \mathfrak{sl}_n$ or $\mathfrak{sp}_{2n}$. Let $\mathcal{N}$ be the set of nilpotent elements in $\mathfrak{g}$. Then*

(1) *There exists a definable set $\mathcal{E}$, such that $\mathcal{E}_F$ is finite for all fields $F$ of sufficiently large residue characteristic, and a definable function $h : \mathcal{N} \to \mathcal{E}$, such that for every $d \in \mathcal{E}$, $h^{-1}(d)$ is an adjoint orbit, and each orbit appears as the fibre of $h$.*
(2) *There exist motivic functions $\mathbf{\Gamma}$ on $\mathcal{E} \times \mathfrak{g}^{\mathrm{rss}}$ and $C$ on $\mathcal{E}$, and a constant $M > 0$, such that for all local fields $F$ of residue characteristic greater than $M$, for every $d \in \mathcal{E}_F$, the function $C^{-1}\Gamma_F(d, \cdot)$ is a representative of the Shalika germ on $\mathfrak{g}^{\mathrm{rss}}$ corresponding to $d$, i.e., coincides with the Shalika germ on some neighbourhood of the origin.*

*Proof.* **(1).** First, note that the set of nilpotent elements $\mathcal{N}$ is, indeed, definable: it is defined by the formula $X^n = 0$ in $\mathfrak{sl}_n$ and by the formula $X^{2n} = 0$ in $\mathfrak{sp}_{2n}$.

For $\mathfrak{g} = \mathfrak{sp}_{2n}$, recall the parametrization of the nilpotent orbits from Theorem 9, and let $\mathcal{E}$ be the set of pairs $(\lambda, \overline{\mathcal{Q}})$ as in Theorem 8. Note that for each pair $(\lambda, \overline{\mathcal{Q}}) \in \mathcal{E}_F$, there is an explicit representative $X_{(\lambda,\overline{\mathcal{Q}})} \in \mathcal{N}_F$. The definition of $X_{(\lambda,\overline{\mathcal{Q}})}$ involves only constant symbols in the extended Denef-Pas language; hence, the orbit of $X_{(\lambda,\overline{\mathcal{Q}})}$ is a definable set, and the map $h$ can be seen explicitly in Theorem 9.

For $\mathfrak{g} = \mathfrak{sl}_n$, the proof is essentially carried out in Diwadkar (2006, Section 6); here, we reinterpret it using the most recent version of motivic integration, and state it more generally. We are assuming that we are working with $\mathbf{G} = \mathbf{SL}_n$, and $n$ is fixed. There is a certain awkwardness to the proof caused by the fact that quotients, even by very nice definable equivalence relations, are not easy (sometimes impossible) to code in a first-order language.

Here we need to make a construction that allows us to handle the quotient $F^\times/(F^\times)^m$, where $F$ is the valued field, and so we use the union of the languages $\mathcal{L}_{\mathrm{DP}m}$ defined above in Section 2.3, as $m$ runs over the divisors of $n$.

More precisely, for every partition $\lambda$ of $n$, add the symbols for constants of the valued field sort $d_{\lambda,1}, \dots, d_{\lambda,m}$, where $m = \gcd(\lambda)$.

Now, define the set $\mathcal{E}$ as the disjoint union over all partitions $\lambda$ of $n$, of sets $\mathcal{E}_\lambda$, defined as follows. Given a partition $\lambda$, let $m = \gcd(\lambda)$ as above. We have $m$ constant symbols corresponding to this partition, $d_{\lambda,1}, \dots, d_{\lambda,m}$, in the language. Recall the formulas $\phi_{\ell,m}$ from (4) in Section 2.3. With this value of $m$, exactly one of the formulas $\psi_{\ell,m} :=$ '$\exists y_1, \dots, y_\ell\ \phi_{\ell,m}(y_1, \dots, y_\ell)$' holds. If $\psi_{\ell,m}$ holds, we interpret the constant symbols $d_{\lambda,1}, \dots, d_{\lambda,\ell}$ as units of the valued field such that $\phi_{\ell,m}(\overline{\mathrm{ac}}(d_{\lambda,1}), \dots, \overline{\mathrm{ac}}(d_{\lambda,\ell}))$ holds. Set the rest of the $d_i$ equal to 1. (Note that this construction of the language is consistent with Section 2.3, but incorporates the union over $\lambda$.) Then let

$$\mathcal{E}_\lambda := \cup_{k=0}^{m-1}\{\varpi^k d_{\lambda,1}, \dots, \varpi^k d_{\lambda,m}\}.$$

This is a definable set since it consists of just the constant symbols in the language.

Now, for every partition $\lambda$, and every $d \in \mathcal{E}_{\geq F}$, we have the elements $X_d$ as defined in Proposition 5. The disjoint union of the orbits of these elements is precisely $\mathcal{N}_F$.

We will also need two observations about the set $\mathcal{E}$ for the proof of the second part of the theorem:

(1) The set $\mathcal{E}^{\geq a}$ parametrizing orbits of dimension at least $a$ is definable for $a = 0, \ldots, n$.
(2) The cardinality of the set $\mathcal{E}^{\geq a}$ is bounded independently of the field.

The first observation holds since the dimension of the orbit depends only on the partition $\lambda$; hence, the set $\mathcal{E}^{\geq a}$ is the disjoint union of $\mathcal{E}_\lambda$ over a prescribed set of partitions $\lambda$ that depends only on $a$. The second observation is immediate from the definition. Indeed, with the above notation, in the case of $\mathfrak{sl}_n$ the upper bound on the size of $\mathcal{E}^{\geq a}$ is given by $N(n, a) := \sum_\lambda \gcd(\lambda)^2$, where the sum is over the partitions $\lambda$ of $n$ that give rise to orbits of dimension at least $a$. In the case of $\mathfrak{sp}_{2n}$, the statement is trivial since the number of orbits of a given dimension is field-independent from the start.

Now we turn our attention to Part (2).

**(2).** First, let us discuss the restriction on the residue characteristic of $F$. We will be using Cluckers et al. (2014a, Corollary 4.4) which states (in our case, without exponentials):

> Given a family of definable test functions $\{f_a\}_{a \in S} \subset C_c^\infty(\mathfrak{g})$ with some definable set $S$, there exists a constant $M$ and a motivic function $h$ on $\mathfrak{g} \times S$ such that for all non-Archimedean local fields $F$ of residue characteristic greater than $M$,
>
> $$\mu_X(f_a) = h_F(X, a).$$

Here we use this corollary with $S = \mathcal{E}$. Recall from Part (1) that there exists a constant $M_0$ such that for all $F$ with residue characteristic greater than $M_0$, for every $d \in \mathcal{E}_F$, we have an element $X_d \in \mathfrak{g}(F)$ (an explicit matrix whose entries are constant symbols in the language $\mathcal{L}_{\mathrm{DP}m}$ for some $m$), and the set $\{X_d\}_{d \in \mathcal{E}_F}$ is a set of representatives of nilpotent orbits in $\mathfrak{g}(F)$. By the matching theorem (Theorem 13 for the case of $\mathfrak{sl}_n$ and Theorem 14 for $\mathfrak{sp}_{2n}$), there exists a unique pair $(\mathcal{F}, v)$ that corresponds to the orbit of $X_d$; in particular, $X_d \in \mathfrak{g}_{\mathcal{F}}$ and $v = X_d \mod \mathfrak{g}_{\mathcal{F}}^+ \in V_{\mathcal{F}}$. Note that $\mathfrak{g}_{\mathcal{F}}^+$ is an open compact subset of $\mathfrak{g}(F)$, and so is its translate $X_d + \mathfrak{g}_{\mathcal{F}}^+$. Let $f_d$ be the characteristic function of the coset $X_d + \mathfrak{g}_{\mathcal{F}}^+$. It is definable by Cluckers et al. (2014a, Lemma 3.2). Thus we have a family of definable test functions $\{f_d\}_{d \in \mathcal{E}}$ indexed by the definable set $\mathcal{E}$. Let $M$ be the maximum of $M_0$ and the constant from the statement of Cluckers et al. (2014a, Corollary 4.4) quoted above, for this specific family. This will be the constant that appears as the restriction on the residue characteristic in our theorem.

Now we are ready to prove the statement of Part (2), for fields $F$ with residue characteristic greater than $M$. The argument proceeds by downward induction on the dimension of the nilpotent orbit. The base case is an orbit of the top dimension, and the idea is to construct a definable test function whose support intersects only this orbit, which allows us to isolate the Shalika germ attached to the chosen orbit. For orbits of smaller dimension it is of course not possible to isolate a single Shalika germ, but it is possible to construct a definable test function whose support intersects only the given orbit and orbits of strictly higher dimension. This is where a theorem of Barbasch and Moy, refined by DeBacker and quoted above as Lemma 11, is needed, and this is how downward induction on the dimension proceeds.

Thus, for the base case, let $\lambda = (n)$ be the partition that gives rise to the orbits of the maximal dimension, which we denote by $a_{\max}$. Let $F$ be a local field with residue characteristic greater than $M$, and let $d \in \mathcal{E}_{\geq F}$. Let $X_d$ be the explicit representative of the corresponding orbit, as above. Let $f_d$ be the corresponding test function constructed above, i.e., the characteristic function of the coset $X_d + \mathfrak{g}_{\mathcal{F}+}$, with $(\mathcal{F}, v)$ the pair corresponding to $X_d$.

By Lemma 11, the orbit of $X_d$ is the unique nilpotent orbit of minimal dimension intersecting $X_d + \mathfrak{g}_{\mathcal{F}}^{+}$; since there are no orbits of dimension greater than $a_{\max}$, the orbit of $X_d$ is the unique nilpotent orbit intersecting the support of the test function $f_d$. Thus, for the test function $f_d$ the Shalika germ expansion has only one term, namely

$$\mu_X(f_d) = \Gamma_{X_d}(X)\mu_{X_d}(f_d),$$

where the expansion holds for $X \in U_{f_d} \cap \mathfrak{g}(F)^{\mathrm{rss}}$, with $U_{f_d}$ some neighbourhood of the origin (which depends on the test function $f_d$). By Cluckers et al. (2014a, Corollary 4.4), $\mu_{X_d}(f_d)$ is a motivic function of $d$ (that is, for a fixed $d$, a motivic constant, which we denote by $C(d)$), and $\mu_X(f_d)$ is a motivic function of $X$ and $d$. More precisely, there exists a motivic function $\Gamma(X, d)$ such that $\Gamma_F(X, d) = \mu_X(f_d)$. (Note that here we are using our definition of the constant $M$, and the assumption that the residue characteristic of $F$ is greater than $M$.) If necessary, we can shrink $U_d$ to make it definable. (Since $U_d$ is open, there exists a lattice of the form $\mathfrak{g}_{x,r}$, i.e., defined entirely by inequalities on the valuations of the entries of $X$, which is contained in $U_d$.) This establishes the base case.

Now, let us assume the statement of the theorem holds for the orbits of dimension at least $a$ (where $a$ is an even integer). Let $F$ be as above, and let $d \in \mathcal{E}_F$ be a point such that the orbit of $X_d$ has dimension $a - 2$. As above, there exists a unique pair $(\mathcal{F}, v)$ that corresponds to the orbit of $X_d$, (i.e. the orbit of $X_d$ is the unique orbit of minimal dimension intersecting $X_d + \mathfrak{g}_{\mathcal{F}}^{+}$). Let $f_d$ be the characteristic function of the coset $X_d + \mathfrak{g}_{\mathcal{F}}^{+}$, as above. Then the intersection of its support with $\mathcal{N}_F$ is the union of its intersection with the orbit of $X_d$, and subsets of orbits of strictly higher dimension, i.e., of dimension at least $a$. Then there exists a neighbourhood of 0, which we will denote by $U_{f_d}$, such that for $X \in U_{f_d}$,

$$\mu_X(f_d) = \Gamma_{X_d}(X)\mu_{X_d}(f_d) + \sum_{d' \in \mathcal{E}_F^{\geq a}} \Gamma_{X_{d'}}(X)\mu_{X_{d'}}(f_d),$$

where the sum runs over the set of representatives of nilpotent orbits of dimension at least $a$. As in the base case, we can shrink $U_{f_d}$ to make it definable. Then, for $X \in U_{f_d}$, we have

$$\Gamma_{X_d}(X)\mu_d(f_d) = \mu_X(f_d) - \sum_{d' \in \mathcal{E}_F^{\geq a}} \Gamma_{X_{d'}}(X)\mu_{X_{d'}}(f_d). \tag{18}$$

The right-hand side of (18) is almost a motivic function. (Almost, because the Shalika germs corresponding to the orbits of greater dimension labelled by the points $d'$ are ratios of motivic functions and motivic constants.) More precisely, by the inductive assumption, for the Shalika germs occurring on the right-hand side of (18), we have $\Gamma_{X_{d'}} = C(d')^{-1}\Gamma_F(X, d')$ in some definable neighbourhood $U_{d'}$ of the origin. Let $U$ be the intersection of $U_{f_d}$ and all the $U_{d'}$ where $d'$ runs over $\mathcal{E}^{\geq a}$.

Clearing denominators on both sides, we see that it remains only to prove that the product of the motivic constants $\prod_{\mathcal{E}^{\geq a}} C(d')$ is itself a motivic constant. In the case of $\mathfrak{sp}_{2n}$ this is clear, since the indexing set in the product is field-independent, so we just have a fixed finite product of motivic constants. In the case of $\mathfrak{sl}_n$, recall the sets $\mathcal{E}^{\geq a}$ defined in Part (1) above. Let $P_a$ be the set of partitions $\lambda$ that give rise to the orbits of dimension at least $a$, so that $\mathcal{E}^{\geq a} = \cup_{\lambda \in P_a} \mathcal{E}_\lambda$. Recall from Part (1) that for each partition $\lambda$ we have the constant symbols $d_{\lambda,1}, \dots, d_{\lambda,m_\lambda}$, where $m_\lambda = \gcd(\lambda)$, and some of these constants specialize to 1 in a given field $F$, depending on the number of roots of unity in $F$. By definition, we have

$$\mathcal{E}^{\geq a} = \bigsqcup_{\lambda \in P_a} \bigsqcup_{j=0}^{m_\lambda - 1} \{\varpi^j d_{\lambda,1}, \dots, \varpi^j d_{\lambda,m_\lambda}\}.$$

Let us define, for each $\lambda \in P_a$ and each $\ell, j$ with $0 \leq j \leq m_\lambda - 1$ and $1 \leq \ell \leq m$, a constant function

$$\varphi_{\ell,j} := \begin{cases} 1 & \text{if } d_{\lambda,\ell} = 1 \\ C(d') & \text{if } d' = \varpi^j d_{\lambda,\ell} \text{ with } d_{\lambda,\ell} \neq 1. \end{cases}$$

Then we can write

$$\prod_{\mathcal{E}^{\geq a}} C(d') = \prod_{\lambda \in P_a} \left( \prod_{\ell=1}^{m_\lambda} \prod_{j=0}^{m_\lambda - 1} \varphi_{\ell,j} \prod_{j=0}^{m_\lambda - 1} C(\varpi^j) \right).$$

(Note that here, because of our convention on the interpretation of the symbols $d_{\lambda,\ell}$ in the given field, the trivial coset of $k^\times/(k^\times)^m$ plays a special role. The functions $\varphi_{\ell,j}$ are defined in order to remove all occurrences of the trivial coset from the product, and the last factor in the product formula above re-introduces the trivial coset, counted with correct multiplicity.) Thus we have represented $\prod_{\mathcal{E}\geq a} C(d')$ as a product of a fixed (i.e. field-independent) number of motivic constants, which proves it is itself a motivic constant, and completes the proof of the induction step. □

We note for future reference that the motivic constants $C(d')$ are positive, since they are obtained as products of volumes of definable sets.

## 6.2 Corollaries

The first consequence of Theorem 15 is an alternate proof that Shalika germ expansion holds in large positive characteristic. (This is already known, thanks to the work of DeBacker.) Indeed, if an equality of motivic functions holds in characteristic zero, it holds in large positive characteristic by the Transfer Principle of Cluckers and Loeser (2008).

However, the results of Appendix B to Shin and Templier 2015, allow us to also prove a different type of corollary. First, we must recall some notation and a theorem of Harish-Chandra, which we quote here from Kottwitz (2005, Theorem 17.9).

For a regular semisimple element $X$ of $\mathfrak{g}$, let $D(X) = \prod_{\alpha\in\Phi} |\alpha(X)|$ be the *Weyl discriminant* of $X$ (cf. Kottwitz 2005, § 7 for alternative definitions). For $d \in \mathcal{E}_F$, let $\overline{\Gamma}_d(X) := |D(X)|^{1/2}\Gamma_d(X)$ be the normalized Shalika germ. Note that here we mean the canonical Shalika germ, not just the provisional Shalika germ considered above in Theorem 15; thus, it is a function defined on the set of all regular semisimple elements in $\mathfrak{g}(F)$. Let $\mathbf{T}$ be a maximal torus of $\mathbf{G}$, and $\mathfrak{t}$ its Lie algebra. Harish-Chandra proved the following result.

**Theorem 16 (Harish-Chandra (1999), Kottwitz (2005, Theorem 17.9)).** *Every normalized Shalika germ $\overline{\Gamma}_d$ is a locally bounded function on $\mathfrak{t}$. (Here the local field $F$ is assumed to have characteristic zero.)*

Now, suppose we fix a definable compact subset in $\mathfrak{t}$ (or, more generally, a family of such definable compact subsets), so that we can vary the local field and still talk about the bound for the normalized Shalika germs, restricted to the specific set. We can ask, how does the bound on $\overline{\Gamma}_d$ depend on the field $F$? (Or, on the compact subset in question?) The next theorem answers both questions. Note that it is more convenient for us to talk about subsets of $\mathfrak{g}$ rather than subsets of $\mathfrak{t}$. Since there are finitely many conjugacy classes of tori (with an upper bound on their number independent of the field), and since Shalika germs are conjugation-invariant, the local boundedness on $\mathfrak{g}^{\mathrm{rss}}$ follows.

We would like to state a general result on the dependence of the bound for Shalika germs restricted to a compact subset of $\mathfrak{g}(F)$ on the field $F$ and on the compact subset in question. Typically, the compact subsets one is interested in are Moy-Prasad filtration lattices or other similar subsets. Since here we are working with explicitly defined Lie algebras, we will say that the compact sets $K_n$ form a *family of congruence lattices* if every set $K_n$ is defined by the formulas

$$\mathrm{ord}(X_{ij}) \geq \alpha_{ij}(n),$$

where $\alpha_{ij}$ are $\mathbb{Z}$-valued Presburger-definable functions of the parameter $n$. We believe most natural situations where questions about uniform bounds arise should satisfy this property; for example, Moy-Prasad filtration lattices in classical Lie algebras satisfy this condition.

**Theorem 17.** *Let $\mathfrak{g} = \mathfrak{sl}_n$ or $\mathfrak{sp}_{2n}$. Let $K_n$ be a family of congruence lattices in $\mathfrak{g}$, indexed by a parameter $n \in \mathbb{Z}$. Then there exists a constant $M > 0$ (that depends only on the formulas defining the family $K_n$), and definable $\mathbb{Z}$-valued functions $a$ and $b$ on $\mathcal{E}$, such that for all local fields $F$ with residue characteristic greater than $M$, for every $d \in \mathcal{E}_F$,*

$$|\overline{\Gamma}_d(X)| \leq q^{a(d)+b(d)n} \text{ for } X \in K_{nF} \cap \mathfrak{g}^{\mathrm{rss}}(F), \tag{19}$$

*where $q$ is the cardinality of the residue field of $F$.*

*Proof.* Note that since $\mathcal{E}_F$ is a finite set with an upper bound on its cardinality independent of $F$, an equivalent formulation would be to demand the existence of constants $a$ and $b$ such that (19) holds, independently of $d$.

Let $U_d$ be the neighbourhood given in the proof of Theorem 15. Recall that this is a definable neighbourhood on which the Shalika germ expansion holds for the specific test function $f_d$ constructed in that proof. Let $U$ be the intersection of the definable sets $U_d$, for $d \in \mathcal{E}_F$. (It is non-empty and definable since the cardinality of $\mathcal{E}_F$ is bounded independently of $F$.) Then on the set $U$, by Theorem 15, we have that

$$\overline{\Gamma}_d(X) = |D(X)|^{1/2} C(d)^{-1} \Gamma_{dF}(X),$$

where $C$ is a motivic function of $d$, and $\Gamma_d$ is a motivic function of $d$ and $X$. The discriminant $D(X)$ is a definable function since it is a polynomial in the entries of $X$ (cf. Kottwitz 2005, § 7.5), hence, $|D(X)|^{1/2}$ is a motivic function in our sense (cf. Cluckers et al. 2014a, §B.3.1). Thus the right-hand side is the ratio of a motivic function of $X$ and $d$, and a motivic function $C(d)$. Since for each $d$, $C(d)$ is a positive motivic constant, i.e., an element of $\mathbb{Z}[q^{-1}, (1-q^i)^{-1}, i > 0]$, and since $\#\mathcal{E}_F$ is bounded independently of $F$, there exist constants $a_1, a_2 \geq 0$ such that $q^{a_2} \geq C(d) \geq q^{-a_1}$ for all $d \in \mathcal{E}_F$. By Harish-Chandra's Theorem, quoted above as Theorem 16, for every local field $F$ of characteristic zero, there exists a constant $A_{dF}$ (that depends on $F$) such that

$$|\overline{\Gamma}_d(X)| \leq A_{dF} \text{ for } X \in U_F.$$

Therefore, for the fields $F$ of characteristic zero, we have an estimate for the motivic function $|D(X)|^{1/2}\Gamma_{dF}(X)$, given by

$$|D(X)|^{1/2}|\Gamma_{dF}(X)| = C(d)|\overline{\Gamma}_d(X)| \leq q^{a_2}A_{dF} \text{ for } X \in U_F.$$

Then by the uniform boundedness principle for motivic functions, Shin and Templier (2015, Theorem 14.6), there exist constants $M$ and $a_d \in \mathbb{Z}$ such that for all local fields $F$ with residue characteristic greater than $M$, we have

$$|D(X)|^{1/2}|\Gamma_{dF}(X)| \leq q^{a_d} \text{ for } X \in U_F.$$

Finally, we obtain, for $X \in U_F$, that $|\overline{\Gamma}_d(X)| \leq q^{a_d+a_1}$. Let $a(d) = a_d + a_1$. Note that since all of the functions involved were motivic functions of $d$, this constant $a(d)$ depends definably on $d$ (though as noted at the beginning of the proof, this seems to be unimportant). Thus, we have proved the theorem for one specific definable open compact set – namely, $U$.

Now we can extend it to an arbitrary family of congruence lattices $\{K_n\}_{n>0}$ using the homogeneity of Shalika germs. Namely, for every $n$ there exists an integer $j(n)$ such that $\varpi^{j(n)}K_n \subset U$, and $j(n)$ is a Presburger-definable function of $n$. Indeed, $U$ has to contain some congruence lattice defined by $\mathrm{ord}(X_{ij}) \geq \beta_{ij}$, with some constants $\beta_{ij} \in \mathbb{Z}$. Then we can take $j(n) := \max_{i,j}(-\alpha_{ij}(n) + \beta_{ij})$, where $\alpha_{ij}$ are the functions in the definition of the family $\{K_n\}_{n>0}$. Then, by definition of the canonical Shalika germs $\Gamma_d$, we have

$$\Gamma_d(X) = |\varpi^{j(n)}|^m \Gamma_d(\varpi^{2j(n)}X),$$

where $m$ is the dimension of the nilpotent orbit with parameter $d$. Note that by definition, for $t \in F^{\times}$, we have $|D(tX)| = |t|^{\dim(\mathfrak{g})-r}|D(X)|$, where $r$ is the rank of $\mathfrak{g}$ (cf. Kottwitz 2005, (17.11.2)). Putting this together, we obtain, for $X \in K_{nF}$,

$$\begin{aligned}\overline{\Gamma}_d(X) &= |D(X)|^{1/2}\Gamma_d(X) = |D(\varpi^{2j(n)}X)|^{1/2}|\varpi^{-j(n)}|^{\dim(\mathfrak{g})-r}\,|\varpi^{j(n)}|^m\,\Gamma_d(\varpi^{2j(n)}X)\\ &= q^{j(n)(\dim(\mathfrak{g})-r-m)}\,\overline{\Gamma}_d(\varpi^{2j(n)}X).\end{aligned}$$

Therefore, we have

$$|\overline{\Gamma}_d(X)| \leq q^{a(d)+j(n)(\dim(\mathfrak{g})-r-m)}.$$

It remains only to observe that since $j(n)$ is a Presburger function on $\mathbb{Z}$, it is piecewise-linear, and the statement follows. □

**Acknowledgements** This paper clearly owes a debt to the ideas of T.C. Hales and to the thesis of Jyotsna Diwadkar. The second author is grateful to Raf Cluckers and Immanuel Halupczok for multiple helpful communications. We thank the organizers of the WIN workshop in Luminy who made this collaboration possible. The second and third authors were supported by NSERC.

## References


Barbasch, D., Moy, A.: Local character expansions. Ann. Sci. École Norm. Sup. (4) **30**(5), 553–567 (1997) (English, with English and French summaries). MR1474804 (99j:22021)

Cluckers, R., Loeser, F.: Constructible motivic functions and motivic integration. Invent. Math. **173**(1), 23–121 (2008)

Cluckers, R., Loeser, F.: Motivic integration in all residue field characteristics for Henselian discretely valued fields of characteristic zero. J. Reine Angew. Math. **701**, pp. 1–31, (2015)

Cluckers, R., Gordon, J., Halupczok, I.: Local integrability results in harmonic analysis on reductive groups in large positive characterisitic. Ann. Sci. Ec. Norm. Sup. http://arxiv.org/abs/1111.7057 **47**(6) pp. 1163–1195 (2014a)

Cluckers, R., Gordon, J., Halupczok, I.: Motivic functions, integrability, and applications to harmonic analysis on p-adic groups. Electron. Res. Announc. (2014b). http://arxiv.org/abs/1111.7057 **21**, pp. 137–152, (2014)

Cluckers, R., Hales, T., Loeser, F.: Transfer principle for the fundamental lemma. In: Clozel, L., Harris, M., Labesse, J.-P., Ngô, B.-C. (eds.) On the Stabilization of the Trace Formula. International Press, Boston (2011b)

Collingwood, D.H., McGovern, W.M.: Nilpotent Orbits in Semisimple Lie Algebras. Van Nostrand Reinhold Mathematics Series. Van Nostrand Reinhold Co., New York (1993). MR1251060 (94j:17001)

DeBacker, S.: Homogeneity results for invariant distributions of a reductive p-adic group. Ann. Sci. École Norm. Sup. (4) **35**(3), 391–422 (2002a) (English, with English and French summaries). MR1914003 (2003i:22019)

DeBacker, S.: Parametrizing nilpotent orbits via Bruhat-Tits theory. Ann. Math. (2) **156**(1), 295–332 (2002b). MR1935848 (2003i:20086)

Diwadkar, J.M.: Nilpotent conjugacy classes of reductive p-adic Lie algebras and definability in Pas's language. Ph.D. Thesis, University of Pittsburgh (2006)

Harish-Chandra: Harmonic analysis on reductive p-adic groups. In: Moore, C.C. (ed.) Harmonic Analysis on Homogeneous Spaces. Proceedings of Symposia in Pure Mathematics, vol. 26, pp. 167–192. American Mathematical Society, Providence (1973). MR0340486 (49 #5238)

Harish-Chandra: Admissible invariant Distributions on Reductive p-Adic Groups. University Lecture Series, vol. 16. American Mathematical Society, Providence (1999). With a preface and notes by Stephen DeBacker and Paul J. Sally, Jr. MR1702257 (2001b:22015)

Kottwitz, R.E.: Harmonic analysis on reductive p-adic groups and Lie algebras. In: Harmonic Analysis, the Trace Formula, and Shimura Varieties, pp. 393–522 (2005). MR2192014, American mathematical Society, Providence, RI for Clay Mathematics Institute, Cambridge, MA, (2006m:22016)

Lam, T.Y.: Introduction to Quadratic Forms Over Fields. Graduate Studies in Mathematics, vol. 67. American Mathematical Society, Providence (2005)

McNinch, G.J.: Nilpotent orbits over ground fields of good characteristic. Math. Ann. **329**(1), 49–85 (2004). MR2052869 (2005j:17018)

Moy, A., Prasad, G.: Unrefined minimal K-types for p-adic groups. Invent. Math. **116**(1–3), 393–408 (1994). MR1253198 (95f:22023)

Nevins, M.: On nilpotent orbits of $SL_n$ and $Sp_{2n}$ over a local non-Archimedean field. Algebra Representation Theory **14**, 161–190 (2011)

Rabinoff, J.: The Bruhat-Tits building of a p-adic Chevalley group and an application to representation theory. Harvard Senior Thesis (2005)

Ranga Rao, R.: Orbital integrals in reductive groups. Ann. Math. (2) **96**, 505–510 (1972). MR0320232 (47 #8771)

Robson, L.: Shalika Germs are Motivic. M.Sc. Essay, University of British Columbia, Vancouver (2012)

Shalika, J.A.: A theorem on semi-simple $\mathcal{P}$-adic groups. Ann. Math. (2) **95**, 226–242 (1972). MR0323957 (48 #2310)

Shin, S.W., Templier, N.: Sato-Tate Theorem for Families and Low-Lying Zeroes of Automorphic L-functions. Preprint. http://arxiv.org/abs/1208.1945. With appendices by R. Kottwitz and J. Gordon, R. Cluckers and I. Halupczok, Invent. Math. (2015)

Waldspurger, J.-L.: Intégrales orbitales nilpotentes et endoscopie pour les groupes classiques non ramifiés. Astérisque **269**, vi+449 (2001)

# The Conjectural Relation Between Generalized Shalika Models on $SO_{4n}(F)$ and Symplectic Linear Models on $Sp_{4n}(F)$: A Toy Example

Agnès David, Marcela Hanzer, and Judith Ludwig

**Abstract** We show that if an irreducible admissible representation of $SO_4(F)$ has a generalized Shalika model, its theta lift to $Sp_4(F)$ is non-zero and has a symplectic linear model.

## 1 Introduction

The recent progress towards proving the Local Langlands Conjectures for classical groups (cf. Arthur 2013; Gan and Takeda 2010a, 2011; Cogdell et al. 2011; Jiang and Soudry 2012; Ginzburg et al. 2001 and many more) increased the interest in understanding characterizations of images of Langlands functorial transfers and the finer structures of $L$- and $A$-packets. One way of distinguishing representations is by the models they have. As an example of how models can be used to characterize images of transfers consider the following situation: Let $F/\mathbb{Q}_p$ be a finite extension and let $\tau$ be an irreducible unitary supercuspidal representation of $GL_{2n}(F)$. Then (cf. Jiang et al. 2010b, Theorem 1.1) $\tau$ is a local Langlands functorial transfer from $SO_{2n+1}(F)$ if and only if $\tau$ has a Shalika model. Furthermore it turns out that the existence of certain models of representations of different groups is very much related through Langlands type correspondences. In this article we investigate how generalized Shalika models on the split group $SO_4(F)$ are related to symplectic linear models on $Sp_4(F)$ via the local theta correspondence.

A. David
Mathematical Research Unit, University of Luxembourg, Luxembourg, Luxembourg

M. Hanzer (✉)
Department of Mathematics, University of Zagreb, Zagreb, Croatia
e-mail: hanmar@math.hr

J. Ludwig
Mathematical Institute, University of Bonn, Bonn, Germany

 
M.J. Bertin et al. (eds.), *Women in Numbers Europe*, Association for Women in Mathematics Series 2, DOI 10.1007/978-3-319-17987-2_4

More precisely in Jiang et al. (2010a) the authors conjecture the following:

*Conjecture 1.1 (Jiang et al. (2010a), p. 542).* Let $\pi$ be an irreducible admissible representation of $SO_{4n}(F)$ which has a generalized Shalika model. Then the representation $\theta(\pi)$ of $Sp_{4n}(F)$ associated with $\pi$ via the local theta correspondence is non-zero and has a symplectic linear model.

Here, and in the remainder of the paper, $F/\mathbb{Q}_p$ is a finite extension. Furthermore $\theta(\pi, m)$ denotes the "small" theta lift of a representation $\pi$ and $\Theta(\pi, m)$ denotes the "big" theta lift of $\pi$ to the symplectic group $Sp_{2m}(F)$ (cf. Kudla 1996, p. 33). If $m$ is understood, we denote $\Theta(\pi, m)$ by $\Theta(\pi)$ and $\theta(\pi, m)$ by $\theta(\pi)$. The dual pair used in this theta correspondence consists of a symplectic and a full orthogonal group and the restriction to the special orthogonal group is explained below.

The goal of this article is to prove

**Theorem (Theorem 5.2).** *Conjecture 1.1 is true for* $n = 1$.

The result that led to the conjecture in the first place and provides evidence for it is

**Theorem 1.1 (Jiang et al. (2010b), Theorems 1.1 and 1.2).** *Let* $\tau$ *be an irreducible unitary supercuspidal representation of* $GL_{2n}(F)$ *which has a Shalika model. Then the Langlands quotient* $\pi$ *of the induced representation* $\tau\nu^{1/2} \rtimes_{SO_{4n}(F)} 1$ *has a generalized Shalika model. The Langlands quotient* $\sigma$ *of the induced representation* $\tau\nu^{1/2} \rtimes_{Sp_{4n}(F)} 1$ *has a symplectic linear model. Furthermore* $\theta(\pi) = \sigma$.

Here $\nu$ denotes the character of $GL_{2n}(F)$ obtained by composing the determinant with the norm on the non-archimedean field $F$ and we may regard any character of $F^*$ as a character of $GL_{2n}(F)$ analogously. Throughout the text we use Zelevinsky's notation for the parabolic induction for the general linear groups and for classical groups as introduced, e.g., in Tadić (1998) Sections 1 and 2.

We will eventually prove Theorem 5.2 by reducing it to a calculation of Jacquet-modules. The following two results make this reduction possible.

**Theorem 1.2 (Jiang et al. (2013), Theorem 1.2).** *Let* $\pi$ *be an irreducible admissible representation of* $SO_{4n}(F)$ *and assume it has a generalized Shalika model. Then there exists an irreducible admissible representation* $\tau$ *of* $GL_{2n}(F)$ *such that* $\pi$ *is a quotient of the induced representation*

$$\tau\nu^{1/2} \rtimes_{SO_{4n}(F)} 1 \twoheadrightarrow \pi.$$

The following theorem is the specialization of Theorem 1.3 in Jiang et al. (2013) to our situation ($n = 1$).[1]

[1]The $\theta_{\mathcal{N}_1}$ model is the Shalika model for $GL_2$, the other one the symplectic model, see the paragraph before Theorem 1.3 in Jiang et al. (2013).

**Theorem 1.3 (Jiang et al. (2013), Theorem 1.3).** *Let $\tau$ be an irreducible admissible representation of $GL_2(F)$. If the induced representation $\tau\nu^{1/2} \rtimes_{SO_4(F)} 1$ has a generalized Shalika model, then $\tau$ either has a symplectic model or a Shalika model.*

The strategy to prove Theorem 5.2 is as follows: The theorems quoted above allow to first study the representations $\tau$. We determine the set of representations $\tau$ that admit a symplectic model or a Shalika model. We then analyze them case by case and study the representations obtained when we parabolically induce $\tau$ to representations of $SO_4(F)$ and $Sp_4(F)$, respectively. We check when these induced representations have the respective models. In particular we need to make sure in any of the cases that the set of $\tau$'s for which the induction to $SO_4(F)$ has a generalized Shalika model agrees with the set of $\tau$'s for which the induction to $Sp_4(F)$ has a symplectic linear model. Finally we verify that the models factor through the relevant quotients of the inductions and that these quotients are related via the theta correspondence.

The key to get our hands on these representations is the following: We can show the existence of some of the models by proving that a certain (twisted) Jacquet-module has a trivial quotient. We prove the existence of these quotients by calculating the Jacquet-modules explicitly using the Geometric Lemma of Bernstein and Zelevinsky (see Theorem 5.1 in Bernstein and Zelevinsky 1977).

As the groups we study have such small rank we can explicitly describe the possible representations $\tau$ and can then do explicit Jacquet-module calculations to find all the information on the models we need. In higher rank we cannot pin down the representations $\tau$ as explicitly. From the general version of Theorem 1.3 we know that they all have a $\theta_{\mathcal{N}_r}$-model, but there is no explicit description of such representations. Therefore our method of analyzing everything explicitly case by case will not be successful and proving Conjecture 1.1 in general will require a different approach.

The plan of the paper is as follows: In Section 2 we recall the various models and determine the set of representations $\tau$ of $GL_2(F)$ that admit a symplectic model or a Shalika model. We also give some background on the theta correspondence. We then prove Theorem 5.2 case by case in Sections 3–5 depending on the properties of $\tau$. Section 3 deals with the square-integrable case. In Section 4 we study the case where $\tau$ is a character. We finish the proof by treating the case where $\tau$ is an irreducible principal series representation in Section 5.

## 2 Notation and Preliminaries

We recall the various models occurring in these notes specialized to the case at hand. For the general definitions we refer to Jiang et al. (2010a) Section 2. Let $F/\mathbb{Q}_p$ be a finite extension and fix a non-trivial additive character $\psi : F \to \mathbb{C}^*$. Let

$$J_n := \begin{pmatrix} & & & 1 \\ & & 1 & \\ & \iddots & & \\ 1 & & & \end{pmatrix} \in \mathrm{GL}_n(F)$$

and set $J := J_4$. In the special orthogonal group $\mathrm{SO}_4$, whose $F$-points are given by

$$\mathrm{SO}_4(F) = \{A \in \mathrm{GL}_4(F) | \, {}^T AJA = J, \det A = 1\},$$

we fix the maximal diagonal torus $T$ and the Borel subgroup $B$ of upper triangular matrices. We let $P = MN$ be the standard maximal parabolic subgroup, whose Levi subgroup $M$ is isomorphic to $\mathrm{GL}_2$.

It is embedded via

$$\iota : \mathrm{GL}_2(F) \hookrightarrow \mathrm{SO}_4(F), \; g \mapsto \begin{pmatrix} g & 0 \\ 0 & J_2{}^T g^{-1} J_2 \end{pmatrix}$$

and the $F$-points of its unipotent radical $N$ are given by all matrices

$$y(X) = \begin{pmatrix} I_2 & X \\ 0 & I_2 \end{pmatrix},$$

such that ${}^T X = -J_2 X J_2$. We refer to $P$ as the Siegel subgroup. The subgroup $\mathcal{H} \subset P(F)$ generated by all $\iota(g)$ for $g \in \mathrm{Sp}_2(F)$ and all $y \in N(F)$ is called the *generalized Shalika subgroup* of $\mathrm{SO}_4(F)$. We extend $\psi$ to a character $\psi_{\mathcal{H}} : \mathcal{H} \to \mathbb{C}^*$ by $\psi_{\mathcal{H}}(y(X)) = \psi\left(tr\left(\begin{pmatrix} -1 & 0 \\ 0 & 1 \end{pmatrix} X\right)\right)$ and by demanding it is trivial on $\iota(\mathrm{Sp}_2(F))$.

**Definition 2.1.** An irreducible admissible representation $\pi$ of $\mathrm{SO}_4(F)$ is said to have a generalized Shalika model if

$$\mathrm{Hom}_{\mathcal{H}}(\pi, \psi_{\mathcal{H}}) \neq 0.$$

**Definition 2.2.** Let $\tau$ be an irreducible admissible representation of $\mathrm{GL}_2(F)$.

- The representation $\tau$ has a Shalika model if

$$\mathrm{Hom}_{\mathcal{S}}(\tau, \psi_{\mathcal{S}}) \neq 0,$$

where $\mathcal{S} = \left\{ \begin{pmatrix} a & x \\ 0 & a \end{pmatrix} \middle| a \in F^*, x \in F \right\} \subset \mathrm{GL}_2(F)$ is the Shalika subgroup and we have extended $\psi$ to a character $\psi_{\mathcal{S}} : \mathcal{S} \to \mathbb{C}^*$, $\psi_{\mathcal{S}}\left(\begin{pmatrix} a & x \\ 0 & a \end{pmatrix}\right) = \psi(x/a)$.

- The representation $\tau$ has a symplectic model if

$$\mathrm{Hom}_{\mathrm{Sp}_2(F)}(\tau, 1_{\mathrm{Sp}_2(F)}) \neq 0.$$

- $\tau$ has a linear model if

$$\mathrm{Hom}_{\mathrm{GL}_1(F)\times\mathrm{GL}_1(F)}(\tau, 1_{\mathrm{GL}_1(F)\times\mathrm{GL}_1(F)}) \neq 0.$$

*Remark.* In the previous definitions, we followed the conventions from Jiang et al. (2010a). As it is clear from the definitions, and will be mentioned below in Lemma 2.2, an irreducible representation of $\mathrm{GL}_2(F)$ having a Shalika model necessarily has trivial central character. There is also a more general definition of the Shalika model, accounting for the representations which do not necessarily have trivial central character (cf. Gan and Takeda 2010b).

**Theorem 2.1 (Jacquet and Rallis (1996), Section 6).** *If an irreducible admissible representation $\tau$ of $\mathrm{GL}_{2n}(F)$ has a Shalika model, then $\tau$ has a linear model.*

In the symplectic group $\mathrm{Sp}_4$, whose $F$-points are given by

$$\mathrm{Sp}_4(F) = \left\{ A \in \mathrm{GL}_4(F) \,\middle|\, {}^T A \begin{pmatrix} 0 & J_2 \\ -J_2 & 0 \end{pmatrix} A = \begin{pmatrix} 0 & J_2 \\ -J_2 & 0 \end{pmatrix} \right\},$$

we fix the maximal diagonal torus $T$ and the Borel subgroup $B$ of upper triangular matrices. We have a standard maximal parabolic subgroup $P = MN$ with Levi $M \cong \mathrm{GL}_2$ embedded via

$$\mathrm{GL}_2(F) \hookrightarrow \mathrm{Sp}_4(F),\ g \mapsto \begin{pmatrix} g & 0 \\ 0 & J_2{}^T g^{-1} J_2 \end{pmatrix}.$$

The group $\mathrm{Sp}_2(F) \times \mathrm{Sp}_2(F)$ injects into $\mathrm{Sp}_4(F)$ via

$$\left( \begin{pmatrix} a & b \\ c & d \end{pmatrix}, \begin{pmatrix} w & x \\ y & z \end{pmatrix} \right) \mapsto \begin{pmatrix} a & & & b \\ & w & x & \\ & y & z & \\ c & & & d \end{pmatrix}. \tag{1}$$

**Definition 2.3.** An irreducible admissible representation $\sigma$ on $\mathrm{Sp}_4(F)$ has a symplectic linear model if

$$\mathrm{Hom}_{\mathrm{Sp}_2(F)\times\mathrm{Sp}_2(F)}(\sigma, 1_{\mathrm{Sp}_2(F)\times\mathrm{Sp}_2(F)}) \neq 0.$$

*Remark.* Note that in the definitions of models all representations are assumed to be irreducible. If the corresponding Hom-space for an admissible not necessarily irreducible representation is non-zero, we speak of *functionals* instead of *models*, so, e.g., if $\pi$ is a possibly reducible admissible representation of $SO_4(F)$ such that $\mathrm{Hom}_{\mathcal{H}}(\pi, \psi_{\mathcal{H}}) \neq 0$, we say that $\pi$ has a non-zero generalized Shalika functional.

**Lemma 2.2.** *Let $\tau$ be an irreducible admissible representation of* $\mathrm{GL}_2(F)$.

1. *If $\tau$ has a symplectic model, then $\tau$ is a character.*
2. *The representation $\tau$ has a Shalika model if and only if $\tau$ is generic with trivial central character.*

*Proof.* For the proof of the first part note that $\mathrm{Sp}_2 = \mathrm{SL}_2$. So $\tau$ has a symplectic model if and only if there exists a non-zero functional $\lambda \in \mathrm{Hom}_{\mathrm{SL}_2(F)}(\tau, 1_{\mathrm{SL}_2})$. Then $V_\tau/ker(\lambda) \cong 1_{\mathrm{SL}_2(F)}$ as a representation of $\mathrm{SL}_2(F)$. The restriction of any smooth irreducible representation of $\mathrm{GL}_2(F)$ to $\mathrm{SL}_2(F)$ decomposes into a finite direct sum of irreducible representations, each occurring with multiplicity one (cf. Labesse and Langlands 1979, Lemmas 2.4 & 2.6). Furthermore the representations occurring in the restriction are permuted by any set $S$ of representatives of $\mathrm{GL}_2(F)/\mathrm{SL}_2(F)F^*$. More precisely for any two irreducible smooth representations $\tau_1$ and $\tau_2$ occurring in $\tau|_{\mathrm{SL}_2(F)}$, there exists $g \in S$ such that $\tau_1 \cong {}^g\tau_2$, where ${}^g\tau_2$ is the representation given by ${}^g\tau_2(h) = \tau_2(g^{-1}hg)$ for $h \in \mathrm{SL}_2(F)$. Therefore if $\tau|_{\mathrm{SL}_2(F)}$ contains the trivial representation as a subrepresentation, $\tau = \chi$ is a character.

For the second part we unravel the definitions to see that any non-zero $\lambda \in \mathrm{Hom}_{\mathcal{S}}(\tau, \psi_{\mathcal{S}})$ is in fact a Whittaker functional on $V_\tau$. Furthermore we have

$$\lambda(\omega_\tau(t)v) = \lambda\left(\tau\left(\begin{pmatrix} t & 0 \\ 0 & t \end{pmatrix}\right)v\right) = \lambda(v)$$

for all $v \in V_\tau$ and $t \in F^*$ if and only if the central character $\omega_\tau$ is trivial. □

So here are the options for $\tau$:

1. The representation $\tau$ is supercuspidal. Note that having trivial central character implies that $\tau$ is unitary and so we are in special case of Theorem 1.1 above, where the implication of Conjecture 1.1 is known to hold.
2. $\tau$ is a generic subquotient of a reducible principal series. Then $\tau$ is an essentially square-integrable representation with trivial central character and in particular $\tau$ is unitary. It follows that

$$\tau \hookrightarrow \chi\nu^{1/2} \times \chi\nu^{-1/2} \cong \chi(\nu^{1/2} \times \nu^{-1/2})$$

and therefore $\tau \cong \chi\,\mathrm{St}_{\mathrm{GL}_2(F)}$, where $\mathrm{St}_{\mathrm{GL}_2(F)}$ denotes the Steinberg representation of $\mathrm{GL}_2(F)$. The condition that the central character is trivial furthermore implies that $\chi^2 = 1$.
3. The representation $\tau$ is a character. Note we can write $\chi = \chi_0\nu^s$, where $\chi_0$ is unitary and $\nu^s = |det|^s$, where $s \in \mathbb{R}$.

4. $\tau$ is an irreducible principal series representation, $\tau \cong \chi_1\nu^{s_1} \times \chi_2\nu^{s_2}$, where $\chi_1, \chi_2$ are unitary characters and $s_1, s_2 \in \mathbb{R}$. The condition that the central character is trivial gives

$$\chi_1\chi_2 = 1 \quad \text{and} \quad s_1 + s_2 = 0.$$

Therefore $\tau \cong \chi\nu^s \times \chi^{-1}\nu^{-s}$ for a unitary character $\chi$.

We briefly explain how to restrict the theta correspondence between symplectic and full orthogonal groups to the correspondence between representations of symplectic and special orthogonal groups. Let $\emptyset_{2n}(F)$ be the element

$$\epsilon = \begin{pmatrix} I_{n-1} & & & \\ & & 1 & \\ & 1 & & \\ & & & I_{n-1} \end{pmatrix}.$$

For an irreducible admissible representation $\tau$ of $\mathrm{SO}_{2n}(F)$, we denote by $\tau^\epsilon$ the representation of $\mathrm{SO}_{2n}(F)$ on the same space, defined by $\tau^\epsilon(g) = \tau(\epsilon g \epsilon^{-1})$. Recall that we can pass between irreducible admissible representations of $\mathrm{O}_{2n}(F)$ and $\mathrm{SO}_{2n}(F)$ as follows:

**Lemma 2.3 (cf. Mœglin et al. (1987) 3.II.5, Lemme).**

1. *Let $\pi$ be an irreducible admissible representation of $\mathrm{O}_{2n}(F)$. Then $\pi|_{\mathrm{SO}_{2n}(F)}$ is irreducible if and only if $\pi \ncong \pi \otimes \det$.*
2. *Let $\tau$ be an irreducible admissible representation of $\mathrm{SO}_{2n}(F)$. Then either (A) $\tau \ncong \tau^\epsilon$, in which case $\mathrm{Ind}_{\mathrm{SO}_{2n}(F)}^{\mathrm{O}_{2n}(F)}(\tau) =: \pi$ is irreducible and satisfies $\pi = \pi \otimes \det$, or (B) $\tau \cong \tau^\epsilon$ in which case $\mathrm{Ind}_{\mathrm{SO}_{2n}(F)}^{\mathrm{O}_{2n}(F)}(\tau)$ is reducible and the direct sum of two non-equivalent irreducible representations $\pi$ and $\pi \otimes \det$.*

We fix a non-trivial additive character $\phi$ of $F$. All Weil representations occurring in this article will be with respect to this character. Furthermore for $n = 1, 2$ we fix the splittings $\mathrm{O}_{2n}(F) \times \mathrm{Sp}_{2n}(F) \to Mp_{4n^2}(F)$ and for later purposes $\mathrm{O}_4(F) \times \mathrm{Sp}_2(F) \to Mp_8(F)$ as described explicitly in Kudla (1994).

*Remark.* Note that we do not demand that $\phi = \psi$, where $\psi$ entered the definition of the generalized Shalika model. The theta correspondence in general does depend on the character $\phi$ of $F$, but in our case, the theta lifts of representations of $\mathrm{SO}_4(F)$ we consider are the same for every choice of $\phi$. This follows from our explicit computations of theta lifts below and can be explained, e.g., by the description of these representations as Langlands quotients, cf. Appendix C of Gan and Ichino (2014).

Using the above lemma we can restrict the theta correspondence, i.e., we can relate the largest $\mathrm{Sp}_{2n}(F)$-invariant quotient which is an isotype of $\tau$ in the appropriate Weil representation with the similar quotient corresponding to $\pi$ as follows: Let $\tau$ be an irreducible admissible representation of $SO_{2n}(F)$. Then

$$\theta(\tau, n) := \theta(\pi, n) \text{ if (A)} \tag{2}$$

$$\theta(\tau, n) := \theta(\pi, n) \oplus \theta(\pi \otimes det, n) \text{ if (B)}$$

*Remark.* We often use the following fact: assume that $\pi$ is an irreducible representation of $O_4(F)$ such that $\pi \cong \pi \otimes \det$. Then, the first occurrence index of $\pi$ in theta correspondence, denoted by $n(\pi)$, is exactly 2, i.e., $\pi$ occurs for the first time in theta correspondence with $Sp_4(F)$. This follows from the general fact (Sun and Zhu 2012):

**Theorem 2.4.** *Assume that $\sigma$ is an irreducible admissible representation of the split $O_{2n}(F)$. Then the following holds:*

$$n(\sigma) + n(\sigma \otimes \det) = 2n.$$

## 3 The Case of Square-Integrable $\tau$

From now on, if $\pi$ is a standard representation of a classical group, we denote by $L(\pi)$ its Langlands quotient.

**Lemma 3.1.** *Let $\chi$ be a quadratic character of $F^*$. Then the representation $\chi\, St_{GL_2(F)}\, \nu^{1/2} \rtimes_{Sp_4(F)} 1$ is of length three if $\chi \neq 1$ and of length two if $\chi = 1$. In the case $\chi = 1$ there is an irreducible, tempered subrepresentation (necessarily generic), and in the case $\chi \neq 1$ two non-equivalent, irreducible, square-integrable subrepresentations. The Langlands quotient $L(\chi\, St_{GL_2(F)}\, \nu^{1/2} \rtimes_{Sp_4(F)} 1)$ has a symplectic linear model.*

*Proof.* The reducibility issues are dealt with in Sally and Tadić (1993), Proposition 5.4 and Theorem 5.2. On the other hand, we know that $\chi\, St_{GL_2(F)}$ has a Shalika model and so by Theorem 2.1 it also has a linear model. Then we reason as in Ginzburg et al. (1999), p. 878 to conclude that $\chi\, St_{GL_2(F)}\, \nu^{1/2} \rtimes_{Sp_4(F)} 1$ has a non-zero symplectic linear functional. Since irreducible generic representations cannot have a symplectic linear model (cf. Ginzburg et al. 1999, Theorem 1), in the case of $\chi = 1$ we see that this functional factors through the Langlands quotient and gives a model. In the case $\chi \neq 1$, for a fixed non-degenerate character $\psi$, one of the two square-integrable representations is $\psi$-generic, and the other is not. It is not difficult to see that the other one is generic with respect to some other generic character, and the conclusion follows. □

**Lemma 3.2.** *Let $\chi$ be a quadratic character of $F^*$. Then the representation $\chi\, St_{GL_2(F)}\, \nu^{1/2} \rtimes_{SO_4(F)} 1$ is of length two and its Langlands quotient $L(\chi\, St_{GL_2(F)}\, \nu^{1/2} \rtimes_{SO_4(F)} 1)$ has a generalized Shalika model.*

*Proof.* In this proof, we use the equality sign to denote the equality of the representations up to the semisimplification. Note that $\chi \rtimes_{O_2(F)} 1$ is reducible. Thus, the same Jacquet module calculation as in (cf. Sally and Tadić 1993, Proposition 5.4) gives us that the representation $\chi\nu^{1/2}\,\mathrm{St}_{\mathrm{GL}_2(F)} \rtimes_{O_4(F)} 1$ is of length three and we have (thus, up to the semisimplification)

$$\chi\nu^{1/2}\,\mathrm{St}_{\mathrm{GL}_2(F)} \rtimes_{O_4(F)} 1 = L(\chi\nu^{1/2}\,\mathrm{St}_{\mathrm{GL}_2(F)} \rtimes 1) + \pi_1 + \pi_2,$$

where the $\pi_i, i = 1, 2$ are (mutually non-isomorphic) square integrable representations. Note that also $\pi_1 \hookrightarrow \chi\nu^1 \rtimes \chi$, and $\pi_2 \hookrightarrow \chi\nu^1 \rtimes \chi \otimes \det$, because $\chi \rtimes 1 = \chi 1_{O_2(F)} \oplus \chi\det_{O_2(F)}$. From this it follows that $\pi_1 \otimes \det \hookrightarrow \chi\nu^1 \rtimes \chi \otimes \det$, and analogously, $\pi_2 \otimes \det \hookrightarrow \chi\nu^1 \rtimes \chi$, so that $\pi_1 \cong \pi_2 \otimes \det$. We conclude that $L(\chi\nu^{1/2}\,\mathrm{St}_{\mathrm{GL}_2(F)} \rtimes_{O_4(F)} 1) \otimes \det \cong L(\chi\nu^{1/2}\,\mathrm{St}_{\mathrm{GL}_2(F)} \rtimes_{O_4(F)} 1)$. Therefore the restriction of $\chi\nu^{1/2}\,\mathrm{St}_{\mathrm{GL}_2(F)} \rtimes_{O_4(F)} 1$ to $SO_4(F)$ decomposes as

$$\begin{aligned}
&\chi\nu^{1/2}\,\mathrm{St}_{\mathrm{GL}_2(F)} \rtimes_{O_4(F)} 1|_{SO_4(F)}\\
&\quad= (\chi\nu^{1/2}\,\mathrm{St}_{\mathrm{GL}_2(F)} \rtimes_{SO_4(F)} 1) \oplus (\chi\nu^{1/2}\,\mathrm{St}_{\mathrm{GL}_2(F)} \rtimes_{SO_4(F)} 1)^{\epsilon}\\
&\quad= L(\chi\nu^{1/2}\,\mathrm{St}_{\mathrm{GL}_2(F)} \rtimes_{O_4(F)} 1)|_{SO_4(F)} + \pi_1|_{SO_4(F)} + \pi_2|_{SO_4(F)}\\
&\quad= L(\chi\nu^{1/2}\,\mathrm{St}_{\mathrm{GL}_2(F)} \rtimes_{SO_4(F)} 1) + (L(\chi\nu^{1/2}\,\mathrm{St}_{\mathrm{GL}_2(F)} \rtimes_{SO_4(F)} 1))^{\epsilon} + \pi_1|_{SO_4(F)}\\
&\quad\quad + \pi_2|_{SO_4(F)}
\end{aligned}$$

where $\pi_1|_{SO_4(F)} = \pi_2|_{SO_4(F)}$ and these two representations are irreducible (as $\pi_1 \not\cong \pi_2$). We conclude

$$\chi\nu^{1/2}\,\mathrm{St}_{\mathrm{GL}_2(F)} \rtimes_{SO_4(F)} 1 = L(\chi\nu^{1/2}\,\mathrm{St}_{\mathrm{GL}_2(F)} \rtimes_{SO_4(F)} 1) + \pi_1|_{SO_4(F)}.$$

The existence of the generalized Shalika model on $L(\chi\nu^{1/2}\,\mathrm{St}_{\mathrm{GL}_2(F)} \rtimes_{SO_4(F)} 1)$ now follows from the general result (Jiang and Qin 2007, Theorem 3.1). Alternatively we will see that we can argue directly: the subquotient $L(\chi\nu^{1/2}\,\mathrm{St}_{\mathrm{GL}_2(F)} \rtimes_{SO_4(F)} 1)$ appears again in Proposition 4.2 and in the proof there we can see the existence of the model directly. □

Next we determine the theta lift of $L(\chi\nu^{1/2}\,\mathrm{St}_{\mathrm{GL}_2(F)} \rtimes_{SO_4(F)} 1)$. As far as we know, unlike in the case of an odd orthogonal-metaplectic pair (cf. Gan and Savin 2012), there still is no explicit description of the theta correspondence for a symplectic-even orthogonal dual pair at the same level. We, therefore, directly calculate the lift by calculating enough bits of one of its Jacquet modules. The main ingredient in these calculations is Kudla's filtration of the Jacquet modules of the Weil representations involved (cf. Kudla 1996, III.8).

For a parabolic subgroup $P$ of a group $G$ and an admissible representation $\pi$ of $G(F)$ we denote by $r_P(\pi)$ the Jacquet module of $\pi$ with respect to $P$. We define $P_1 = M_1N_1$ to be the standard parabolic subgroup of $\mathrm{Sp}_4$ with Levi $M_1$ isomorphic to $\mathrm{GL}_1 \times \mathrm{SL}_2$ and we define $Q_1 = M_1'N_1'$ to be the standard parabolic of $\mathrm{O}_4$ with Levi $M_1'$ isomorphic to $\mathrm{GL}_1 \times \mathrm{O}_2$ . The Jacquet-module functor $r_{P_1}$ induces a functor from $Rep(\mathrm{O}_4(F) \times \mathrm{Sp}_4(F))$ to $Rep(\mathrm{O}_4(F) \times M_1)$, which we will again denote by $r_{P_1}$.

For $n = 1, 2$ we denote by $\omega_{n,n}$ the Weil representation of the double cover $Mp_{4n^2}(F)$ of $\mathrm{Sp}_{4n^2}(F)$ viewed as a representation of a dual pair $(\mathrm{Sp}_{2n}(F), \mathrm{O}_{2n}(F))$ (inside $\mathrm{Sp}_{4n^2}(F)$). The Weil representation of $Mp_8(F)$ viewed as a representation of $(\mathrm{SL}_2(F), \mathrm{O}_4(F))$ will be denoted by $\omega_{1,2}$.

The next proposition could be derived from the much more general results of Gan and Ichino (2014, Proposition C.4(ii) in Appendix C), but we would have to adapt their argument a little bit since, in the notation of Gan and Ichino (2014), in our case $l = 1$ but Proposition C.4(ii) in loc. cit. describes the case of $l = -1$. So we decided to give a direct proof.

**Proposition 3.3.** *Assume $\chi^2 = 1$. Then the theta lift of the representation $L(\chi\nu^{1/2}\,\mathrm{St}_{\mathrm{GL}_2(F)} \rtimes_{\mathrm{SO}_4(F)} 1)$ to $\mathrm{Sp}_4(F)$ is*

$$\theta(L(\chi\nu^{1/2}\,\mathrm{St}_{\mathrm{GL}_2(F)} \rtimes_{\mathrm{SO}_4(F)} 1)) = L(\chi\nu^{1/2}\,\mathrm{St}_{\mathrm{GL}_2(F)} \rtimes_{\mathrm{Sp}_4(F)} 1).$$

*Proof.* Theorem 2.4 guarantees that

$$n(L(\chi\nu^{1/2}\,\mathrm{St}_{\mathrm{GL}_2(F)} \rtimes_{\mathrm{SO}_4(F)} 1)) = 2$$

with $\theta(L(\chi\nu^{1/2}\,\mathrm{St}_{\mathrm{GL}_2(F)} \rtimes_{\mathrm{SO}_4(F)} 1), 2) = \theta(L(\chi\nu^{1/2}\,\mathrm{St}_{\mathrm{GL}_2(F)} \rtimes_{\mathrm{O}_4(F)} 1), 2)$ by the calculation in the proof of Lemma 3.2 and (2). To determine the theta lift $\theta(L(\chi\nu^{1/2}\,\mathrm{St}_{\mathrm{GL}_2(F)} \rtimes_{\mathrm{O}_4(F)} 1), 2)$ we compute a part of its $r_{P_1}$-Jacquet module. We have

$$\omega_{2,2} \twoheadrightarrow L(\chi\nu^{1/2}\,\mathrm{St}_{\mathrm{GL}_2(F)} \rtimes_{\mathrm{O}_4(F)} 1) \otimes \theta(L(\chi\nu^{1/2}\,\mathrm{St}_{\mathrm{GL}_2(F)} \rtimes_{\mathrm{O}_4(F)} 1), 2),$$

so that there is a non-zero intertwining operator, say $T$, of $\mathrm{O}_4(F) \times \mathrm{GL}_1(F) \times \mathrm{SL}_2(F)$-modules such that

$$T : r_{P_1}(\omega_{2,2}) \twoheadrightarrow L(\chi\nu^{1/2}\,\mathrm{St}_{\mathrm{GL}_2(F)} \rtimes_{\mathrm{O}_4(F)} 1) \otimes r_{P_1}(\theta(L(\chi\nu^{\frac{1}{2}}\,\mathrm{St}_{\mathrm{GL}_2(F)} \rtimes_{\mathrm{O}_4(F)} 1), 2)).$$

Now, $r_{P_1}(\omega_{2,2})$ has a filtration

$$\{0\} \subset J_1^{(1)} \subset J_1^{(0)} = r_{P_1}(\omega_{2,2}),$$

such that $J_1^{(1)} \cong \mathrm{Ind}_{\mathrm{GL}_1(F) \times \mathrm{SL}_2(F) \times \mathrm{GL}_1(F) \times \mathrm{O}_2(F)}^{\mathrm{GL}_1(F) \times \mathrm{SL}_2(F) \times \mathrm{O}_4(F)}(\sigma_1 \otimes \omega_{1,1})$, where $\sigma_1$ is the representation of $\mathrm{GL}_1(F) \times \mathrm{GL}_1(F)$ by left and right translations on the space of smooth, compactly supported functions on $\mathrm{GL}_1(F)$ (denoted by $S(\mathrm{GL}_1(F))$). Furthermore we know that $J_1^{(0)}/J_1^{(1)} \cong \nu^0 \otimes \omega_{1,2}$. We have $T|_{J_1^{(1)}} \neq 0$, as otherwise we would have

$$\nu^0 \otimes \omega_{1,2} \twoheadrightarrow L(\chi\nu^{1/2}\,\mathrm{St}_{\mathrm{GL}_2(F)} \rtimes_{\mathrm{O}_4(F)} 1) \otimes r_{P_1}(\theta(L(\chi\nu^{1/2}\,\mathrm{St}_{\mathrm{GL}_2(F)} \rtimes_{\mathrm{O}_4(F)} 1), 2))$$

which would mean $n(L(\chi\nu^{1/2}\,\mathrm{St}_{\mathrm{GL}_2(F)} \rtimes_{\mathrm{O}_4(F)} 1)) \leq 1$, and that is impossible. The second Frobenius reciprocity gives us that the space of $\mathrm{GL}_1(F) \times \mathrm{SL}_2(F) \times \mathrm{GL}_1(F) \times O_2(F)$-intertwining operators from $\sigma_1 \otimes \omega_{1,1}$ to

$$\overline{r_{Q_1}(L(\chi\nu^{1/2}\,\mathrm{St}_{\mathrm{GL}_2(F)} \rtimes_{\mathrm{O}_4(F)} 1))} \otimes r_{P_1}(\theta(L(\chi\nu^{1/2}\,\mathrm{St}_{\mathrm{GL}_2(F)} \rtimes_{\mathrm{O}_4(F)} 1), 2))$$

is non-zero. Now, we use the fact that

$$r_{Q_1}(L(\chi\nu^{\frac{1}{2}}\,\mathrm{St}_{\mathrm{GL}_2(F)} \rtimes_{\mathrm{O}_4(F)} 1)) = \chi\nu^0 \otimes \chi(\nu^1 \rtimes 1)$$

(we shall prove this in Lemma 3.4). This means that there is an $\mathrm{GL}_1(F)$-epimorphism from $\sigma_1$ to $\chi\nu^0$. But the maximal isotypic component of $\chi\nu^0$ in $S(\mathrm{GL}_1(F))$ (when we view it as a $\mathrm{GL}_1(F) \times \mathrm{GL}_1(F)$-module) is again $\chi\nu^0$. This means that $r_{P_1}(\theta(L(\chi\nu^{1/2}\,\mathrm{St}_{\mathrm{GL}_2(F)} \rtimes_{\mathrm{O}_4(F)} 1), 2))$ has an irreducible subquotient of the form $\chi\nu^0 \otimes *$, where $*$ is some representation of $\mathrm{SL}_2(F)$.

Now we settle the case of $\chi \neq 1$. Then, we can read off the cuspidal support of

$$\theta(L(\chi\nu^{1/2}\,\mathrm{St}_{\mathrm{GL}_2(F)} \rtimes_{\mathrm{SO}_4(F)} 1), 2)$$

(e.g. Kudla 1996); it is a subquotient of $\chi\nu^1 \times \chi\nu^0 \rtimes_{\mathrm{Sp}_4(F)} 1$. The representation $\chi\nu^1 \times \chi\nu^0 \rtimes_{\mathrm{Sp}_4(F)} 1$ is of length six (cf. Proposition 5.4 of Sally and Tadić 1993), and has analogous subquotients as a representation $\chi\nu^1 \times \chi\nu^0 \rtimes_{\mathrm{O}_4(F)} 1$. This means that

$$r_{P_1}(L(\chi\nu^{1/2}\,\mathrm{St}_{\mathrm{GL}_2(F)} \rtimes_{\mathrm{Sp}_4(F)} 1)) = \chi\nu^0 \otimes \chi\nu^1 \rtimes_{\mathrm{SL}_2(F)} 1,$$

and the representation $L(\chi\nu^{1/2}\,\mathrm{St}_{\mathrm{GL}_2(F)} \rtimes_{\mathrm{Sp}_4(F)} 1)$ comes with the multiplicity two in $\chi\nu^1 \times \chi\nu^0 \rtimes_{\mathrm{Sp}_4(F)} 1$. Now, from the expression for $r_{P_1}(\chi\nu^1 \times \chi\nu^0 \rtimes_{\mathrm{Sp}_4(F)} 1)$ in the proof of Lemma 3.4, we see that $L(\chi\nu^{\frac{1}{2}}\,\mathrm{St}_{\mathrm{GL}_2(F)} \rtimes_{\mathrm{Sp}_4(F)} 1)$ is the only subquotient of $\chi\nu^1 \times \chi\nu^0 \rtimes_{\mathrm{Sp}_4(F)} 1$ having $\chi\nu^0 \otimes *$ in its $r_{P_1}$-Jacquet module and the conclusion follows.

Now let $\chi = 1$. We continue with the analysis of the Jacquet module $r_{P_1}(\omega_{2,2})$. Since $\Theta(1_{\mathrm{SL}_2}, 2) = \nu^1 \rtimes_{\mathrm{O}_2(F)} 1$, we have a non-zero intertwining from $\nu^0 \otimes \nu^0 \otimes \omega_{1,1} \to \nu^0 \otimes \nu^0 \otimes 1_{\mathrm{SL}_2(F)} \otimes \nu^1 \rtimes_{\mathrm{O}_2(F)} 1$. Therefore,

$$r_{P_1}(\theta(L(\nu^{1/2}\,\mathrm{St}_{\mathrm{GL}_2(F)} \rtimes_{\mathrm{O}_4(F)} 1), 2)) \geq \nu^0 \otimes 1_{\mathrm{SL}_2(F)}$$

in the appropriate Grothendieck group. We conclude that

$$\theta(L(\nu^{1/2}\,\mathrm{St}_{\mathrm{GL}_2(F)} \rtimes_{\mathrm{O}_4(F)} 1), 2) \in \{L(\nu^{1/2}\,\mathrm{St}_{\mathrm{GL}_2(F)} \rtimes_{\mathrm{Sp}_4(F)} 1), L(\nu^1 \times \nu^0 \rtimes_{\mathrm{SL}_2(F)} 1)\}.$$

This follows from the fact that

$$\nu^0 \rtimes 1_{\mathrm{SL}_2(F)} = L(\nu^{1/2}\,\mathrm{St}_{\mathrm{GL}_2(F)} \rtimes_{\mathrm{Sp}_4(F)} 1) \oplus L(\nu^1 \times \nu^0 \rtimes 1),$$

and these two representations are the only irreducible subquotients of $\nu^1 \times \nu^0 \rtimes_{\mathrm{Sp}_4(F)} 1$ with $\nu^0 \otimes 1_{\mathrm{SL}_2(F)}$ in their Jacquet module.

Assume now that $\theta(L(\nu^{1/2}\,\mathrm{St}_{\mathrm{GL}_2(F)} \rtimes_{\mathrm{O}_4(F)} 1), 2) = L(\nu^1 \times \nu^0 \rtimes 1)$. Then, we have an epimorphism

$$r_{P_1}(\omega_{2,2}) \twoheadrightarrow L(\nu^{1/2}\,\mathrm{St}_{\mathrm{GL}_2(F)} \rtimes_{\mathrm{O}_4(F)} 1) \otimes r_{P_1}(L(\nu^1 \times \nu^0 \rtimes 1)),$$

and since we have an epimorphism $r_{P_1}(L(\nu^1 \times \nu^0 \rtimes 1)) \to \nu^{-1} \otimes \nu^0 \rtimes_{\mathrm{SL}_2(F)} 1$, there is a non-zero epimorphism, say $T$

$$r_{P_1}(\omega_{2,2}) \twoheadrightarrow L(\nu^{1/2}\,\mathrm{St}_{\mathrm{GL}_2(F)} \rtimes_{\mathrm{O}_4(F)} 1) \otimes \nu^{-1} \otimes \nu^0 \rtimes_{\mathrm{SL}_2(F)} 1.$$

Now we analyze the restrictions of $T$ to the terms of the filtration of $r_{P_1}(\omega_{2,2})$. We get that there is a non-zero $\mathrm{GL}_1(F) \times \mathrm{GL}_1(F) \times \mathrm{SL}_2(F) \times \mathrm{O}_2(F)$-intertwining

$$\sigma_1 \otimes \omega_{1,1} \to \nu^0 \otimes \nu^1 \rtimes_{\mathrm{O}_2(F)} 1 \otimes \nu^{-1} \otimes \nu^0 \rtimes_{\mathrm{SL}_2(F)} 1,$$

which is impossible. This proves the proposition. □

**Lemma 3.4.** *Assume* $\chi^2 = 1$. *Then*

$$r_{Q_1}(L(\chi\nu^{1/2}\,\mathrm{St}_{GL_2(F)} \rtimes_{O_4(F)} 1)) = \chi\nu^0 \otimes \chi\nu^1 \rtimes 1.$$

*Proof.* We use the structure formula (*) on page 2 of Ban (1999) to compute the Jacquet module of the induced representation $\pi := \chi\nu^1 \times \chi\nu^0 \rtimes_{O_4(F)} 1$ with respect to $Q_1$. We get that

$$\begin{aligned} r_{Q_1}(\pi) &= \chi\nu^1 \otimes \chi\det + \chi\nu^1 \otimes \chi 1_{O_2(F)} + \chi\nu^{-1} \otimes \chi\det \\ &\quad + \chi\nu^{-1} \otimes \chi 1_{O_2(F)} + 2\chi\nu^0 \otimes \chi\nu^1 \rtimes 1. \end{aligned}$$

Since the multiplicity of $L(\chi\nu^{1/2}\,\mathrm{St}_{GL_2(F)} \rtimes_{O_4(F)} 1)$ in $\pi$ is two the lemma follows. □

## 4 The Case of the Representation $\chi\nu^s \rtimes 1$

In this section, we consider the case when the representation $\tau$ is a character of $\mathrm{GL}_2(F)$. We write $\tau$ as $\chi\nu^s$, with $\chi$ a unitary character and $s$ in $\mathbb{R}$.

## *4.1* $SO_4(F)$*-Side*

First of all we prove that for any unitary character $\chi$ and any $s \in \mathbb{R}$, the representation $\pi := \chi\nu^s \rtimes 1$ of $SO_4(F)$ has a generalized Shalika model by showing that a certain twisted Jacquet-module of $\pi$ has a trivial quotient.

The calculation is done by applying the Geometric Lemma and we adapt the notation as follows: As before let $\mathcal{H}$ be the Shalika subgroup of $SO_4(F)$. We let $P := P(F) = MV$ denote the $F$-points of the Siegel subgroup, and we write $\mathcal{H} = NV$, where $N \cong SL_2(F)$. We form the twisted Jacquet module $r_{V,\psi}(\pi)$ of $\pi$ with respect to the group $V$ and the character $\psi := \psi_{\mathcal{H}}|_V$. Recall that it is defined as the $SL_2(F)$-module given by the quotient of $\pi$ by the space

$$\pi_{V,\psi} := \mathrm{span}_{\mathbb{C}}\{\pi(X)f - \psi(X)f : X \in V, f \in \pi\}.$$

**Lemma 4.1.** *For any unitary character $\chi$ of $GL_2(F)$ and any $s$ in $\mathbb{R}$, the representation $\chi\nu^s \rtimes_{SO_4(F)} 1$ has a non-zero generalized Shalika functional.*

*Proof.* It follows from the definitions that if $r_{V,\psi}(\chi\nu^s \rtimes_{SO_4(F)} 1)$ has a trivial quotient, $\chi\nu^s \rtimes_{SO_4(F)} 1$ has a non-zero generalized Shalika functional. The geometric lemma Bernstein and Zelevinsky (1977) gives a description of the composition of functors $F := r_{V,\psi} \circ i_{P,SO_4(F)}$ from $GL_2(F)$-representations to $SL_2(F)$-representations. Here $i_{P,SO_4(F)}$ denotes the functor of normalized parabolic induction. In order to apply it note the following: Firstly, under the action of $\mathcal{H}$ by right translation, the space $P \setminus SO_4(F)$ decomposes into two orbits $P \setminus SO_4(F) = P \cup Pw_1\mathcal{H}$ (we easily get that $Pw_1P = Pw_1\mathcal{H}$), where $w_1 = \begin{bmatrix} 0 & I_2 \\ I_2 & 0 \end{bmatrix}$. Furthermore all the requirements such as good decomposition, etc. of Section 5.1 of Bernstein and Zelevinsky (1977) for the triples $(P, M, V)$ and $(\mathcal{H}, N, V)$ are satisfied. Abbreviate $Y := Pw_1\mathcal{H}$. There is an $SL_2(F)$-invariant subspace $\tau_Y$ of $\chi\nu^s \rtimes 1$ which consists of all functions in $\chi\nu^s \rtimes 1$ which vanish outside of $Pw_1\mathcal{H}$. We apply the Jacquet functor $r_{V,\psi}$ to the filtration of $SL_2(F)$-representations

$$\{0\} \subset \tau_Y \subset \chi\nu^s \rtimes 1.$$

Then Theorem 5.2 of Bernstein and Zelevinsky (1977) implies that $r_{V,\psi}(\chi\nu^s \rtimes 1/\tau_Y)$ is the $SL_2(F)$-module given by the restriction of $\chi\nu^s$ from $GL_2(F)$ to $SL_2(F)$. We get the same result for $r_{V,\psi}(\tau_Y)$. We conclude that the $SL_2(F)$–representation $r_{V,\psi}(\chi\nu^s \rtimes 1)$ has length two and each subquotient is isomorphic to the trivial representation. We conclude that $r_{V,\psi}(\chi\nu^s \rtimes 1)$ has a trivial quotient. □

We remind the reader that we need to determine the irreducible quotients of $\chi\nu^s \rtimes 1$ and decide when they have a generalized Shalika model.

**Proposition 4.2.** *Whenever the representation $\chi\nu^s \rtimes 1$ of $SO_4(F)$ is irreducible it has a generalized Shalika model.*

1. *Assume that* $\chi^2 \neq 1$. *Then the representation* $\chi\nu^s \rtimes 1$ *of* $\mathrm{SO}_4(F)$ *is irreducible.*
2. *Assume that* $\chi^2 = 1$. *The representation* $\chi\nu^s \rtimes_{\mathrm{SO}_4(F)} 1$ *is reducible if and only if* $s = \pm\frac{1}{2}$. *Then, in the appropriate Grothendieck group,* $\chi\nu^{\frac{1}{2}} \rtimes_{\mathrm{SO}_4(F)} 1 = L(\chi\,\mathrm{St}_{\mathrm{GL}_2(F)}\,\nu^{\frac{1}{2}} \rtimes_{\mathrm{SO}_4(F)} 1) + L(\chi\nu^1 \rtimes \chi)$. *The representation* $L(\chi\,\mathrm{St}_{\mathrm{GL}_2(F)}\,\nu^{\frac{1}{2}} \rtimes_{\mathrm{SO}_4(F)} 1)$ *has a generalized Shalika model and* $L(\chi\nu^1 \rtimes \chi)$ *does not admit one.*

*Proof.* Note that if the representation $\chi\nu^s \rtimes 1$ of $\mathrm{O}_4(F)$ is irreducible, then the representation $\chi\nu^s \rtimes_{\mathrm{SO}_4(F)} 1$ of $\mathrm{SO}_4(F)$ is irreducible, since we saw that $\chi\nu^s \rtimes 1|_{\mathrm{SO}_4(F)} = \chi\nu^s \rtimes_{\mathrm{SO}_4(F)} 1 + (\chi\nu^s \rtimes_{\mathrm{SO}_4(F)} 1)^{\epsilon}$. So, the reducibility points for $\chi\nu^s \rtimes_{\mathrm{SO}_4(F)} 1$ are among the reducibility points for $\chi\nu^s \rtimes 1$. A necessary condition for reducibility here is $\chi^2 = 1$. In that case, using the spinor norm, we have $\chi\nu^s \rtimes 1 \cong \chi(\nu^s \rtimes 1)$. We can extend, as we already saw, the Jacquet module calculations from the case of $\mathrm{Sp}_4(F)$ to the $\mathrm{O}_4(F)$ case if the rank-one reducibilities are the same (i.e., in the case when $\chi \rtimes 1$ reduces in $\mathrm{SL}_2(F)$). We conclude (cf. Sally and Tadić 1993, Proposition 5.4.) that the only cases of reducibility are $s = \pm\frac{1}{2}$, and the length of the representation $\chi\nu^s \rtimes 1$ is three. Now, analogously as in the case of the representation $\chi\nu^{\frac{1}{2}}\,\mathrm{St}_{\mathrm{GL}_2(F)} \rtimes 1$, when restricting to $\mathrm{SO}_4(F)$, we obtain that the length of $\chi\nu^{\frac{1}{2}} \rtimes_{\mathrm{SO}_4(F)} 1$ is two. Namely, in $\mathrm{O}_4(F)$

$$\chi\nu^{\frac{1}{2}} \rtimes 1 = L(\chi\nu^{\frac{1}{2}}\,\mathrm{St}_{\mathrm{GL}_2(F)} \rtimes 1) + L(\chi\nu^1 \rtimes \chi) + L(\chi\nu^1 \rtimes \chi \otimes \det).$$

We get that $L(\chi\nu^{\frac{1}{2}}\,\mathrm{St}_{\mathrm{GL}_2(F)} \rtimes 1) \cong L(\chi\nu^{\frac{1}{2}}\,\mathrm{St}_{\mathrm{GL}_2(F)} \rtimes 1) \otimes \det$ and $L(\chi\nu^1 \rtimes \chi)|_{\mathrm{SO}_4(F)} = L(\chi\nu^1 \rtimes \chi \otimes \det)|_{\mathrm{SO}_4(F)}$, so that

$$\chi\nu^{\frac{1}{2}} \rtimes_{\mathrm{SO}_4(F)} 1 = L(\chi\nu^{\frac{1}{2}}\,\mathrm{St}_{\mathrm{GL}_2(F)} \rtimes_{\mathrm{SO}_4(F)} 1) + L(\chi\nu^1 \rtimes \chi)|_{\mathrm{SO}_4(F)}.$$

We already know that $L(\chi\nu^{\frac{1}{2}}\,\mathrm{St}_{\mathrm{GL}_2(F)} \rtimes_{\mathrm{SO}_4(F)} 1)$ has a generalized Shalika model. Take $\chi = 1$ for a moment. Then, it is easy to see that $L(\nu^1 \rtimes 1)|_{\mathrm{SO}_4(F)}$ is actually the trivial representation of $\mathrm{SO}_4(F)$ and it does not admit a generalized Shalika model, since $\psi$ is a non-trivial character of $\mathcal{H}$. Similarly, if $\chi \neq 1$, then $\chi$ composed with the spinor norm is equal to one, since the spinor norm on the Shalika subgroup is trivial (cf. Lemma 2.2 on p. 79 of Kudla 1996 and Zassenhaus 1962 for the unipotent elements in $\mathcal{H}$). We can thus also directly see that $L(\chi\nu^{\frac{1}{2}}\,\mathrm{St}_{\mathrm{GL}_2(F)} \rtimes_{\mathrm{SO}_4(F)} 1)$ has a generalized Shalika model, since $\chi\nu^{-\frac{1}{2}} \rtimes_{\mathrm{SO}_4(F)} 1$ has one and $L(\chi\nu^1 \rtimes \chi)|_{\mathrm{SO}_4(F)}$ does not, and $L(\chi\nu^{\frac{1}{2}}\,\mathrm{St}_{\mathrm{GL}_2(F)} \rtimes_{\mathrm{SO}_4(F)} 1)$ is a quotient of $\chi\nu^{-\frac{1}{2}} \rtimes_{\mathrm{SO}_4(F)} 1$. □

## 4.2 $\mathrm{Sp}_4(F)$-*Side*

We now analyze the representation $\chi\nu^s \rtimes_{\mathrm{Sp}_4(F)} 1$, where as above $\chi$ is unitary and $s \in \mathbb{R}$. In the following $\mathcal{R}(G)$ denotes the category of smooth representations of the topological group $G$.

Let $S(\mathrm{SL}_2(F))$ denote the space of smooth, compactly supported functions on $\mathrm{SL}_2(F)$. It comes equipped with an action $R$ of the group $\mathrm{SL}_2(F) \times \mathrm{SL}_2(F)$ given by $R(g_1, g_2)\phi(x) = \phi(g_1^{-1}xg_2)$, for $\phi \in S(\mathrm{SL}_2(F))$, $(g_1, g_2) \in \mathrm{SL}_2(F) \times \mathrm{SL}_2(F)$. In the study of symplectic linear functionals on $\chi\nu^s \rtimes_{\mathrm{Sp}_4(F)} 1$ the following lemma will be very useful.

**Lemma 4.3.** *We have an exact sequence of* $\mathrm{SL}_2(F) \times \mathrm{SL}_2(F)$*-representations*

$$0 \to S(\mathrm{SL}_2(F)) \to \chi\nu^s \rtimes 1|_{\mathrm{SL}_2(F)\times\mathrm{SL}_2(F)} \to \chi\nu^{s+1/2} \rtimes 1 \otimes \chi\nu^{s+1/2} \rtimes 1 \to 0. \qquad (3)$$

*Proof.* Again, we use Theorem 5.2 of Bernstein and Zelevinsky (1977) and adapt our notation. So, $P = MU$ denotes the Siegel parabolic subgroup of $\mathrm{Sp}_4(F)$ (with our previous choice of the Borel subgroup). We let $Q = NV$ with $Q = N = \mathrm{SL}_2(F) \times \mathrm{SL}_2(F)$ and $V = \{e\}$. We decompose $r_{V,1} \circ i_{P,\mathrm{Sp}_4(F)}(\chi\nu^s)$. Here $r_{V,1}$ turns out to be just the restriction to $\mathrm{SL}_2(F) \times \mathrm{SL}_2(F)$. To meet the requirements of the geometric lemma (decomposability with respect to $Q$), we have to take $M = P$ and $U = \{e\}$. We describe $P\backslash\mathrm{Sp}_4(F)/Q$ using (Kudla 1996, Chapter 4, Proposition 2.1). We have just two orbits of $Q$-action on $P \backslash \mathrm{Sp}_4(F)$; there is an open orbit $Pw^{-1}Q$, where $w = \begin{bmatrix} I_2 & 0 \\ -I_2 & I_2 \end{bmatrix}$, and (closed) orbit $PQ$. We have the following filtration ($\tau = \chi\nu^s \rtimes 1$)

$$0 \subset \tau_1 \subset \tau,$$

where $\tau_1$ is a subset of functions in $\tau$ vanishing outside of $PwQ$. In the notation of Bernstein and Zelevinsky (1977) we get that the two subgroups which we use to decompose $r_{V,1} \circ i_{P,\mathrm{Sp}_4(F)}$ are

$$M' = \left\{ \begin{bmatrix} a & b & 0 & -b \\ c & d & -c & 0 \\ 0 & 0 & a & -b \\ 0 & 0 & -c & d \end{bmatrix} : \begin{bmatrix} a & b \\ c & d \end{bmatrix} \in \mathrm{SL}_2(F) \right\}$$

and

$$N' = wM'w^{-1} = \left\{ \begin{bmatrix} a & 0 & 0 & -b \\ 0 & d & -c & 0 \\ 0 & -b & a & 0 \\ -c & 0 & 0 & d \end{bmatrix} : \begin{bmatrix} a & b \\ c & d \end{bmatrix} \in \mathrm{SL}_2(F) \right\}.$$

Then, restricted to $\tau_1$, $r_{V,1}$ acts as $i_{N',Q} \circ w \circ r_{V',1}(\chi\nu^s)$. Note that the restriction $r_{V',1} : \mathcal{R}(P) \to \mathcal{R}(M')$ applied to $\chi\nu^s$ gives the trivial character of $M'$, and with conjugation by $w$, it again gives the trivial character of $N'$, so we have $i_{N',Q}(1)$ (compact induction). Note that $N'$ is isomorphic to $\mathrm{SL}_2(F)$, so that this isomorphism gives an embedding $\mathrm{SL}_2(F) \to \mathrm{SL}_2(F) \times \mathrm{SL}_2(F)$ with $g \mapsto (g, J_2 g J_2)$. By Proposition 2.3 in Chapter 4 of Kudla (1996) $i_{N',Q}(1) \cong S(\mathrm{SL}_2(F))$. The intertwining operator from $i_{N',Q}(1)$ to $S(\mathrm{SL}_2(F))$ is given by $T(f)(x) = f(1, J_2 x J_2)$.

On the other hand, $r_{V,1}$ on $\tau/\tau_1$ is composed of $i_{N',Q} \circ e \circ r_{V',1} : \mathcal{R}(P) \to \mathcal{R}(Q)$, where now $N' = M' = P \cap Q$, so that $r_{V',1}$ denotes the restriction of the representation of $P$ to representation of $P \cap Q$ and $i_{N',Q}$ is compact induction from representations of $P \cap Q$ to representations of $Q$. It is easy to see that $P \cap Q$ consists of matrices of the form

$$\begin{bmatrix} a_1 & 0 & 0 & b_1 \\ 0 & a_2 & b_2 & 0 \\ 0 & 0 & a_2^{-1} & 0 \\ 0 & 0 & 0 & a_1^{-1} \end{bmatrix}.$$

Note that, in the case of the normalized induction, $\chi\nu^s \rtimes 1$ as a representation of $\mathrm{Sp}_4(F)$ is actually induced from the representation $\chi\nu^s\delta_P^{1/2}$, with our choice of $M = P$ (so that $U = \{e\}$). When we restrict to $P \cap Q$, we get $(\chi\nu^{s+3/2} \otimes \chi\nu^{s+3/2})\left(\begin{bmatrix} a_1 & b_1 \\ 0 & a_1^{-1} \end{bmatrix}, \begin{bmatrix} a_2 & b_2 \\ 0 & a_2^{-1} \end{bmatrix}\right)$. When we then induce to $\mathrm{SL}_2(F) \times \mathrm{SL}_2(F)$, we get $\chi\nu^{s+3/2} \rtimes' 1 \otimes \chi\nu^{s+3/2} \rtimes' 1$ as a representations of $\mathrm{SL}_2(F) \times \mathrm{SL}_2(F)$ (the prime denotes the unnormalized induction, so in our usual notation of the normalized induction, we get $\chi\nu^{s+1/2} \rtimes 1 \otimes \chi\nu^{s+1/2} \rtimes 1$). □

We now analyze (3) to see if $\chi\nu^s \rtimes 1|_{\mathrm{SL}_2(F)\times\mathrm{SL}_2(F)}$ has a trivial quotient. We use the following fact: for any irreducible smooth representation $\sigma$ of $\mathrm{SL}_2(F)$, the largest $\sigma$-isotypic component of $S(\mathrm{SL}_2(F))|_{R_{\mathrm{SL}_2(F)\times 1}}$ is $\sigma \otimes \tilde{\sigma}$, as an $\mathrm{SL}_2(F) \times \mathrm{SL}_2(F)$-module.

**Proposition 4.4.** *Let $\chi$ be a unitary character and $s \in \mathbb{R}$. Assume $(\chi, s) \neq (1, -\frac{3}{2})$. Then the representation $\chi\nu^s \rtimes_{\mathrm{Sp}_4(F)} 1$ has a symplectic linear model. The representation $\nu^{-3/2} \rtimes_{\mathrm{Sp}_4(F)} 1$ has the trivial representation as a subquotient, and the trivial representation obviously has a symplectic linear model.*

*Proof.* Assume $\chi$ is a ramified character. Then, using the Bernstein center decomposition, we get that in that case the epimorphism $S(\mathrm{SL}_2(F)) \to 1 \otimes 1$ can be extended to $\chi\nu^s \rtimes 1|_{\mathrm{SL}_2(F)\times\mathrm{SL}_2(F)}$, since it is non-zero anyway only on the Bernstein component in which $1 \otimes 1$, as a representation of $\mathrm{SL}_2(F) \times \mathrm{SL}_2(F)$, lies. This component is different from the component in which $\chi\nu^{s+1/2} \rtimes 1 \otimes \chi\nu^{s+1/2} \rtimes 1$ lies. This means that $\chi\nu^s \rtimes 1$ has a non-zero linear symplectic model.

Assume that $\chi = 1$ and $s = \frac{1}{2}$, so that $\chi\nu^{s+1/2} \rtimes 1 \otimes \chi\nu^{s+1/2} \rtimes 1$ has $1 \otimes 1$ as a trivial quotient. Then obviously, $\chi\nu^s \rtimes 1|_{\mathrm{SL}_2(F)\times\mathrm{SL}_2(F)}$ has a trivial quotient.

Assume that $\chi$ is unramified, but $\chi\nu^{s+1/2} \neq \nu^{\pm 1}$. For a smooth representation $\pi$ of $SL_2(F) \times SL_2(F)$ we denote by $r_{V_1,V_2}(\pi)$ the Jacquet module of $\pi$ with respect to the upper-triangular unipotent subgroup of the first and, then, of the second copy of $SL_2(F)$ (we can view it as $r_{V_1}(r_{V_2}(\pi))$). We get then a smooth $GL_1(F) \times GL_1(F)$ module. We apply this Jacquet functor, which is exact, on the exact sequence (3). Since $S(SL_2(F)) \twoheadrightarrow 1_{SL_2(F)} \times 1_{SL_2(F)}$, we have $r_{V_1,V_2}(S(SL_2(F)) \twoheadrightarrow \nu^{-1} \otimes \nu^{-1}$, so that $\nu^{-1} \otimes \nu^{-1}$ is a subquotient of $r_{V_1,V_2}(\chi\nu^s \rtimes 1|_{SL_2(F)\times SL_2(F)})$. Projectivity of cuspidal representations (Lemma 26. of Bernstein 1992) gives us an epimorphism

$$r_{V_1,V_2}(\chi\nu^s \rtimes 1|_{SL_2(F)\times SL_2(F)}) \twoheadrightarrow \nu^{-1} \otimes \nu^{-1}.$$

Actually, Lemma 26 of Bernstein (1992) requires finite-length representations, but we can just apply it to the representation $\chi\nu^s \rtimes 1/W$, where $W$ is a subspace of $S(SL_2(F))$ such that $S(SL_2(F))/W \cong 1_{SL_2(F)} \times 1_{SL_2(F)}$, i.e., we apply it on $r_{V_1,V_2}(\chi\nu^s \rtimes 1/W)$ which is then clearly of finite length (as a $GL_1(F) \times GL_1(F)$-module). Frobenius reciprocity then gives rise to an $SL_2(F) \times GL_1(F)$-intertwining operator, say $T$,

$$T : r_{V_2}(\chi\nu^s \rtimes 1) \to \nu^{-1} \rtimes 1 \otimes \nu^{-1}.$$

If the image of this operator is $\nu^{-1} \rtimes 1 \otimes \nu^{-1}$, then, there would exist an epimorphism

$$T_1 : r_{V_2}(\chi\nu^s \rtimes 1) \twoheadrightarrow St_{SL_2(F)} \otimes \nu^{-1}.$$

If $T_1|_{r_{V_2}(S(SL_2(F))} = 0$, this would give us an epimorphism

$$r_{V_2}(\chi\nu^{s+1/2} \rtimes 1 \otimes \chi\nu^{s+1/2} \rtimes 1) = \chi\nu^{s+1/2} \rtimes 1 \otimes r_{V_2}(\chi\nu^{s+1/2} \rtimes 1) \twoheadrightarrow St_{SL_2(F)} \otimes \nu^{-1},$$

which is impossible, by the requirement $\chi\nu^{s+1/2} \neq \nu^{-1}$. So, we conclude that $T_1|_{r_{V_2}(S(SL_2(F))} \neq 0$, so that we have an epimorphism

$$r_{V_2}(S(SL_2(F)) \twoheadrightarrow St_{SL_2(F)} \otimes \nu^{-1}.$$

By the Frobenius reciprocity, this would give us a non-zero intertwining map

$$S(SL_2(F)) \to St_{SL_2(F)} \otimes \nu^{-1} \rtimes 1.$$

The image of this intertwiner is $St_{SL_2(F)} \otimes \nu^{-1} \rtimes 1$ or $St_{SL_2(F)} \otimes 1_{SL_2(F)}$. Note that the maximal isotypic quotient of $St_{SL_2(F)}$ in $S(SL_2(F))$ is $\Theta(St_{SL_2(F)}) = St_{SL_2(F)}$, so we would have an epimorphism $St_{SL_2(F)} \twoheadrightarrow \nu^{-1} \rtimes 1$ or $St_{SL_2(F)} \twoheadrightarrow 1_{SL_2(F)}$, which is impossible in both of the cases.

We conclude that we have an epimorphism $r_{V_2}(\chi\nu^s \rtimes 1) \to 1_{SL_2(F)} \otimes \nu^{-1}$.

Now we continue analogously: the Frobenius isomorphism gives us a non-zero $SL_2(F) \times SL_2(F)$-intertwining operator, say $T_2$, $\chi\nu^s \rtimes 1 \to 1_{SL_2(F)} \otimes \nu^{-1} \rtimes 1$.

If the image of this intertwiner were to be $1_{\mathrm{SL}_2(F)} \otimes \nu^{-1} \rtimes 1$, we would have an epimorphism, say $T_3$, from $\chi\nu^s \rtimes 1$ to $1_{\mathrm{SL}_2(F)} \otimes \mathrm{St}_{\mathrm{SL}_2(F)}$. If $T_3$ restricted to $S(\mathrm{SL}_2(F))$ is zero, this would give an epimorphism

$$\chi\nu^{s+1/2} \rtimes 1 \otimes \chi\nu^{s+1/2} \rtimes 1 \twoheadrightarrow 1_{\mathrm{SL}_2(F)} \otimes \mathrm{St}_{\mathrm{SL}_2(F)},$$

which is impossible by our choice of $\chi\nu^s$. So, $T_3$ restricted to $S(\mathrm{SL}_2(F))$ is non-zero, but this is impossible with the form of the isotypic component, recalled above. Thus, the image of $T$ is $1_{\mathrm{SL}_2(F)} \otimes 1_{\mathrm{SL}_2(F)}$, which is what we wanted. □

Using Propositions 4.2 and 4.4, we conclude

**Proposition 4.5.** *For any irreducible subquotient $\pi$ of $\chi\nu^s \rtimes_{\mathrm{SO}_4(F)} 1$ having a generalized Shalika model, its ("small") theta-lift to $\mathrm{Sp}_4(F)$ is non-zero and has a linear symplectic model.*

*Proof.* Assume $\chi^2 \neq 1$. Then, the representations $\chi\nu^s \rtimes_{\mathrm{SO}_4(F)} 1$ and $\chi\nu^s \rtimes_{\mathrm{Sp}_4(F)} 1$ are irreducible. From Theorem 2.4 we get that $n(\chi\nu^s \rtimes_{\mathrm{SO}_4(F)} 1) = 2$. Moreover, in the same way as in Proposition 3.3, we get that $\theta(\chi\nu^s \rtimes_{\mathrm{SO}_4(F)} 1, 2) = \chi\nu^s \rtimes_{\mathrm{Sp}_4(F)} 1$ and then apply Propositions 4.2 and 4.4. Assume that $\chi^2 = 1$. Then, if $\chi \neq 1$ then for $s \neq \pm\frac{1}{2}$ the representations $\chi\nu^s \rtimes_{\mathrm{SO}_4(F)} 1$ and $\chi\nu^s \rtimes_{\mathrm{Sp}_4(F)} 1$ are both irreducible and the conclusion follows as previously. For $s = \frac{1}{2}$, we know that $L(\chi\nu^{\frac{1}{2}} \mathrm{St}_{\mathrm{GL}_2(F)} \rtimes_{\mathrm{SO}_4(F)} 1)$ is a subquotient of $\chi\nu^s \rtimes_{\mathrm{SO}_4(F)} 1$ and has a non-zero generalized Shalika model, and in Proposition 3.3 we have already proved that $\theta(L(\chi\nu^{\frac{1}{2}} \mathrm{St}_{\mathrm{GL}_2(F)} \rtimes_{\mathrm{SO}_4(F)} 1)) = L(\chi\nu^{\frac{1}{2}} \mathrm{St}_{\mathrm{GL}_2(F)} \rtimes_{\mathrm{Sp}_4(F)} 1)$. The other subquotients of $\chi\nu^s \rtimes_{\mathrm{SO}_4(F)} 1$ do not have the generalized Shalika models (Proposition 4.2). We have also proved that $\theta(L(\nu^{\frac{1}{2}} \mathrm{St}_{\mathrm{GL}_2(F)} \rtimes_{\mathrm{SO}_4(F)} 1)) = L(\nu^{\frac{1}{2}} \mathrm{St}_{\mathrm{GL}_2(F)} \rtimes_{\mathrm{Sp}_4(F)} 1)$ (the case of $\chi = 1$) in (Proposition 3.3), so the conclusion is the same. Note that in the case $s = \frac{3}{2}$ the small theta lift of $\nu^{3/2} \rtimes_{\mathrm{SO}_4(F)} 1$ is the trivial representation of $\mathrm{Sp}_4(F)$ (cf. Theorem 5.1 (ii) of Kudla 1996), and all the cases are covered. □

# 5 Final Case and Proof of the Main Theorem

## 5.1 The Case of Irreducible Principal Series

In this section, we consider the case where $\tau$ is an irreducible principal series of $\mathrm{GL}_2(F)$ with trivial central character. The representation $\tau$ is of the form $\chi\nu^s \times \chi^{-1}\nu^{-s}$, with $\chi$ a unitary character and $s$ in $\mathbb{R}$. The irreducibility condition for $\tau$ is: $(\chi^2, |s|) \neq (1, \frac{1}{2})$.

**Proposition 5.1.** *Let $\chi$ be a unitary character of $\mathrm{GL}_2(F)$ and $s$ in $\mathbb{R}$, with $(\chi^2, |s|) \neq (1, \frac{1}{2})$, and let $\tau$ be the representation $\chi\nu^s \times \chi^{-1}\nu^{-s}$.*

*Then, the representation* $\tau\nu^{\frac{1}{2}} \rtimes_{SO_4(F)} 1$ *has* $\left(\chi\nu^s \rtimes_{SO_4(F)} 1\right)^{\epsilon}$ *as unique irreducible quotient. Its theta lift to* $Sp_4(F)$ *is:*

*(i)* $\chi\nu^s \rtimes_{Sp_4(F)} 1$ *if* $(\chi, s) \neq (1, \pm\frac{3}{2})$;
*(ii)* $1_{Sp_4(F)}$ *if* $(\chi, s) = (1, \pm\frac{3}{2})$.

*This theta lift has a symplectic model.*

*Proof.* We first note that the map sending $f$ to $s \mapsto f(\epsilon s \epsilon)$ gives an isomorphism:

$$
\begin{aligned}
(\tau\nu^{\frac{1}{2}} \rtimes_{SO_4(F)} 1)^{\epsilon} &= \left(\left(\chi\nu^{s+\frac{1}{2}} \times \chi^{-1}\nu^{-s+\frac{1}{2}}\right) \rtimes_{SO_4(F)} 1\right)^{\epsilon} \\
&\cong \left(\chi\nu^{s+\frac{1}{2}} \times \chi\nu^{s-\frac{1}{2}}\right) \rtimes_{SO_4(F)} 1 = \chi\nu^{s}(\nu^{\frac{1}{2}} \times \nu^{-\frac{1}{2}}) \rtimes_{SO_4(F)} 1.
\end{aligned}
$$

The last representation has a unique irreducible quotient, namely $\chi\nu^s \rtimes_{SO_4(F)} 1$. This implies that the representation $\tau\nu^{\frac{1}{2}} \rtimes_{SO_4(F)} 1$ has a unique irreducible quotient, $\left(\chi\nu^s \rtimes_{SO_4(F)} 1\right)^{\epsilon}$.

Since $\left(\chi\nu^s \rtimes_{SO_4(F)} 1\right)^{\epsilon}$ and $\chi\nu^s \rtimes_{SO_4(F)} 1$ are non-isomorphic, these two representations have the same theta lift to $Sp_4(F)$ (Lemma 2.3 and Relation (2)). Proposition 4.5 and its proof then give the desired result. □

## 5.2 Main Theorem

**Theorem 5.2.** *Let* $\pi$ *be an irreducible smooth admissible representation of* $SO_4(F)$ *with a generalized Shalika model. Then* $\theta(\pi)$ *is non-zero and has a symplectic linear model.*

*Proof.* By Theorem 1.2 the representation $\pi$ is a quotient of $\tau\nu^{1/2} \rtimes_{SO_4(F)} 1$ for an irreducible admissible representation $\tau$ of $GL_2(F)$. If $\tau$ is supercuspidal, everything is known by Theorem 1.1. So we may assume that $\tau$ is not supercuspidal.

Lemma 2.2 combined with Lemma 3.2, Propositions 4.2 and 5.1 implies that $\pi$ must then be one of the representations in the first column of the following table.

| $\pi$ | $\theta(\pi)$ |
|---|---|
| $L(\chi St_{GL_2(F)}\nu^{1/2} \rtimes_{SO_4(F)} 1)$ | $L(\chi St_{GL_2(F)}\nu^{1/2} \rtimes_{Sp_4(F)} 1)$ |
| $\chi\nu^s \rtimes_{SO_4(F)} 1$, s.t. $(\chi^2, s) \neq (1, \pm\frac{1}{2})$ and $(\chi, s) \neq (1, \pm\frac{3}{2})$ | $\chi\nu^s \rtimes_{Sp_4(F)} 1$ |
| $\nu^{\pm 3/2} \rtimes_{SO_4(F)} 1$ | $1_{Sp_4(F)}$ |
| $\left(\chi\nu^s \rtimes_{SO_4(F)} 1\right)^{\epsilon}$, s.t. $(\chi^2, \lvert s\rvert) \neq (1, 1/2)$ | $\chi\nu^s \rtimes_{Sp_4(F)} 1$ |

Note that in the first column, the first three entries indeed all have a generalized Shalika model. The last entry of the first column might have a generalized Shalika model. Note furthermore that all entries in the second column are non-zero. We have shown in Lemma 3.1 and Proposition 4.4 that all the representations in the second column have a symplectic linear model. Finally by Propositions 3.3 and 4.5 a representation in the second column is indeed the theta lift of the representation in the same line in the first column, which completes the proof. □

**Acknowledgements** This project started at the WIN-Europe conference in October 2013. We would like to thank the organizers of the conference and the CIRM in Luminy for providing such excellent working conditions. We are grateful to the referee for several helpful comments. MH has been supported in part by the Croatian Science Foundation under the project 9364. JL would like to thank Imperial College London for providing financial support in form of a Doris Chen Mobility Award.

## References

Arthur, J.: The Endoscopic Classification of Representations. American Mathematical Society Colloquium Publications, vol. 61. American Mathematical Society, Providence (2013). Orthogonal and symplectic groups

Ban, D.: Parabolic induction and Jacquet modules of representations of $O(2n, F)$. Glas. Mat. Ser. III **34**(54), 147–185 (1999)

Bernstein, I.N., Zelevinsky, A.V.: Induced representations of reductive p-adic groups I. Ann. Sci. École Norm. Sup. **10**(4), 441–472 (1977)

J. Bernstein, Draft of: Representations of p-adic groups, preprint, 1992, available at http://www.math.tau.ac.il/~bernstei

Cogdell, J.W., Piatetski-Shapiro, I.I., Shahidi, F.: Functoriality for the quasisplit classical groups. In: On Certain $L$-Functions. Clay Mathematics Institute, vol. 13, pp. 117–140. American Mathematical Society, Providence (2011)

Gan, W.T., Ichino, A.: Formal degrees and local theta correspondence. Invent. Math. **195**, 509–672 (2014)

Gan, W.T., Savin, G.: Representations of metaplectic groups I: epsilon dichotomy and local Langlands correspondence. Compos. Math. **148**, 1655–1694 (2012)

Gan, W.T., Takeda, S.: The local Langlands conjecture for Sp(4). Int. Math. Res. Not. IMRN **15**, 2987–3038 (2010a)

Gan, W.T., Takeda, S.: On Shalika periods and a theorem of Jacquet-Martin. Am. J. Math. **132**, 475–528 (2010b)

Gan, W.T., Takeda, S.: The local Langlands conjecture for GSp(4). Ann. Math. **173**(2), 1841–1882 (2011)

Ginzburg, D., Rallis, S., Soudry, D.: On a correspondence between cuspidal representations of $GL_{2n}$ and $\widetilde{Sp}_{2n}$. J. Am. Math. Soc. **12**, 849–907 (1999)

Ginzburg, D., Rallis, S., Soudry, D.: Generic automorphic forms on $SO(2n + 1)$: functorial lift to $GL(2n)$, endoscopy, and base change. Int. Math. Res. Not. **14**, 729–764 (2001)

Jacquet, H., Rallis, S.: Uniqueness of linear periods. Compos. Math. **102**, 65–123 (1996)

Jiang, D., Qin, Y.: Residues of Eisenstein series and generalized Shalika models for $SO_{4n}$. J. Ramanujan Math. Soc. **22**, 101–133 (2007)

Jiang, D., Soudry, D.: Appendix: On the local descent from $GL(n)$ to classical groups [appendix to mr2931222]. Am. J. Math. **134**, 767–772 (2012)

Jiang, D., Nien, C., Qin, Y.: Symplectic supercuspidal representations and related problems. Sci. China Math. **53**, 533–546 (2010a)

Jiang, D., Nien, C., Qin, Y.: Symplectic supercuspidal representations of $GL(2n)$ over $p$-adic fields. Pac. J. Math. **245**, 273–313 (2010b)

Jiang, D., Nien, C., Qin, Y.: Generalized Shalika models of $p$-adic $SO_{4n}$ and functoriality. Israel J. Math. **195**, 135–169 (2013)

Kudla, S.S.: Notes on the local theta correspondence (lectures at the European School in Group Theory), preprint, 1996, available at http://www.math.utoronto.ca/~skudla/castle.pdf

Kudla, S.: Notes on the local theta correspondence (lectures at the European School in Group Theory) (1996)

Labesse, J.-P., Langlands, R.P.: $L$-indistinguishability for SL(2). Can. J. Math. **31**, 726–785 (1979)

Mœglin, C., Vignéras, M.-F., Waldspurger, J.-L.: Correspondances de Howe sur un corps $p$-adique. Lecture Notes in Mathematics, vol. 1291. Springer, Berlin (1987)

Sally, Jr., P.J., Tadić, M.: Induced representations and classifications for $GSp(2, F)$ and $Sp(2, F)$. Mém. Soc. Math. France (N.S.) **52**, 75–133 (1993)

Sun, B., Zhu, C.-B.: Conservation relations for local theta correspondence. http://arxiv.org/pdf/1204.2969v2.pdf (2012)

Tadić, M.: On reducibility of parabolic induction. Israel J. Math. **107**, 29–91 (1998)

Zassenhaus, H.: On the spinor norm. Arch. Math. **13**, 434–451 (1962)

# Bad Reduction of Genus Three Curves with Complex Multiplication

**Irene Bouw, Jenny Cooley, Kristin Lauter, Elisa Lorenzo García, Michelle Manes, Rachel Newton, and Ekin Ozman**

**Abstract** Let $C$ be a smooth, absolutely irreducible genus 3 curve over a number field $M$. Suppose that the Jacobian of $C$ has complex multiplication by a sextic CM-field $K$. Suppose further that $K$ contains no imaginary quadratic subfield. We give a bound on the primes $\mathfrak{p}$ of $M$ such that the stable reduction of $C$ at $\mathfrak{p}$ contains three irreducible components of genus 1.

**MSC 2010:** 11G15, 14K22, 15B33

## 1 Introduction

In Goren and Lauter (2007), Goren and Lauter study genus 2 curves whose Jacobians are absolutely simple and have complex multiplication (CM) by the ring of integers $\mathcal{O}_K$ of a quartic CM-field $K$, and they show that if such a curve has bad reduction to characteristic $p$ then there is a solution to the embedding problem, formulated as follows Goren and Lauter (2007):

I. Bouw
Institute of Pure Mathematics, University of Ulm, Ulm, Germany

J. Cooley
Mathematics Institute, University of Warwick, Coventry, UK

K. Lauter (✉)
Microsoft Research, Redmond, WA, USA
e-mail: klauter@microsoft.com

E. Lorenzo García
Mathematisch Instituut, Universiteit Leiden, Leiden, The Netherlands

M. Manes
Department of Mathematics, University of Hawaii, Honolulu, HI, USA

R. Newton
Max Planck Institute for Mathematics, Bonn, Germany

E. Ozman
Mathematics Department, University of Texas at Austin, Austin, TX, USA

Department of Mathematics, Bogazici University, Istanbul, Turkey

 
M.J. Bertin et al. (eds.), *Women in Numbers Europe*, Association for Women in Mathematics Series 2, DOI 10.1007/978-3-319-17987-2_5

Let $K$ be a quartic CM-field which does not contain a proper CM-subfield, and let $p$ be a prime. The embedding problem concerns finding a ring embedding $\iota : \mathcal{O}_K \hookrightarrow \mathrm{End}(E_1 \times E_2)$, such that the Rosati involution coming from the product polarization induces complex conjugation on $\mathcal{O}_K$, and $E_1, E_2$ are supersingular elliptic curves over $\overline{\mathbb{F}}_p$.

In this paper, we consider genus 3 curves whose Jacobians have CM by a sextic CM-field that does not contain a proper CM-subfield. By analogy with Goren and Lauter (2007), we formulate an embedding problem for the genus 3 case as follows.

**Problem 6.3 (The Embedding Problem).** Let $\mathcal{O}$ be an order in a sextic CM-field $K$, and let $p$ be a prime number. The *embedding problem* for $\mathcal{O}$ and $p$ is the problem of finding elliptic curves $E_1, E_2, E_3$ defined over $\overline{\mathbb{F}}_p$, and a ring embedding

$$i : \mathcal{O} \hookrightarrow \mathrm{End}(E_1 \times E_2 \times E_3)$$

such that the Rosati involution on $\mathrm{End}(E_1 \times E_2 \times E_3)$ induces complex conjugation on $\mathcal{O}$. We call such a ring embedding a *solution to the embedding problem* for $\mathcal{O}$ and $p$.

In this paper, we prove the following result on solutions to the embedding problem. We refer to Section 6.3 for the precise statement.

**Theorem 6.9.** *Let $K$ be a sextic CM-field such that $K$ does not contain a proper CM-subfield. Let $\mathcal{O}$ be an order in $K$. There exists an explicit bound on the rational primes $p$ for which the embedding problem has a solution, and this bound depends only on the order $\mathcal{O}$.*

As in the genus 2 case, Theorem 6.9 yields a bound on certain primes of bad reduction for the curve $C$. However, the result is not as strong as in the genus 2 case, since there are more possibilities for the reduction of $C$. We discuss the statement of the result.

Let $C$ be a smooth, absolutely irreducible genus 3 curve over a number field $M$ whose Jacobian has CM by an order $\mathcal{O}$ in a sextic CM-field $K$. We say that $C$ has bad reduction at a rational prime $p$ if there exists a prime $\mathfrak{p}$ of $M$ above $p$ at which $C$ has bad reduction. In Corollary 4.3, we observe that if $C$ has bad reduction at a prime $\mathfrak{p}$, there are two possibilities for the stable reduction $\overline{C}_{\mathfrak{p}}$ of $C$ at $\mathfrak{p}$. Either $\overline{C}_{\mathfrak{p}}$ contains three irreducible components of genus 1 or $\overline{C}_{\mathfrak{p}}$ contains one irreducible component of genus 1 and one of genus 2.

In this paper, we restrict our attention to the first of these two possibilities. In Proposition 6.5, we show that if $C$ has bad reduction at a prime $\mathfrak{p}$ above $p$ and the stable reduction contains three genus 1 curves, then the embedding problem for $\mathcal{O}$ and $p$ has a solution. Theorem 6.9 therefore yields the following result on the primes of bad reduction of $C$.

**Theorem 6.8.** *Let $C$ be a genus 3 curve whose Jacobian has CM by an order $\mathcal{O}$ in a sextic CM-field $K$ that does not contain a proper CM-subfield. There exists an explicit bound on the primes $p$ where the stable reduction contains three irreducible components of genus* 1.

We do not consider all primes of bad reduction of $C$ in Theorem 6.8 for the following reason. If the stable reduction of $C$ at $\mathfrak{p}$ contains three irreducible components of genus 1, then the reduction $\overline{J}_{\mathfrak{p}}$ of the Jacobian $J$ of $C$ is isomorphic to the product $E_1 \times E_2 \times E_3$ of elliptic curves as polarized abelian varieties (Proposition 4.2). This yields a ring embedding

$$\iota : \mathcal{O} = \mathrm{End}(J) \hookrightarrow \mathrm{End}(\overline{J}_{\mathfrak{p}}) = \mathrm{End}(E_1 \times E_2 \times E_3),$$

which has the property that the Rosati involution on $\mathrm{End}(E_1 \times E_2 \times E_3)$ restricts to complex conjugation on the image of $\mathcal{O}$ (Section 4.3). This is precisely the statement that $\iota$ is a solution to the embedding problem for $\mathcal{O}$ and $p$.

Consider a prime $\mathfrak{p}$ where the curve $C$ has bad reduction, but the stable reduction $\overline{C}_{\mathfrak{p}}$ contains an irreducible component $E$ of genus 1 and an irreducible component $D$ of genus 2 (Corollary 4.3). In this case — an example of which is described in Section 5.2 — the reduction $\overline{J}_{\mathfrak{p}}$ of the Jacobian of $C$ is the product of $E$ with the Jacobian of $D$ as polarized abelian varieties. The abelian variety $\overline{J}_{\mathfrak{p}}$ is still isogenous to a product of elliptic curves (Theorem 4.5), but $\overline{J}_{\mathfrak{p}}$ is not isomorphic to a product of elliptic curves as polarized abelian varieties. This suggests that a different formulation of the embedding problem would be needed to draw conclusions for such primes $\mathfrak{p}$. We do not discuss the correct formulation of the embedding problem for this case in the present paper, but leave it as a direction for future work.

The assumption that the CM-field $K$ does not contain a proper CM-field is also present in the genus 2 case in Goren and Lauter (2007). However, in the genus 2 case, this assumption is equivalent to the assumption that the CM-type of the Jacobian $J$ is primitive. We refer to Section 3.4 for more details. In characteristic zero, the condition that the CM-type corresponding to $J$ is primitive is equivalent to the assumption that $J$ is absolutely simple (Theorem 3.2).

In the genus 3 case, the assumption that the CM-field $K$ does not contain a proper CM-subfield still implies that the CM-type of the Jacobian $J$ is primitive. However, the converse does not hold. Even in the case that the sextic CM-field $K$ contains a proper CM-subfield there exist primitive CM-types (Section 3). In Section 6.4, we discuss why the embedding problem needs to be formulated differently for such CM-fields. We show that, in the case where $K$ contains a proper CM-subfield, the embedding problem as we have formulated it has solutions for any prime $p$ and some order $\mathcal{O}$ of $K$.

Finally, we have not included the condition that the elliptic curves $E_i$ are supersingular in the formulation of the embedding problem, in contrast to the formulation in genus 2, because for a set of Dirichlet density $1/2$, the elliptic curves $E_i$ are ordinary.

## 1.1 Relation to a Result of Gross and Zagier

One of the motivations of Goren and Lauter for studying solutions of the embedding problem in genus 2 was generalizing a result of Gross and Zagier on singular moduli of elliptic curves in Gross and Zagier (1985). Recall that singular moduli are values $j(\tau)$ of the modular function $j$ at imaginary quadratic numbers $\tau$. Gross and Zagier define the product

$$J(d_1, d_2) = \left( \prod_{[\tau_1],[\tau_2]} (j(\tau_1) - j(\tau_2)) \right)^{4/w_1 w_2},$$

where the product runs over equivalence classes of imaginary quadratic numbers $\tau_i$ with discriminants $d_i$, where the $d_i$ are assumed to be relatively prime. Here $w_i$ denotes the number of units in $\mathbb{Q}(\tau_i)$. The function $J$ is closely related to the value of the Hilbert class polynomial of an imaginary quadratic field at a point $\tau$ corresponding to a different imaginary quadratic field.

Under some assumptions, Gross and Zagier show that $J(d_1, d_2)$ is an integer, and their main result gives a formula for the factorization of this integer. The result of Gross and Zagier may be reinterpreted as a formula for the number of isomorphisms between the reductions of the elliptic curves $E_i$ corresponding to the $\tau_i$ at all rational primes $p$. This problem is equivalent to counting embeddings of $\mathrm{End}(E_2)$ into the endomorphism ring of the reduction of $E_1$ at $p$.

Goren and Lauter (2007, Corollary 5.1.3) prove a generalization of the result of Gross and Zagier. They consider curves of genus 2 with CM by a quartic CM-field. In their result, the function $j$ is replaced by suitable Siegel modular functions $f/\Theta^k$. Here $f$ is a Siegel modular form of weight $10k$ with values in a number field and $\Theta$ is a concrete Siegel modular form of weight 10. The modular function $f/\Theta^k$ has the property that for any $\tau$ in the Siegel upper half plane the genus 2 curve corresponding to $\tau$ has bad reduction at the primes dividing the denominator of $(f/\Theta^k)(\tau)$. (See Goren and Lauter 2007, Corollary 5.1.2 for the precise statement.)

The Igusa class polynomials are an analog of the Hilbert class polynomials for quartic CM-fields, where the $j$-invariant is replaced by the absolute Igusa invariants. Goren and Lauter and collaborators (see, for example, Goren and Lauter 2007, 2013; Lauter and Viray 2012) deduce results on the denominators of the coefficients of the Igusa class polynomials from results on the embedding problem for quartic CM-fields. A different approach to generalize the results of Gross and Zagier using arithmetic intersection theory can be found in the work of Bruinier, Kudla, Yang, and collaborators (see, for example, Bruinier and Yang 2006).

The embedding problem for curves of genus 3 studied in this paper does not immediately yield a statement analogous to that of Gross and Zagier. One of the ingredients that is missing is finding good coordinates for the moduli space of curves of genus 3, analogous to the absolute Igusa invariants in genus 2.

In this paper, we discuss several differences between the reduction of CM-curves in genus 2 and in genus 3. The embedding problem in the formulation of Problem 6.3 does not cover all types of bad reduction. Also, in the case that the sextic CM-field $K$ contains a proper CM-subfield the embedding problem should be adapted. It would be interesting to study the implication of these differences for a possible analog of the Igusa class polynomials for sextic CM-fields.

## *1.2 Outline*

The structure of this paper is as follows. Section 2 gives the possibilities for the Galois group of the Galois closure of a sextic CM-field, following work of Dodson (Dodson 1984). Section 3 describes the possible CM-types for a sextic CM-field. We note which of the CM-types are primitive, meaning that they can arise as the CM-type of a simple abelian variety. In Section 4, we describe the possibilities for the reduction of a genus 3 curve and its Jacobian to characteristic $p > 0$. We also give some properties of the Rosati involution attached to a polarized abelian variety, which will be used in Section 6. In Section 5, we give various examples of genus 3 curves with CM; we calculate their CM-types and the reductions of the curves and their Jacobians to characteristic $p > 0$. In Section 6, we consider a genus 3 curve $C$ over a number field $M$ such that its Jacobian has CM by a sextic CM-field $K$ with no proper CM-subfield. We prove a bound on primes such that there exists a solution to the embedding problem, and we use that to give a bound on the primes $p$ such that the stable reduction of $C$ at $p$ contains three elliptic curves. We show that if we drop the assumption that $K$ has no proper CM-subfield, then the embedding problem as stated cannot be used to give a bound on the primes $p$ as above.

We include as an appendix a collection of conditions that a solution to the embedding problem must satisfy, written as equations in the entries of certain matrices in the image of the embedding. These equations may be useful for future work. A refinement of the embedding problem (for example, a version which includes conditions pertaining to the CM-type) would result in extra equations in addition to those in the appendix. It is to be hoped that studying this larger set of equations would yield an explicit bound on the primes for which they have a solution. This would give a bound on the primes $p$ such that the stable reduction of $C$ at $p$ contains three curves of genus 1, even in the case where the CM-field $K$ contains a proper CM-subfield.

## *1.3 Notation and Conventions*

We set the following notation, to be used throughout.

- $\mathbb{F}_p$ is the finite field with $p$ elements.
- $\zeta_N$ is a primitive $N$th root of unity.

- For a field $k$, $\overline{k}$ is an algebraic closure.
- $K$ is a sextic CM-field, i.e., $K$ is a totally imaginary quadratic extension of $K^+$, where $K^+$ is a totally real cubic extension of $\mathbb{Q}$.
- $\mathcal{O}$ is an order of $K$.
- $F$ and $L$ are Galois closures of $K/\mathbb{Q}$ and $K^+/\mathbb{Q}$, respectively, with $G = \mathrm{Gal}(F/\mathbb{Q})$ and $G^+ = \mathrm{Gal}(L/\mathbb{Q})$.
- $\psi$ is a complex embedding $K \hookrightarrow \mathbb{C}$, and $\rho$ is complex conjugation. Hence $\{\psi, \rho \circ \psi\}$ is a conjugate pair of embeddings.
- $(K, \varphi)$ is a CM-type, i.e., a choice of one embedding from each pair of complex conjugate embeddings.
- $A$ is an abelian variety, $\mathrm{End}(A)$ is the endomorphism ring of $A$, and $\mathrm{End}^0(A)$ is $\mathrm{End}(A) \otimes \mathbb{Q}$.
- For $f \in \mathrm{End}(A)$, $f^\vee \in \mathrm{End}(A^\vee)$ is the dual isogeny. The Rosati involution associated with a fixed polarization is denoted by $f \mapsto f^*$, $\mathrm{End}^0(A) \to \mathrm{End}^0(A)$.
- $E$ is an elliptic curve, $j(E)$ is the $j$-invariant of $E$.
- We denote an isomorphism between two abelian varieties over an algebraic closure of the field of definition by $\simeq$.
- We denote an isogeny between two abelian varieties over an algebraic closure of the field of definition by $\sim$.
- $M$ is a number field, $\nu$ (or $\mathfrak{p}$) is a finite place of $M$, $\mathcal{O}_\nu$ is the valuation ring of $\nu$, and $k_\nu$ is the residue field.
- $C$ is a curve over a number field with Jacobian $J$ and genus $g = g(C)$. A curve $C$ is always assumed to be smooth, projective, and absolutely irreducible, unless explicitly mentioned otherwise.
- $B_{p,\infty}$ is the quaternion algebra ramified at $p$ and $\infty$, and $R$ is a maximal order of $B_{p,\infty}$.
- For a matrix $T$, $\mathbf{Tr}(T)$ denotes the sum of its diagonal entries, the trace.
- $\mathbf{Tr}_{K/K_1}$ denotes the trace of a field extension $K/K_1$.
- For an element of a central simple algebra, Nrd denotes the reduced norm.
- $\mathbf{Nm}_{K/K_1}$ denotes the norm of a field extension $K/K_1$; we use $\mathbf{Nm}$ when the extension is clear.

## 2 The Galois Group of the Galois Closure of a Sextic CM-Field

Let $K$ be a sextic CM-field, i.e., $K$ is a totally imaginary quadratic extension of a totally real field $K^+$ with $[K^+ : \mathbb{Q}] = 3$. We denote the Galois closure of $K^+/\mathbb{Q}$ by $L$ and the Galois closure of $K/\mathbb{Q}$ by $F$. We write $G = \mathrm{Gal}(F/\mathbb{Q})$ and $G^+ = \mathrm{Gal}(L/\mathbb{Q})$. The following proposition lists the possibilities for $G$.

**Proposition 2.1.** *Let $K$ be a sextic CM-field, and let $G$ be the Galois group of the Galois closure of $K/\mathbb{Q}$. Then $G$ is one of the following groups:*

*(1)* $C_2 \times C_3 \simeq C_6$,
*(2)* $C_2 \times S_3 \simeq D_{12}$,
*(3)* $(C_2)^3 \rtimes G^+$ *with* $G^+ \in \{C_3, S_3\}$ *acting by permutations on the three copies of* $C_2$.

*In particular, if $K/\mathbb{Q}$ is Galois, then the Galois group $G = \mathrm{Gal}(K/\mathbb{Q}) \simeq C_6$ is cyclic.*

*Proof.* This is proved in Section 5.1.1 of Dodson (1984), for example. □

In the rest of this section, we sketch the proof of Proposition 2.1, following Dodson. Since we restrict to the case of sextic CM-fields, the presentation can be simplified. In the course of the proof, we also give more details on the structure of the extensions $F/\mathbb{Q}$ and $K^+/\mathbb{Q}$ in the different cases. In particular, we show that Case 3 is precisely the case where $K$ does not contain an imaginary quadratic subfield.

Galois theory implies that we have the following exact sequence of groups:

$$1 \to \mathrm{Gal}(F/L) \to G \to G^+ \to 1.$$

**Lemma 2.2.** *We have*

$$\mathrm{Gal}(F/L) \simeq (C_2)^v, \qquad 1 \leq v \leq 3$$

*and*

$$G^+ \in \{C_3, S_3\}.$$

*Proof.* This lemma is a special case of the proposition in Section 1.1 of Dodson (1984). We give the proof here for convenience.

We first remark that $K = K^+(\sqrt{-\delta})$ for some totally positive square-free $\delta \in K^+$. We write $\delta_1 := \delta, \delta_2, \ldots, \delta_r$ for the $G^+$-conjugates of $\delta$. It follows that

$$F = L(\sqrt{-\delta_1}, \ldots, \sqrt{-\delta_r}).$$

Every element $h \in \mathrm{Gal}(F/L)$ sends $\sqrt{-\delta_i}$ to $\pm\sqrt{-\delta_i}$. Moreover, $h$ is determined by its action on these elements. It follows that $\mathrm{Gal}(F/L) \simeq (C_2)^v$ is an elementary abelian 2-group.

Since $\delta \in K^+$ it follows that $[\mathbb{Q}(\delta) : \mathbb{Q}]$ divides 3. We conclude that the number of $G^+$-conjugates of $\delta$ is at most 3.

The statement on $G^+$ immediately follows from the fact that $[K^+ : \mathbb{Q}] = 3$. This proves the lemma. □

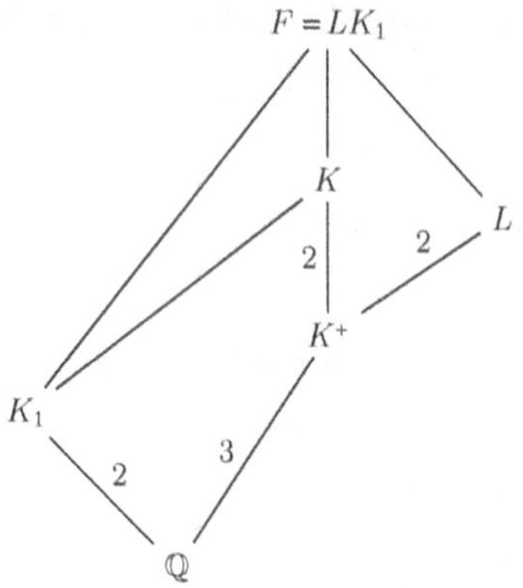


**Fig. 1** Field extensions in Case 2

*Proof of Proposition 2.1.* We start the classification. Note that $\mathrm{Gal}(K/K^+)$ is generated by complex conjugation. It follows that complex conjugation is also an element of $G$. This element, which we denote by $\rho$, is an element of the center of $G$.

**Case I**: $K/\mathbb{Q}$ Galois.
Since $K/\mathbb{Q}$ is Galois, $G = \mathrm{Gal}(K/\mathbb{Q})$ is a group of order 6, hence either cyclic or $S_3$. Since the Galois closure $L$ of $K^+/\mathbb{Q}$ is a totally real subfield of $K$, it follows that $K^+ = L$. This implies that $\mathrm{Gal}(K/K^+)$ is a normal subgroup of $G$ which has order 2. It follows that $G \simeq C_6$ is cyclic. Note that $K$ contains the imaginary quadratic subfield $K_1 := K^{C_3}$ and $K = K_1K^+$. This corresponds to Case 1 of Proposition 2.1.

**Case II**: $K/\mathbb{Q}$ is not Galois and $K$ contains an imaginary quadratic field $K_1$.
Since $K$ contains an imaginary quadratic field $K_1$, we have $F = LK_1$ and $G \simeq C_2 \times G^+$. If $G^+ \simeq C_3$, then $L = K^+$ and $K/\mathbb{Q}$ is Galois, which contradicts our assumption. It follows that $G^+ \simeq S_3$ and $G \simeq C_2 \times S_3$. This is Case 2 of Proposition 2.1. We obtain the field diagram in Figure 1.

**Case III**: $K/\mathbb{Q}$ is not Galois and $K$ does not contain an imaginary quadratic subfield.
This case corresponds to Case 3 of Proposition 2.1. In this case the integer $v$ from Lemma 2.2 is not equal to 1, i.e., we have $v = 2$ or 3. The following claim completes the proof of Proposition 2.1.

**Claim.** The case $v = 2$ does not occur. This claim is a special case of the second proposition in Section 5.1.1 of Dodson (1984). We give the proof here for completeness.

Recall that $\rho \in \mathrm{Gal}(F/L)$ denotes complex conjugation and is contained in the center of $G$. Let $\sigma \in G^+$ be an element of order 3. Then $\sigma$ acts on $\mathrm{Gal}(F/L) = (C_2)^v$ by conjugation. This action has two orbits of length 1, corresponding to the identity element and $\rho$. All other orbits have length 3. It follows that $3 \mid (2^v - 2)$. The claim follows. □

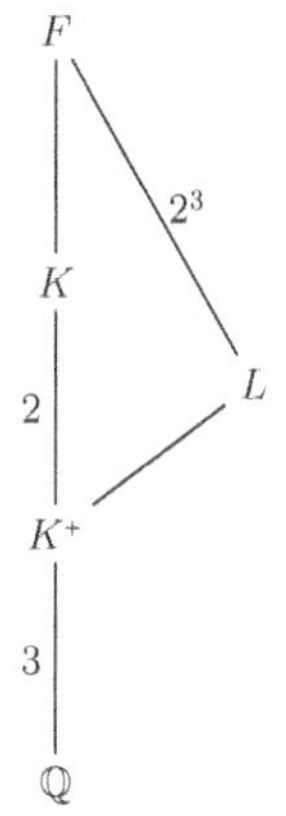


**Fig. 2** Field extensions in Case 3

Of primary interest to us in the rest of this paper is Case 3 of Proposition 2.1, in which $K$ does not contain an imaginary quadratic subfield. We have seen that $G \simeq (C_2)^3 \rtimes G^+$ with $G^+ \in \{C_3, S_3\}$. Figure 2 describes the field extensions in Case 3.

## 3 Primitive CM-Types

Let $K$ be a sextic CM-field. As in Section 2, we write $K^+$ for the totally real cubic subfield of $K$. The complex embeddings $K \hookrightarrow \mathbb{C}$ come in pairs $\{\psi, \rho \circ \psi\}$, where $\rho$ denotes complex conjugation. Recall that a CM-type $(K, \varphi)$ is a choice of one embedding from each of these pairs. The goal of this section is to determine the primitive CM-types. We start by recalling the definition from Milne (2006), Section 1.1. For examples we refer to Section 5.

**Definition 3.1.** Let $(K, \varphi)$ and $(K_1, \varphi_1)$ be CM-types. We say that $(K, \varphi)$ is *induced* from $(K_1, \varphi_1)$ if $K_1$ is a subfield of $K$ and the restriction of $\varphi$ to $K_1$ coincides with $\varphi_1$. A CM-type is called *primitive* if it is not induced from a CM-type on any proper CM-subfield of $K$.

Let $A$ be an abelian variety and let $K$ be a CM-field with $[K : \mathbb{Q}] = 2\dim(A)$. We say that $A$ has *complex multiplication (CM) by K* if the endomorphism algebra $\mathrm{End}^0(A) = \mathrm{End}(A) \otimes \mathbb{Q}$ contains $K$. We say that a curve $C$ has *CM by K* if its Jacobian has CM by $K$. We say that $A$ (or $C$) has CM if there exists a CM-field $K$ such that $A$ (or $C$) has CM by $K$. If $\mathrm{End}(A)$ is an order $\mathcal{O}$ in a CM-field $K$ with $[K : \mathbb{Q}] = 2\dim(A)$, we say that $A$ has *CM by* $\mathcal{O}$.

The following theorem gives a geometric interpretation of what it means for the CM-type of a CM-abelian variety to be primitive in characteristic zero. For convenience, we say that an abelian variety $A$ defined over a field $M$ is simple if it is absolutely simple, meaning that $A \otimes_M \overline{M}$ is not isogenous to a product of abelian varieties of lower dimension. Similarly, we say that two abelian varieties $A_1, A_2$ defined over $M$ are isogenous if there exists an isogeny $\varphi : A_1 \to A_2$ defined over the algebraic closure of $M$.

**Theorem 3.2.** *Let $A$ be an abelian variety defined over a field of characteristic zero. Suppose that $A$ has CM with CM-type $(K, \varphi)$. Then the CM-type $(K, \varphi)$ is primitive if and only if the abelian variety $A$ is simple.*

*Proof.* This is proved in Theorem 3.5 of Chapter 1 of Lang (1983). See also Remark 1.5.4.2 of Chai et al. (2014). □

We refer to Section 1.5.5 of Chai et al. (2014) for an explanation of why we need to assume that $A$ is defined over a field of characteristic zero in Theorem 3.2.

The following result gives a useful criterion for determining whether a given CM-type is primitive. For a proof, we refer to Theorem 3.6 of Chapter 1 of Lang (1983). For a CM-type $(K, \varphi)$ and $h \in \mathrm{Aut}(K)$, we write

$$\varphi h = \{\varphi_i \circ h \mid \varphi_i \in \varphi\}.$$

**Proposition 3.3.** *Let $(K, \varphi)$ be a CM-type. We write $(F, \Phi)$ for the induced CM-type of the Galois closure of $K/\mathbb{Q}$. Let*

$$H_\Phi = \{h \in G = \mathrm{Gal}(F/\mathbb{Q}) \mid \Phi h = \Phi\}.$$

*Then $(K, \varphi)$ is primitive if and only if*

$$K = F^{H_\Phi}.$$

We now determine the primitive sextic CM-types in each of the cases of Proposition 2.1. We first consider Case 3. Recall that in the proof of Proposition 2.1 we showed that Case 3 is precisely the case where $K$ does not contain an imaginary quadratic subfield.

**Corollary 3.4.** *Suppose that we are in Case 3 of Proposition 2.1, i.e., $K$ does not contain an imaginary quadratic field. Then every CM-type $(K, \varphi)$ is primitive.*

*Proof.* Suppose that $(K, \varphi)$ is not primitive. Then $K$ contains a proper CM-subfield $K_1$. Since $K$ is sextic, $K_1$ is an imaginary quadratic field. This yields a contradiction. □

### 3.1 Primitive Types in Case 1

We now consider Case 1 from Proposition 2.1. This is the case in which $K/\mathbb{Q}$ is Galois, with Galois group $G \simeq C_6$. We choose a generator $\sigma$ of $G$. Note that complex conjugation corresponds to $\sigma^3$. Up to replacing $\varphi$ by its complex conjugate, every CM-type $(K, \varphi)$ may be written as

$$\varphi_{a,b} = \{1, \sigma^a, \sigma^b\}, \qquad 0 < a, b < 6, \quad a \equiv 1 \pmod 3,\ b \equiv 2 \pmod 3.$$

We find 4 cases:

$$\{a, b\} \in \{\{1,2\}, \{1,5\}, \{4,2\}, \{4,5\}\}.$$

Note that changing the generator $\sigma$ of $G$ to $\sigma^{-1}$ changes $\{4,5\}$ to $\{1,2\}$, therefore we do not have to consider the choice $\{4,5\}$.

We write $H_{a,b}$ for the subgroup fixing the CM-type as in Proposition 3.3. Then $H_{1,2} = H_{1,5} = \{1\}$ and $H_{4,2} = \langle\sigma^2\rangle \simeq C_3$. Note that $K_1 := K^{H_{4,2}}$ is the imaginary quadratic subfield of $K$, which is a CM-field. We conclude that $\varphi_{4,2}$ is induced from $K_1$, and hence imprimitive. The other CM-types are primitive.

## 3.2 *Primitive Types in Case 2*

We now consider Case 2 from Proposition 2.1. We refer to Section 2 for a description of the fields involved. Recall that $K = K_1K^+$. Therefore, an embedding $\psi : K \hookrightarrow \mathbb{C}$ corresponds to an ordered pair $(\psi_1, \psi^+)$, where $\psi_1 : K_1 \hookrightarrow \mathbb{C}$ is an embedding of $K_1$ and $\psi^+ : K^+ \hookrightarrow \mathbb{C}$ is an embedding of $K^+$. Since $K^+$ is totally real, the image of $\psi^+$ is contained in $\mathbb{R}$. We denote the three possible complex embeddings of $K^+$ by $\chi_i$ for $i = 1, 2, 3$. We fix a complex embedding of $K_1$ and denote it by 1. We denote the other complex embedding of $K_1$ by $-1$.

A CM-type $(K, \varphi)$ consists of a triple of these ordered pairs in which no two of the pairs are complex conjugates. Since $\mathrm{Gal}(K_1/\mathbb{Q})$ is generated by complex conjugation, we simply choose one of the two complex embeddings of $K_1$ for each embedding $\chi_i$ of $K^+$. This means that we may write

$$\varphi = \{(\epsilon_i, \chi_i) \mid i = 1, 2, 3\}, \qquad \epsilon_i \in \{\pm 1\}.$$

Identifying $\varphi$ with its complex conjugate yields four different CM-types.

We determine the imprimitive types. The only CM-field properly contained in $K$ is the imaginary quadratic field $K_1$. The restriction of the embedding $(\epsilon_i, \chi_i)$ to $K_1$ is just $\epsilon_i$. Therefore, the CM-type $\varphi = \{(\epsilon_i, \chi_i)\}$ is imprimitive if and only if $\epsilon_i$ is independent of $i$. We conclude that there is a unique imprimitive CM-type. The other three are primitive.

## 3.3 Examples of CM-Types

We give examples of CM-types illustrating each of the three cases of Proposition 2.1.

*Example 3.5 ($K/\mathbb{Q}$ is Galois with Galois Group $G \simeq C_6$).* Let $K$ be $\mathbb{Q}(\zeta_7)$ where $\zeta_7$ is a primitive seventh root of unity. The maximal totally real subfield of $K$ is $K^+ = \mathbb{Q}(\zeta_7 + \zeta_7^{-1})$, which has degree three over $\mathbb{Q}$ (the minimal polynomial of $\zeta_7 + \zeta_7^{-1}$ over $\mathbb{Q}$ is $x^3 + x^2 - 2x - 1$). The field $K$ is a totally imaginary quadratic extension of $K^+$.

The automorphism $\sigma$ which maps $\zeta_7$ to $\zeta_7^5$ generates $\mathrm{Gal}(K/\mathbb{Q})$. The fixed field of $\langle\sigma^2\rangle$ is $\mathbb{Q}(\zeta_7^4+\zeta_7^2+\zeta_7) = \mathbb{Q}(\sqrt{-7})$. This is the unique imaginary quadratic extension of $\mathbb{Q}$ contained in $\mathbb{Q}(\zeta_7)$. Therefore, the only imprimitive CM-type admitted by $K$ is $\varphi_{2,4} = \{1, \sigma^2, \sigma^4\}$; the CM-types $\varphi_{a,b} = \{1, \sigma^a, \sigma^b\}$ for $\{a, b\} \neq \{4, 2\}$ with $a \equiv 1 \pmod 3$, $b \equiv 2 \pmod 3$ are all primitive.

The following examples have been taken from the database of Klüners and Malle (2011).

*Example 3.6 (The Galois Closure of $K/\mathbb{Q}$ is $D_{12}$).* Let $K$ be the sextic field obtained by adjoining a root of the irreducible polynomial $f(x) = x^6 - 3x^5 + x^4 + 10x^2 - 9x + 3$. Then $K$ is a totally imaginary quadratic extension of the totally real cubic field $K^+ = \mathbb{Q}(\alpha)$ where the minimal polynomial of $\alpha$ is $g(x) = x^3 - 7x^2 + 12x - 3$. The Galois closure $F$ of $K/\mathbb{Q}$ is the compositum of the Galois closure of $K^+$ with the unique imaginary quadratic subfield $K_1$ of $K$, given by the minimal polynomial $x^2 + 3x + 3$. The Galois group of $F$ is isomorphic to $S_3 \times C_2 \simeq D_{12}$. Denote the roots of $g(x)$ by $\alpha_1 := \alpha, \alpha_2, \alpha_3$.

Let $\chi_i : \alpha_1 \mapsto \alpha_i$ denote the three real embeddings of $K^+$ and $\pm 1$ denote the two complex embeddings of $K_1$. Then the CM-type $\varphi = \{(1, \chi_1), (1, \chi_2), (1, \chi_3)\}$ of $K$ is imprimitive, since its restriction to the quadratic imaginary subfield $K_1$ is also a CM-type. The remaining three CM-types of $K$ are primitive. For clarity, the primitive CM-types are as follows: $\{(1, \chi_1), (-1, \chi_2), (-1, \chi_3)\}, \{(1, \chi_1), (1, \chi_2), (-1, \chi_3)\}, \{(1, \chi_1), (-1, \chi_2), (1, \chi_3)\}$.

*Example 3.7 (The Galois Closure of $K/\mathbb{Q}$ is $(C_2)^3 \rtimes C_3$).* Let $K = \mathbb{Q}(\beta)$ be the degree 6 extension of $\mathbb{Q}$ where the minimal polynomial of $\beta$ is $f(x) = x^6 - 2x^5 + 5x^4 - 7x^3 + 10x^2 - 8x + 8$. Let $F$ be the Galois closure of $K$. Then $\mathrm{Gal}(F/\mathbb{Q})$ is $(C_2)^3 \rtimes C_3$. Moreover, $K$ is a CM-field since $K$ is a totally imaginary quadratic extension of $K^+ = \mathbb{Q}(\alpha)$ where the minimal polynomial of $\alpha$ over $\mathbb{Q}$ is $g(x) = x^3 - 7x^2 + 14x - 7$. Note that $K$ contains no quadratic subfield, hence every CM-type is primitive.

## 3.4 Comparison with the Genus 2 Case

The following proposition characterizes primitive CM-types for quartic CM-fields.

**Proposition 3.8.** *Let $K$ be a quartic CM-field. The following are equivalent.*

*(1) The CM-type is primitive.*
*(2) The CM-field $K$ does not contain an imaginary quadratic subfield.*

*Proof.* We recall the argument from Example 8.4.(2) of Shimura (1998) in which we find a classification of the possible Galois groups of quartic CM-fields $K$ together with the possible CM-types. It follows from this classification that if $K$ contains a proper CM-subfield $K_1 \neq \mathbb{Q}$ then $K/\mathbb{Q}$ is Galois with Galois group $G \simeq C_2 \times C_2$. Moreover, in this case all CM-types are imprimitive. Namely, denoting again complex conjugation by $\rho$, we may write $G = \{1, \rho, \sigma, \rho\sigma\}$. Then the possible CM-types are $\{1, \sigma\}$ and $\{1, \rho\sigma\}$, which are fixed by $\langle\sigma\rangle$ and $\langle\rho\sigma\rangle$, respectively. Therefore, the statement follows from Proposition 3.3. □

Proposition 3.8 explains why Goren and Lauter (2006) and Goren and Lauter (2007) restrict to the case where the quartic CM-field does not contain an imaginary quadratic subfield. For quartic CM-fields, this is equivalent to requiring that the CM-type is primitive. However, as we have seen in our discussion of the primitive types in Cases 1 and 2 of Proposition 2.1, these two properties are not equivalent for sextic CM-fields.

We give two concrete examples of genus 2 curves with CM to illustrate Proposition 3.8. These are similar to the genus 3 examples given in Section 5.1. We consider two smooth projective curves defined by the following affine equations

$$D_1 : \quad y^5 = x(x-1),$$
$$D_2 : \quad y^8 = x(x-1)^4.$$

One easily verifies that both curves have genus 2.

The curve $D_1$ has CM by $K_1 := \mathbb{Q}(\zeta_5)$ with CM-type $(1, 2)$ in the notation of Section 5.1. The Galois group of $K_1/\mathbb{Q}$ is cyclic of order 4, hence its unique subgroup of order 2 is generated by complex conjugation, which cannot fix the CM-type. Indeed, the Jacobian of $D_1$ is simple. In the genus 2 case, all CM-types of a cyclic CM-field are primitive. We have already seen that this does not hold in general for genus $g \geq 3$.

The curve $D_2$ has CM by $K_2 := \mathbb{Q}(\zeta_8)$. The corresponding Galois group is isomorphic to $C_2 \times C_2$, hence the CM-type is imprimitive. Indeed, the CM-type is $(1, 3)$ which is fixed by $\langle 3 \rangle \subset (\mathbb{Z}/8\mathbb{Z})^*$. The CM-type $(1, 3)$ is induced from the CM-type of the elliptic curve $E := D_2/\langle\tau\rangle$, where $\tau(x, y) = (1/x, y^3/x(x-1))$ is an automorphism of order 4.

# 4 Reduction of CM-Curves and Their Jacobians

Our main result (Theorem 6.8) deals with curves $C$ of genus 3 defined over some number field whose Jacobians have CM by a sextic CM-field $K$. In this section, we describe the possibilities for the reduction of these curves and their Jacobians to characteristic $p > 0$.

## 4.1 The Theorem of Serre–Tate

Let $C$ be a curve of genus $g \geq 2$ defined over a number field $M$, and let $J := \mathrm{Jac}(C)$ be its Jacobian. In the course of our arguments, we allow ourselves to replace $M$ by a finite extension, which we still denote by $M$. Let $\nu$ be a finite place of $M$. We write $\mathcal{O}_\nu$ for the valuation ring of $\nu$ and $k_\nu$ for its residue field. We write $\overline{k_\nu}$ for an algebraic closure of $k_\nu$.

Recall that the abelian variety $J$ has *good reduction* at $\nu$ if there exists an abelian scheme $\mathcal{J}$ over $\mathcal{O}_\nu$ with $\mathcal{J} \otimes_{\mathcal{O}_\nu} M \simeq J$. This implies that the reduction $\overline{J} := \mathcal{J} \otimes_{\mathcal{O}_\nu} \overline{k_\nu}$ is an abelian variety. We say that $J$ has *potentially good reduction* at $\nu$ if there exists a finite extension $M'/M$ and an extension $\nu'$ of $\nu$ such that $J \otimes_M M'$ has good reduction at $\nu'$.

The following theorem is Theorem 6 of Serre and Tate (1968).

**Theorem 4.1 (Serre–Tate).** *Let $J$ be an abelian variety with CM defined over a number field $M$. Let $\nu$ be a finite place of $M$. Then $J$ has potentially good reduction at $\nu$.*

Since there are at most finitely places where $J$ does not have good reduction, there exists a finite extension of $M$ over which $J$ has good reduction everywhere.

## 4.2 Reduction of Genus 3 Curves with CM

We now describe the restrictions imposed by Theorem 4.1 on the reduction of the curve $C$.

Recall that $C$ is a curve of genus $g(C) \geq 2$ defined over a number field $M$. We say that $C$ has *good reduction* at a finite place $\nu$ of $M$ if there exists a model $\mathcal{C}$ over $\mathcal{O}_\nu$ with $\mathcal{C} \otimes_{\mathcal{O}_\nu} M \simeq C$ such that the reduction $\overline{C} := \mathcal{C} \otimes_{\mathcal{O}_\nu} \overline{k_\nu}$ is smooth. Similarly, $C$ has *potentially good reduction* at $\nu$ if it has good reduction over a finite extension of $M$.

We say that $C$ has *semistable reduction* at $\nu$ if there exists a model $\mathcal{C}$ over $\mathcal{O}_\nu$ with $\mathcal{C} \otimes_{\mathcal{O}_\nu} M \simeq C$ such that the reduction $\overline{C}$ is semistable. This means that $\overline{C}$ is reduced and has at most ordinary double points as singularities. The corresponding model $\mathcal{C} = \mathcal{C}_\nu$ is called a *semistable model* of $C$ at $\nu$. The Stable Reduction Theorem

(Deligne and Mumford 1969, Corollary 2.7) states that every curve $C$ admits a semistable model at $\nu$ after replacing $M$ by a finite extension. Since we assume that $g(C) \geq 2$, there exists a unique minimal semistable model, which is called the *stable model* at $\nu$. Its special fiber $\overline{C}$ is called the *stable reduction* of $C$ at $\nu$. The minimality of the stable model implies that $C$ has potentially good reduction if and only if the stable reduction $\overline{C}$ is smooth. If the finite place $\nu$ is fixed, we usually omit it.

We now turn to our situation of interest, namely that of a genus 3 curve whose Jacobian has CM by a sextic CM-field. The following proposition is a consequence of Theorem 4.1.

We say that $C$ has *bad reduction* at $\nu$ if it does not have potentially good reduction at $\nu$. This is equivalent to the stable reduction $\overline{C}$ having singularities. We say that the reduction $\bar{C}$ of $C$ is *tree-like* if the intersection graph of the irreducible components of $\overline{C}$ is a tree. Note that we always consider the reduction $\overline{J}$ (resp. $\overline{C}$) as an abelian variety (resp. curve) defined over the algebraically closed field $\overline{k_\nu}$ for convenience.

**Proposition 4.2.** *Let $C$ be a curve of genus 3 defined over a number field $M$ such that its Jacobian $J = \mathrm{Jac}(C)$ has CM. Let $\nu$ be place of $M$ where $C$ has bad reduction. Then*

*(a) the stable reduction $\overline{C}$ of $C$ is tree-like, and*
*(b) the reduction $\overline{J}$ of $J$ is the product of the Jacobians of the irreducible components of $\overline{C}$ (as polarized abelian varieties).*

*Proof.* Let $\nu$ be a finite place of $M$. After replacing $M$ by a finite extension and choosing an extension of $\nu$, we may assume that $C$ has stable reduction at $\nu$. Let $\mathcal{C}$ be the stable model of $C$. Set $S = \mathrm{Spec}(\mathcal{O}_\nu)$, and define $\mathrm{Pic}^0(\mathcal{C}/S)$ to be the identity component of the Picard variety. Since the stable reduction $\overline{C}$ of $C$ is reduced, Theorem 1 in Section 9.5 of Bosch et al. (1990) states that $\mathrm{Pic}^0(\mathcal{C}/S)$ is a Néron model of $J$.

Theorem 4.1 implies that $J$ has potentially good reduction, i.e., there exists an abelian variety $\mathcal{J}$ over $S$ with generic fiber $J$. Proposition 8 of Section 1.2 in Bosch et al. (1990) shows that $\mathcal{J}/S$ is a Néron model. Since two different Néron models are canonically isomorphic, it follows that $\mathrm{Pic}^0(\mathcal{C}/S) \simeq_S \mathcal{J}$. In particular, it follows that the special fiber $\mathrm{Pic}^0(\mathcal{C}/S) \otimes_{\mathcal{O}_\nu} \overline{k_\nu} \simeq \mathrm{Pic}^0(\overline{C})$ is an abelian variety.

Example 8 of Section 9.2 in Bosch et al. (1990) shows that $\mathrm{Pic}^0(\overline{C})$ is given by an exact sequence

$$1 \to T \to \mathrm{Pic}^0(\overline{C}) \to B := \prod_i \mathrm{Jac}(\tilde{C}_i) \to 1, \tag{4.1}$$

where $B$ is an abelian variety and $T$ is a torus. The product on the right-hand side is taken over the irreducible components of $\overline{C}$. We denote the normalization of an irreducible component $C_i$ of $\overline{C}$ by $\tilde{C}_i$. The torus $T$ satisfies

$$T \simeq \mathbb{G}^t_{m,\overline{k_\nu}}$$

for some $t \geq 0$. The torus $\mathbb{G}_m$ is not compact, and hence not an abelian variety. Since $\mathrm{Pic}^0(\overline{C})$ is an abelian variety, the exact sequence (4.1) implies that $t = 0$, i.e., $\mathrm{Pic}^0(\overline{C})$ contains no torus. By Corollary 12.b of Bosch et al. (1990), this means that the intersection graph of the irreducible components of $\overline{C}$ is a tree. Both statements of the proposition follow from this. □

The corollary below follows immediately from Proposition 4.2. In Section 5, we give examples of each of the cases.

**Corollary 4.3.** *Let $C$ be a genus 3 curve with CM defined over a number field $M$, and let $v$ be a finite place of $M$. One of the following three possibilities holds for the irreducible components of $\overline{C}$ of positive genus:*

*(i) (good reduction) $\overline{C}$ is a smooth curve of genus* 3,
*(ii) $\overline{C}$ has three irreducible components of genus* 1,
*(iii) $\overline{C}$ has an irreducible component of genus* 1 *and one of genus* 2.

Note that the stable reduction $\overline{C}$ may contain irreducible components of genus 0. This happens for the stable reduction $\overline{C}_1$ to characteristic 3 of the curve $C_1$ from Lemma 5.3, for example. One may show that $\overline{C}_1$ has four irreducible components: one of genus 0 and three of genus 1. The three elliptic curves each intersect the genus 0 curve in one point but do not intersect each other. Since the irreducible components of genus 0 do not contribute to the Jacobian, we have not listed them in Corollary 4.3.

*Remark 4.4.* Let $C$ be a curve of genus 3 with CM, defined over a number field $M$. Suppose that $C$ has bad reduction at a finite place $v$ of $M$. In Case (ii) of Corollary 4.3, the reduction $\overline{C}$ of $C$ contains three irreducible components $E_i$ of genus 1. Proposition 4.2 implies that

$$\overline{J} \simeq E_1 \times E_2 \times E_3$$

as polarized abelian varieties, i.e., the polarization on $\overline{J}$ is the product polarization.

In Case (iii) of Corollary 4.3, $\overline{C}$ contains an irreducible component $E$ of genus 1 and an irreducible component $D$ of genus 2. In this case, we have

$$\overline{J} \simeq E \times \mathrm{Jac}(D)$$

and the polarization on $\overline{J}$ is the product polarization induced by the principal polarizations on the components. We show below that in this case $\overline{J}$ is still isogenous to a product of elliptic curves (Theorem 4.5). However, it is **not** true that the polarization of $\overline{J}$ is induced by polarization on the three elliptic curves as we had in Case (ii).

Even in the case where $C$ has good reduction (Case (i) of Corollary 4.3), the reduction $\overline{J}$ of the Jacobian need not be simple even if $J$ is. In this case, the polarization of $\overline{J}$ is induced by the embedding of $\overline{C}$ in its Jacobian and hence is not a product polarization.

The following theorem is a generalization of Theorem 3.2 to positive characteristic.

**Theorem 4.5.** *Let $J$ be an abelian variety of dimension 3 with CM, defined over a number field $M$. Suppose that the reduction $\overline{J}$ of $J$ at a finite place of $M$ is not simple. Then $\overline{J}$ is isogenous to the product of three copies of the same elliptic curve $E$.*

*Proof.* The result is essentially a special case of Theorem 1.3.1.1 of Chai et al. (2014). For the convenience of the reader we sketch a direct proof in our situation.

By assumption, $J$ has CM by the sextic CM-field $K$. This implies that we have an embedding

$$K \hookrightarrow \mathrm{End}^0(\overline{J}).$$

Decompose $\overline{J}$ into isotypic components: $\overline{J} \sim \prod_i A_i^{n_i}$ where the $A_i$ are simple and $A_i \not\sim A_j$ for $i \neq j$. Since $\overline{J}$ is not simple by assumption, for dimension reasons there exists $j$ such that $A_j = E$ is an elliptic curve. We have $K \hookrightarrow \mathrm{End}^0(\overline{J}) = \prod_i M_{n_i}(\mathrm{End}^0(A_i))$. Projecting this ring homomorphism on the $j$th factor gives an injection $K \hookrightarrow M_{n_j}(\mathrm{End}^0(E))$. A dimension argument shows that $n_j = 3$ and therefore $\overline{J} \sim E^3$. □

**Proposition 4.6.** *Let $C$ be a genus 3 curve with CM, defined over a number field $M$. Suppose that $C$ has bad reduction at a finite place $\nu$ of $M$. Then the reduction $\overline{J}$ of the Jacobian $J$ of $C$ is supersingular or $K$ contains an imaginary quadratic field $K_1$.*

*Proof.* Let $C$ and $\overline{J}$ be as in the statement of the proposition. Since $C$ has bad reduction at $\nu$, Corollary 4.3 shows that $\overline{C}$ has an irreducible component $E_1$ of genus 1. It follows that we may regard $E_1$ as abelian subvariety of $\overline{J}$. (This is slightly weaker than the statement in Remark 4.4.) In particular, $\overline{J}$ is not simple. Theorem 4.5 implies therefore that $\overline{J}$ is isogenous to the product of three copies of an elliptic curve $E$. Note that $\overline{J}$ is supersingular if and only if $E$ is.

We assume that $E$ is ordinary. Since $E$ may be defined over a finite field, it has CM and $K_1 := \mathrm{End}^0(E)$ is an imaginary quadratic field contained in the center of $\mathrm{End}^0(E^3) = M_3(K_1)$. Since $\overline{J}$ is isogenous to $E^3$, we obtain an embedding

$$K = \mathrm{End}^0(J) \hookrightarrow \mathrm{End}^0(\overline{J}) \simeq \mathrm{End}^0(E^3) = M_3(K_1).$$

Theorem 1.3.1.1 of Chai et al. (2014) states that $K$ is its own centralizer in $M_3(K_1)$. Since the center of $M_3(K_1)$ is $K_1$, we conclude that $K_1$ is contained in $K$ and the result follows. □

The following corollary summarizes the results so far in the case that the CM-field $K$ does not contain an imaginary quadratic subfield $K_1$.

**Corollary 4.7.** *Let $C$ be a genus 3 curve with CM by $K$, defined over a number field $M$. Suppose that $K$ does not contain an imaginary quadratic subfield. Then the following holds:*

*(a) the CM-type $(K, \varphi)$ of $J$ is primitive, and $J$ is absolutely simple,*
*(b) if $C$ has bad reduction at a finite place $v$, then the reduction of $J$ at $v$ is supersingular.*

*Proof.* Part (a) follows from Corollary 3.4 and Theorem 3.2. Part (b) follows from Proposition 4.6. □

## 4.3 Polarizations and the Rosati Involution

In the rest of this section, we recall some results on the Rosati involution following Sections 20 and 21 of Mumford (1970) and Section 17 of Milne (2008). For precise definitions and more details, we refer to these sources. Let $A$ be an abelian variety and $\lambda : A \to A^\vee$ be a polarization associated with an ample line bundle $\mathcal{L}$ on $A$. The polarization $\lambda$ is an isogeny and therefore has an inverse $\frac{1}{\deg \lambda}\lambda^\vee = \lambda^{-1} \in \mathrm{Hom}(A^\vee, A) \otimes_{\mathbb{Z}} \mathbb{Q}$.

The Rosati involution on $\mathrm{End}^0(A) = \mathrm{End}(A) \otimes \mathbb{Q}$ is defined by

$$f \mapsto f^* = \lambda^{-1} \circ f^\vee \circ \lambda.$$

It satisfies

$$(f+g)^* = f^* + g^*, \quad (fg)^* = g^* f^*, \quad a^* = a$$

for $f, g \in \mathrm{End}^0(A)$ and $a \in \mathbb{Q}$. In the case where $\lambda$ is a principal polarization, i.e., $\deg(\lambda) = 1$, the Rosati involution acts as an involution on $\mathrm{End}(A)$. This is because $\lambda^{-1}$ is in $\mathrm{Hom}(A^\vee, A)$ and not just in $\mathrm{Hom}(A^\vee, A) \otimes_{\mathbb{Z}} \mathbb{Q}$. The natural polarization on a Jacobian is a principal polarization.

The Rosati involution is a positive involution Theorem 1 of Section 21 in Mumford (1970). This means that

$$(f, g) \mapsto \mathbf{Tr}(f \cdot g^*), \qquad \mathrm{End}^0(A) \to \mathbb{Q}$$

defines a positive definite quadratic form on $\mathrm{End}^0(A)$. We refer to Section 21 of Mumford (1970) for the precise definition of the trace. In the case that $A = E$ is an elliptic curve, we choose the polarization $\lambda$ defined as

$$\lambda : E \to \mathrm{Pic}^0(E), \quad P \mapsto [P] - [O].$$

The corresponding Rosati involution sends an isogeny $f$ to its dual isogeny $f^\vee$ and $\mathbf{Tr}(f \cdot f^\vee)$ is $\deg(f)$, the degree of the endomorphism $f$.

**Proposition 4.8.** *Let $A$ be a simple abelian variety defined over a field of characteristic zero with principal polarization $\lambda$. Assume that $A$ has CM by a field $K$. Then the Rosati involution associated with $\lambda$ induces complex conjugation on the CM-field $K$.*

*Proof.* Since $A$ is simple, the endomorphism algebra $\mathrm{End}^0(A)$ equals $K$ and the proposition is proved, for example, in Lemma 1.3.5.4 of Chai et al. (2014). □

*Remark 4.9.* Let $A$ be a simple abelian variety with $\mathrm{End}^0(A) = K$ as in the statement of Proposition 4.8. Let $M$ be a number field over which $A$ can be defined, and let $\mathfrak{p}$ be a prime of $M$ at which $A$ has good reduction. Write $\overline{A}$ for the reduction. We obtain an embedding

$$K \hookrightarrow \mathrm{End}^0(\overline{A}).$$

The Rosati involution on $\mathrm{End}^0(\overline{A})$ is an extension of the Rosati involution on $\mathrm{End}^0(A) = K$, which is complex conjugation by Proposition 4.8.

The following proposition was used in the proofs of Goren and Lauter (2007) but not stated there explicitly.

**Proposition 4.10.** *Suppose that $A = E^n$ is a product of elliptic curves as polarized abelian varieties. Then the Rosati involution acts as*

$$M_n(\mathrm{End}(E)) \to M_n(\mathrm{End}(E)), \qquad (f_{i,j}) \mapsto (f_{j,i}^\vee).$$

*Proof.* The result is well known to the experts. We sketch the argument. The proof we present here is a variant of the proof of Proposition 11.28 (ii) of van der Geer and Moonen (2011).

Let $A = E^n$ be as in the statement of the lemma, and write $p_i : A \to E$ for the projection on the $i$th coordinate. Then any line bundle $\mathcal{L}$ on $A$ satisfies $\mathcal{L} = p_1^*\mathcal{L}_1 \otimes \cdots \otimes p_n^*\mathcal{L}_n$ for suitable line bundles $\mathcal{L}_i$ on $E$.

Consider the natural map $A^\vee = \mathrm{Pic}^0(A) \to (\mathrm{Pic}^0(E))^n = (E^\vee)^n$ which sends a line bundle $\mathcal{L} \in \mathrm{Pic}^0(A)$ to the $n$-tuple $(\mathcal{L}|_{E_i})_i \in (\mathrm{Pic}^0(E))^n$ of the restrictions of $\mathcal{L}$ to the $i$th copy $E_i := (\cdots, 0, E, 0, \cdots)$ of $E$. One shows that this map is an isomorphism. The product polarization $\lambda_A : A \to A^\vee = (E^\vee)^n$ is induced by the natural polarization $\lambda : E \to \mathrm{Pic}^0(E)$ on $E$. In particular, it is also a principal polarization.

Using this identification, it suffices to prove the proposition in the case that $f \in \mathrm{End}(A)$ corresponds to an $n \times n$ matrix with an endomorphism $\alpha \in \mathrm{End}(E)$ as $(j, i)$th component and zeros everywhere else. The endomorphism $f^\vee : A^\vee \to A^\vee$ induced by $f$ sends a line bundle $\mathcal{L}$ on $A$ to $p_i^*(\alpha^*\mathcal{L}_i)$. We conclude that $f^* : A \to A = E^n$ corresponds to the matrix with the dual isogeny $\alpha^\vee$ in the $(i, j)$th coordinate and zeros elsewhere. This proves the proposition. □

# 5 Examples

In this section we discuss some examples of genus 3 curves with CM.

## 5.1 Cyclic Covers

The first type of examples we consider are $N$-cyclic covers of the projective line branched at exactly three points, see also Sections 1.6 and 1.7 of Chapter 1 of Lang (1983). More precisely, let $C$ be a smooth projective curve defined over a field of characteristic zero which admits a Galois cover $\pi : C \to \mathbb{P}^1$ whose Galois group is cyclic of order $N$ such that $\pi$ is branched exactly at three points. We may assume the three branch points to be $0, 1, \infty \in \mathbb{P}^1$.

Kummer theory implies the existence of integers $0 < a_1, a_2 < N$ with $\gcd(N, a_1, a_2) = 1$ such that the extension of function fields corresponding to $\pi$ is

$$\mathbb{Q}(x) \subset \mathbb{Q}(x)[y]/(y^N - x^{a_1}(x-1)^{a_2}).$$

The Galois group of $\pi$ is generated by $\alpha(x, y) = (x, \zeta_N y)$, where $\zeta_N$ is a primitive $N$th root of unity.

Define $0 < a_3 < N$ by $a_1 + a_2 + a_3 \equiv 0 \pmod{N}$. Then a chart at $\infty$ may be given by

$$w^N = z^{a_3}(z-1)^{a_2},$$

where $z = 1/x$. The condition that $\pi$ is branched at $\infty$ is therefore equivalent to $a_3 \equiv -(a_1 + a_2) \not\equiv 0 \pmod{N}$. The Riemann–Hurwitz formula shows that

$$2g(C) - 2 = -2N + \sum_{i=1}^{3} (N - \gcd(N, a_i)).$$

In Lemma 5.1 below, we show that the endomorphism ring of $\mathrm{Jac}(C)$ contains $\mathbb{Q}(\zeta_N)$. Therefore $\mathrm{Jac}(C)$ has CM by $\mathbb{Q}(\zeta_N)$ if and only if $2g(C) = \varphi(N)$, where $\varphi$ denotes Euler's totient function. For example, this condition is satisfied if $N$ is an odd prime. This case is discussed by Lang (Section 1.7 of Chapter 1 of Lang 1983).

The condition $2g(C) = \varphi(N)$ is satisfied for exactly three curves $C_i$, up to isomorphism over $\mathbb{C}$. Kummer theory implies that two tuples $(N, a_1, a_2, a_3)$ and $(M, b_1, b_2, b_3)$ define isomorphic curves if and only if $N = M$ and there exists an integer $c$ with $\gcd(c, N) = 1$ and a permutation $\sigma \in S_3$ such that $b_i \equiv c a_{\sigma(i)} \pmod{N}$ for all $i$. This is similar to the argument in Section 1.7 of Chapter 1 of Lang (1983).

The three curves satisfying this property are:

$$C_1: \quad y^9 = x(x-1)^3,$$
$$C_2: \quad y^7 = x(x-1)^2,$$
$$C_3: \quad y^7 = x(x-1).$$

An alternative equation for $C_1$ is

$$y^3 = z^4 - z, \qquad \text{where } z^3 = x. \tag{5.1}$$

We put $K_{N_i} = \mathbb{Q}(\zeta_{N_i})$ and $G_{N_i} = (\mathbb{Z}/N_i\mathbb{Z})^*$. In the three cases we consider in Lemma 5.1, we have $G_{N_i} \simeq C_6$. For $j \in (\mathbb{Z}/N_i\mathbb{Z})^*$, we denote the corresponding element of $\mathrm{Gal}(K_{N_i}/\mathbb{Q})$ by

$$\sigma_j : \zeta_{N_i} \mapsto \zeta_{N_i}^j,$$

or also by $j$ when no confusion can arise.

The following lemma summarizes the properties of the curves $C_i$.

**Lemma 5.1.** *(a) The curve $C_1$ has CM by $\mathbb{Q}(\zeta_9)$. The CM-type is $(1, 2, 4)$. This type is primitive.*
*(b) The curve $C_2$ has CM by $\mathbb{Q}(\zeta_7)$ and CM-type $(1, 2, 4)$. This type is imprimitive.*
*(c) The curve $C_3$ has CM by $\mathbb{Q}(\zeta_7)$ and CM-type $(1, 2, 3)$. This type is primitive.*

*Proof.* It is easy to check that the automorphism $\alpha$ of $C_i$ has a fixed point. Using this point to embed the curve $C_i$ in its Jacobian, we see that $\alpha$ induces an endomorphism $\alpha \in \mathrm{End}(\mathrm{Jac}(C_i))$ of multiplicative order $N_i$.

We may regard $\alpha \in \mathrm{End}(\mathrm{Jac}(C_i))$ as a primitive $N_i$th root of unity. In all three cases, we have $2g(C_i) = 6 = \varphi(N_i) = [\mathbb{Q}(\zeta_{N_i}) : \mathbb{Q}]$. It follows that $C_i$ has CM by $K_{N_i}$.

To calculate the CM-type of $C_i$ we follow the strategy of Section 1.7 of Chapter 1 of Lang (1983), and identify the cohomology group $H^0(C_i, \Omega)$ of holomorphic differentials with the tangent space of $\mathrm{Jac}(C)$. It suffices to find a basis of $H^0(C_i, \Omega)$ consisting of eigenvectors of $\alpha^*$, the map induced by $\alpha$ on $H^0(C_i, \Omega)$. Such a basis is computed in Section 1.7 of Chapter 1 of Lang (1983). The statement on the CM-type easily follows from this. (The fact that the action of $\langle \alpha \rangle$ on $H^0(C_i, \Omega)$ does not factor through the action of a quotient group provides a second proof that $\alpha$ defines an endomorphism of order $N_i$ of $\mathrm{Jac}(C_i)$.)

We explain what happens for $C_1$. We use a slightly different notation from Theorem 1.7.1 of Chapter 1 of Lang (1983). A basis of $H^0(C_1, \Omega)$ is given by

$$\omega_1 = \frac{y\,\mathrm{d}x}{x(x-1)}, \qquad \omega_2 = \frac{y^2\,\mathrm{d}x}{x(x-1)}, \qquad \omega_4 = \frac{y^4\,\mathrm{d}x}{x(x-1)^2}.$$

Note that $\alpha^*\omega_i = \zeta_9^i\omega_i$. The statement on the CM-type of $\mathrm{Jac}(C_1)$ follows. Primitivity is shown in Section 3.1.

In Example 3.5 we have determined all primitive CM-types for $\mathbb{Q}(\zeta_7)$. The statements on the (im)primitivity of the CM-types of $C_2$ and $C_3$ follow from this.

□

*Remark 5.2.* Lemma 5.1.(b) implies that $\mathrm{Jac}(C_2)$ is not simple. We may also check this directly. The curve $C_2$ admits an automorphism

$$\beta(x, y) = \left(\frac{1}{1-x}, \frac{y^2}{1-x}\right).$$

The curve $E := C_2/\langle\beta\rangle$ has genus 1. This curve has CM by the field $K_1 = \mathbb{Q}(\zeta_7)^{\langle\sigma_2\rangle} = \mathbb{Q}(\sqrt{-7})$.

One checks that $\langle\alpha, \beta\rangle \simeq \mathbb{Z}/7\mathbb{Z}\rtimes\mathbb{Z}/3\mathbb{Z}$ is a non-abelian group. Using the method of Kani and Rosen (1989) or Paulhus (2008), one may also deduce from this that

$$\mathrm{Jac}(C_2) \sim E^3.$$

Our next goal is to describe the reduction behavior of the curves $C_1$ and $C_3$.

**Lemma 5.3.** *(a) The curve $C_1$ has bad reduction at $p = 3$ and good reduction at all other primes.*

*(b) The reduction $\overline{J}_{1,p}$ of the Jacobian $J_1$ of $C_1$ to characteristic $p$ is ordinary if and only if $p \equiv 1 \pmod 9$ and supersingular if and only if $p = 3$ or $p \equiv 2 \pmod 3$.*

*(c) If $p \equiv 4, 7 \pmod 9$, then the abelian variety $\overline{J}_{1,p}$ is simple.*

*Proof.* It is easy to see that $C_1$ has good reduction to characteristic $p \neq 3$. Indeed, (5.1) still defines a smooth projective curve in characteristic $p \neq 3$. We consider the reduction at $p = 3$. In this case, the extension of function fields

$$\mathbb{F}_3(z) \subset \mathbb{F}_3(z)[y]/(y^3 - z(z^3 - 1))$$

defines a purely inseparable field extension. This implies that $\mathbb{F}_3(z)[y]/(y^3-z(z^3-1))$ is the function field of a curve of genus 0. This does not imply that $C_1$ has bad reduction to characteristic 3, since there could be a different model.

We claim that there does not exist a curve of genus 3 in characteristic 3 with an automorphism of order 9. This claim implies that $C$ has bad reduction to characteristic 3. Indeed, if $C$ has potentially good reduction, then the automorphism group $\mathrm{Aut}(\overline{C})$ of the reduction $\overline{C}$ of $C$ contains $\mathrm{Aut}(C)$. Hence, in particular, $\mathrm{Aut}(\overline{C})$ contains an automorphism of order 9.

To obtain a contradiction, we assume that $X$ is a curve of genus 3 in characteristic 3 with an automorphism $\gamma$ of order 9. We consider the Galois cover

$$X \to X/\langle\gamma\rangle.$$

This cover is wildly ramified of order 9 above at least one point. We apply the Riemann–Hurwitz formula to this cover. It follows from Theorem 1.1 of Obus and Pries (2010) that the contribution of a wild ramification point with ramification index 9 to $2g(X) - 2$ in the Riemann–Hurwitz formula is at least $2 \cdot (9-1) + 5 \cdot (3-1) = 26$, which contradicts the assumption that $X$ has genus 3. This proves (a).

We have shown that $C_1$ has bad reduction to characteristic 3. Let $\overline{C}_{1,3}$ be the stable reduction of $C_1$ to characteristic 3. Then $\overline{C}_{1,3}$ contains at least 2 irreducible components of positive genus (Corollary 4.3). Furthermore, there is an automorphism of order 9 acting on $\overline{C}_{1,3}$. The only way this is possible is if $\overline{C}_{1,3}$ contains three irreducible components of positive genus, which are then elliptic curves, each with an automorphism of order 3. The automorphism of order 9 permutes these components. There is a unique elliptic curve with an automorphism of order 3, namely the elliptic curve with $j = 0$. In characteristic 3 this curve may be given by

$$w^3 - w = v^2. \tag{5.2}$$

This curve is supersingular by the Deuring–Shafarevich formula (Crew 1984). We conclude that the reduction $\overline{J}_{1,3}$ of the Jacobian of $C_1$ to characteristic 3 is supersingular. Proposition 4.2.(b) implies that $\overline{J}_{1,3}$ is in fact superspecial: the Jacobian $\overline{J}_{1,3}$ is isomorphic to three copies of the supersingular elliptic curve (5.2) as a polarized abelian variety.

The rest of (b) may be deduced from Yui (1980). For $p \equiv 4, 7 \pmod 9$, Yui's results (Yui 1980) imply that $\overline{J}$ is neither ordinary nor supersingular. In fact, her results imply that $\overline{J}$ has $p$-rank zero, but is not supersingular. Theorem 4.5 therefore implies that $\overline{J}$ is simple. □

The situation for $C_3$ is similar but somewhat easier.

**Lemma 5.4.** *(a) The curve $C_3$ has good reduction at $p \neq 7$ and potentially good reduction at $p = 7$.*
*(b) The reduction $\overline{J}_{3,p}$ of the Jacobian $J_3$ of $C_3$ to characteristic $p$ is ordinary if and only if $p \equiv 1 \pmod 7$ and supersingular if and only if $p = 7$ or $p \equiv -1, 3, 5 \pmod 7$.*

*Proof.* The fact that $C_3$ has good reduction to characteristic $p \neq 7$ follows as in the proof of Lemma 5.3. The curve $C_3$ has potentially good reduction to characteristic 7 as well, see Example 3.8 of Bouw and Wewers (2012). The curve $C_3$ does not have good reduction over $\mathbb{Q}_7$ but acquires good reduction over the extension $\mathbb{Q}_7(\zeta_7)$ of $\mathbb{Q}_7$.

Statement (b) for $p \neq 7$ follows from Yui (1980). We consider the reduction $\overline{C}_{3,7}$ of $C$ to characteristic 7. In characteristic 7, the reduction $\overline{C}_{3,7}$ is given by

$$w^7 - w = v^2,$$

by Example 3.8 of Bouw and Wewers (2012). By the Deuring–Shafarevich formula, it follows that the Jacobian $\overline{J}_{3,7}$ of $\overline{C}_{3,7}$ has $p$-rank 0. To show that it is supersingular, it suffices to find an elliptic quotient of the curve $\overline{C}_{3,7}$.

The curve $\overline{C}_{3,7}$ admits an extra automorphism of order 3 given by

$$\beta(v, w) = (\zeta_3 v, \zeta_3^2 w),$$

where $\zeta_3 \in \mathbb{F}_7^\times$ is an element of order three. The automorphism $\beta$ has exactly two fixed points, namely the points with $w = 0, \infty$. It follows that $E_{3,7} := \overline{C}_{3,7}/\langle\beta\rangle$ is an elliptic curve. This shows that $\overline{J}_{3,7}$ is supersingular. □

## 5.2 A Picard Curve Example

We end this section by considering Example 3 from Section 5 of Koike and Weng (2005), wherein Koike and Weng study Picard curves with CM. We show that the curve in the aforementioned example has bad reduction to characteristic $p = 5$, and that the stable reduction consists of an elliptic curve and a curve of genus 2. We will show that the Jacobian has superspecial reduction in this case. This is an example where the reduction $\overline{J}$ of the Jacobian is isomorphic to $E^3$, but the polarization is neither that of a smooth curve nor the product polarization $E \times \{0\} \times \{0\} + \{0\} \times E \times \{0\} + \{0\} \times \{0\} \times E$.

A Picard curve is a curve of genus 3 given by an equation

$$y^3 = f(x),$$

where $f(x) \in \mathbb{C}[x]$ is a polynomial of degree 4 with simple roots. Every Picard curve admits an automorphism $\alpha(x, y) = (x, \zeta_3 y)$. Therefore, the endomorphism ring of the Jacobian contains $\mathbb{Q}(\zeta_3)$.

Let $C_4$ be the smooth projective curve defined by

$$y^3 = f(x) := x^4 - 13 \cdot 2 \cdot 7^2 \cdot x^2 + 2^3 \cdot 13 \cdot 5 \cdot 47 \cdot x - 5^2 \cdot 31 \cdot 13^2.$$

Koike and Weng show that the Jacobian of $C_4$ has CM by the field $K = K^+K_1$ with $K_1 = \mathbb{Q}(\zeta_3)$ and $K^+ = \mathbb{Q}[t]/(t^3 - t^2 - 4t - 1)$. The CM-field $K$ is Galois over $\mathbb{Q}$, hence we are in Case 1 of Proposition 2.1. One may show that the corresponding CM-type is primitive. For example, one may check using Bouw (2001) that the reduction $\overline{J}_{4,7}$ of the Jacobian $J_4$ of $C_4$ to characteristic 7 has $p$-rank 1, and hence is neither ordinary nor supersingular. It follows from this that the Jacobian $J_4$ is simple. The primitivity of the CM-type follows from this, by Theorem 3.2.

We now consider the reduction of $C_4$. The discriminant of $f$ is $2^{12} \cdot 5^6 \cdot 13^4$ which shows that $C_4$ has good reduction for $p \neq 2, 3, 5, 13$. One may check that $C_4$ also has good reduction at $p = 2, 13$. We do not consider what happens for $p = 3$.

We determine the reduction at $p = 5$. Note that

$$f(x) \equiv x^2(x+2)(x-2) = x^4 + x^2 \pmod{5}. \tag{5.3}$$

Therefore, the stable reduction of $C_4$ contains an irreducible component $\overline{D}$ of genus 2 given by the equation

$$\bar{y}^3 = \bar{x}^2(\bar{x}^2 + 1). \tag{5.4}$$

The reason that this curve has genus 2 rather than 3 is that the 3-cyclic cover $(\bar{x}, \bar{y}) \mapsto \bar{x}$ has only 4 branch points in characteristic 5, and not 5 branch points as it had in characteristic zero. It follows that the curve $C_4$ has bad reduction to characteristic 5, and the reduction of $C_4$ consists of the curve $\overline{D}$ of genus 2 intersecting with an elliptic curve. (We do not actually have to compute the elliptic component to conclude this.) The reduction $\overline{J}_4$ of the Jacobian of $C_4$ is therefore isogenous to the product of an elliptic curve and the abelian surface $\mathrm{Jac}(\overline{D})$. To determine the reduction type of $\overline{J}_4$, we first consider the Jacobian $\mathrm{Jac}(\overline{D})$ of the curve $\overline{D}$ given by Equation (5.4).

One may show by computing the Hasse–Witt matrix of $\overline{D}$ that the Jacobian $J(\overline{D})$ is supersingular. This is a similar calculation to the one we did in Section 5.1. However, since $\overline{D}$ has genus 2, it suffices to compute the $p$-rank. In fact, the Hasse–Witt matrix is identically zero, which shows that $J(\overline{D})$ is superspecial, i.e., isomorphic to the product of two supersingular elliptic curves.

Alternatively, we may note that $\overline{D}$ has additional automorphisms given by

$$\tau(\bar{x}, \bar{y}) = (-\bar{x}, \bar{y}), \qquad \rho(\bar{x}, \bar{y}) = \left(-\frac{1}{\bar{x}}, \frac{\bar{y}}{\bar{x}^2}\right), \qquad \tau \circ \rho(\bar{x}, \bar{y}) = \left(\frac{1}{\bar{x}}, \frac{\bar{y}}{\bar{x}^2}\right).$$

Note that $\tau$ fixes the two points with $\bar{x} = 0, \infty$ and $\rho$ fixes the two points with $\bar{x}^2 = -1$. The quotients $\overline{C}_4/\langle\tau\rangle$ and $\overline{C}_4/\langle\rho\rangle$ are elliptic curves, each with an automorphism of order 3. In particular, these elliptic curves have $j = 0$. Since $p = 5 \equiv 2 \pmod{3}$, they are supersingular. Theorem 4.5 implies that $\overline{J}$ is isogenous to $E_0^3$, where $E_0$ denotes the supersingular elliptic curve over $\overline{\mathbb{F}_5}$ with $j = 0$.

*Remark 5.5.* The examples we discussed in this section all have the property that the CM-field $K$ contains a CM-subfield $K_1$ with $\mathbb{Q} \subsetneq K_1 \subsetneq K$. In Section 6.4, we will show that this implies that the embedding problem, which we formulate in Section 6, has degenerate solutions for every prime. This explains why we exclude this case in Theorem 6.8.

*Remark 5.6.* Let $K$ be a sextic CM-field. It is known how to construct genus 3 curves $C$ in characteristic 0 with CM by $K$ (Shimura 1998, Sections 6.2 and 14.3). We sketch the construction.

We fix a CM-type $(K, \varphi)$. Let $\delta_{K/\mathbb{Q}}$ be the different. For any ideal $\mathfrak{a}$ of $\mathcal{O}_K$ we consider the lattice $\varphi(\mathfrak{a}) = (\varphi_1(\mathfrak{a}), \varphi_2(\mathfrak{a}), \varphi_3(\mathfrak{a}))$. Then

$$A := \mathbb{C}^3/\varphi(\mathfrak{a})$$

is an abelian variety with CM by $(K, \varphi)$. Shimura (Theorem 3 of Section 6.2 in Shimura 1998) shows that all CM abelian varieties occur in this way.

In Section 14.3 of Shimura (1998), Shimura also describes all Riemann forms defining principal polarizations on $A$. Such a Riemann form exists if the following two conditions are satisfied.

- The ideal $\delta_{K/\mathbb{Q}}\mathfrak{a}\bar{\mathfrak{a}} = (a)$ is principal.
- There exists a unit $u \in \mathcal{O}_K$ such that $ua$ is totally imaginary and the imaginary part of $\varphi_i(ua)$ is negative for all $i$.

Every principally polarized abelian variety of dimension 3 is isomorphic to the Jacobian of a (possibly singular) genus 3 curve $C$ by Theorem 4 of Oort and Ueno (1973). More precisely, Oort and Ueno show that the curve $C$ is of compact type, meaning that $A$ is isomorphic to the product of the Jacobians of the irreducible components of positive genus of $C$. (This notion is essentially the same as the notion "tree-like" that we used in Section 4.2.) In our situation, the abelian variety $A$ is simple, and it follows that the curve $C$ is smooth.

# 6 Embedding Problem

## *6.1 Formulation of the Embedding Problem*

Let $C$ be a genus 3 curve defined over some number field $M$. We assume that the Jacobian $J = \mathrm{Jac}(C)$ has CM by a sextic CM-field $K$. After replacing $M$ by a finite extension if necessary, we may assume that $J$ has good reduction (Theorem 4.1) and that $C$ has stable reduction at all finite places of $M$.

In this section, we make the following important assumption.

**Assumption 6.1.** We assume that $K$ does not contain an imaginary quadratic subfield.

Recall that Assumption 6.1 implies that the CM-type of $C$ is primitive (Corollary 4.7). The reason for making this assumption is discussed in Section 6.4.

Let $\mathfrak{p}$ be a finite prime of $M$ where the curve $C$ has bad reduction. We write $\bar{k}$ for the algebraic closure of the residue field at $\mathfrak{p}$ and let $p$ denote the residue characteristic. We want to bound these primes $p$. (See Theorem 6.8 for the precise statement of our result.) Recall from Corollary 4.3 that there are two possibilities for the reduction $\overline{C}$ of $C$. In this section, we only deal with the case where $\overline{C}$ has three irreducible components of genus 1 and postpone the other case for future work. To summarize, we make the following assumption on the prime $\mathfrak{p}$.

**Assumption 6.2.** Let $\mathfrak{p}$ be a finite prime of $M$, such that the stable reduction $\overline{C} = \overline{C}_{\mathfrak{p}}$ of $C$ at $\mathfrak{p}$ contains three elliptic curves as irreducible components (Case (ii) of Corollary 4.3).

Let $\mathfrak{p}$ be as in Assumption 6.2. We write $E_1, E_2, E_3$ for the three elliptic curves that are the irreducible components of $\overline{C}$. We write $\overline{J}$ for the reduction of $J$ at $\mathfrak{p}$. Recall from Remark 4.4 that we have an isomorphism

$$\overline{J} \simeq E_1 \times E_2 \times E_3$$

as polarized abelian varieties, i.e., the polarization on $\overline{J}$ is the product polarization. Corollary 4.7 implies that the $E_i$ are supersingular. In particular, they are isogenous. (This also follows from Theorem 4.5.)

Let $\mathrm{End}(J) = \mathcal{O} \subset \mathcal{O}_K$. Reduction at the prime $\mathfrak{p}$ gives an injective ring homomorphism

$$\mathcal{O} \hookrightarrow \mathrm{End}(\overline{J}) \simeq \mathrm{End}(E_1 \times E_2 \times E_3).$$

**Problem 6.3 (The Embedding Problem).** Let $\mathcal{O}$ be an order in a sextic CM-field $K$, and let $p$ be a prime number. The *embedding problem* for $\mathcal{O}$ and $p$ is the problem of finding elliptic curves $E_1, E_2, E_3$ defined over a field of characteristic $p$, and a ring embedding

$$i : \mathcal{O} \hookrightarrow \mathrm{End}(E_1 \times E_2 \times E_3)$$

such that the Rosati involution on $\mathrm{End}(E_1 \times E_2 \times E_3)$ induces complex conjugation on $\mathcal{O}$. We call such a ring embedding a *solution to the embedding problem* for $\mathcal{O}$ and $p$.

The following result states that if we have a solution to the embedding problem then the elliptic curves $E_i$ are automatically isogenous. The proof we give here works directly with the abelian variety $E_1 \times E_2 \times E_3$ without considering it as the reduction of an abelian variety in characteristic zero. However the proof is essentially the same as the proofs of Theorem 4.5 and Proposition 4.6.

**Lemma 6.4.** *Let $K$ be a sextic CM-field. Suppose that there exist elliptic curves $E_1, E_2, E_3$ defined over a field of characteristic $p > 0$ and an injective $\mathbb{Q}$-algebra homomorphism*

$$i : K \hookrightarrow \mathrm{End}^0(E_1 \times E_2 \times E_3).$$

*Then the elliptic curves $E_1$, $E_2$, and $E_3$ are all isogenous. Furthermore, if $K$ contains no imaginary quadratic subfield, then the $E_i$ are supersingular.*

*Proof.* First suppose that no two of the elliptic curves $E_1$, $E_2$, $E_3$ are isogenous. Then

$$i: K \hookrightarrow \mathrm{End}^0(E_1 \times E_2 \times E_3) = \begin{pmatrix} \mathrm{End}^0 E_1 & 0 & 0 \\ 0 & \mathrm{End}^0 E_2 & 0 \\ 0 & 0 & \mathrm{End}^0 E_3 \end{pmatrix}$$
$$= \mathrm{End}^0 E_1 \times \mathrm{End}^0 E_2 \times \mathrm{End}^0 E_3.$$

Projecting on the factor $\mathrm{End}^0 E_i$ gives a ring homomorphism $K \hookrightarrow \mathrm{End}^0 E_i$. Since $K$ is a field, this ring homomorphism must be injective. But $\mathrm{End}^0 E_i$ is either an imaginary quadratic field or a quaternion algebra, neither of which can contain a sextic field.

Now suppose that exactly two of the elliptic curves are isogenous. Without loss of generality, we may assume that $E_1 \sim E_2$ and $E_1 \not\sim E_3$. Then

$$i: K \hookrightarrow \mathrm{End}^0(E_1 \times E_2 \times E_3) = \begin{pmatrix} \mathrm{End}^0 E_1 & \mathrm{End}^0 E_1 & 0 \\ \mathrm{End}^0 E_1 & \mathrm{End}^0 E_1 & 0 \\ 0 & 0 & \mathrm{End}^0 E_3 \end{pmatrix}$$
$$= M_2(\mathrm{End}^0 E_1) \times \mathrm{End}^0 E_3.$$

Again, projecting on the factor $\mathrm{End}^0 E_3$, we see that $K \hookrightarrow \mathrm{End}^0 E_3$. This is impossible for dimension reasons. Thus, we have proved that all three elliptic curves are isogenous.

Now suppose that $K$ contains no imaginary quadratic subfield and that the elliptic curves $E_i$ are ordinary. Then $\mathrm{End}^0 E_1 = K_1$ for some imaginary quadratic field $K_1$ and

$$i: K \hookrightarrow \mathrm{End}^0(E_1 \times E_2 \times E_3) = M_3(K_1).$$

Let $\beta$ be a generator for $K$ over $\mathbb{Q}$ and let $f$ be its minimal polynomial, which has degree 6. The matrix $i(\beta) \in M_3(K_1)$ has a minimal polynomial of degree at most 3 over $K_1$. Since $i$ is an injective $\mathbb{Q}$-algebra homomorphism, this means that $f$ splits over $K_1$. Since $K_1$ is quadratic, this implies that $K_1 \hookrightarrow K$, contradicting the assumption that $K$ contains no imaginary quadratic subfield. □

**Proposition 6.5.** *Let $C$ be a genus 3 curve such that $\mathcal{O} := \mathrm{End}(\mathrm{Jac}(C))$ is an order in a sextic CM-field $K$ satisfying Assumption 6.1. Let $M$ be a number field over which $C$ is defined, and let $\mathfrak{p}$ be a prime of bad reduction of $C$ such that Assumption 6.2 is satisfied. Write $p$ for the residue characteristic of $\mathfrak{p}$. Then there exists a solution to the embedding problem for $\mathcal{O}$ and $p$. Moreover, in this situation the three elliptic curves are supersingular.*

*Proof.* Let $C$ be as in the statement of the proposition. Then the CM-type of its Jacobian $J$ is primitive (Corollary 4.7.(a)). Therefore the Rosati involution acts as complex conjugation on $\mathrm{End}^0(J) = K$ by Proposition 4.8. The canonical polarization on the Jacobian $J$ is a principal polarization, therefore the Rosati involution also acts on $\mathrm{End}(J) = \mathcal{O}$.

Assumption 6.2 implies that the reduction $\overline{J}$ of the Jacobian at $\mathfrak{p}$ is isomorphic to a product of three elliptic curves $E_i$ as polarized abelian varieties. These elliptic curves are supersingular (Corollary 4.7.(b)). Remark 4.9 shows that we obtain a solution to the embedding problem. □

## 6.2 Endomorphisms of $\overline{J}$ as $3 \times 3$ Matrices

In this section we describe the ring $\mathrm{End}(E_1 \times E_2 \times E_3)$ from the embedding problem (Problem 6.3). Recall that we may assume that the $E_i$ are isogenous (Lemma 6.4). We recall from Proposition 4.10 the description of the Rosati involution corresponding to the product polarization on $E_1 \times E_2 \times E_3$.

We can view an element $f \in \mathrm{End}(E_1 \times E_2 \times E_3)$ as a matrix

$$f = \begin{pmatrix} f_{1,1} & f_{1,2} & f_{1,3} \\ f_{2,1} & f_{2,2} & f_{2,3} \\ f_{3,1} & f_{3,2} & f_{3,3} \end{pmatrix},$$

where $f_{i,j} \in \mathrm{Hom}(E_j, E_i)$. Given two endomorphisms $f, g$ the composition $f \circ g$ corresponds to multiplication of matrices. Since the polarization on $\overline{J} = E_1 \times E_2 \times E_3$ is the product polarization, the Rosati involution $f \mapsto f^*$ sends $f$ to

$$\begin{pmatrix} f_{1,1}^\vee & f_{2,1}^\vee & f_{3,1}^\vee \\ f_{1,2}^\vee & f_{2,2}^\vee & f_{3,2}^\vee \\ f_{1,3}^\vee & f_{2,3}^\vee & f_{3,3}^\vee \end{pmatrix}$$

where $f_{i,j}^\vee$ denotes the dual isogeny of $f_{i,j}$.

For $i = 2, 3$, let $\psi_i : E_1 \to E_i$ be an isogeny of degree $\delta_i$. The composition

$$E_1 \times E_1 \times E_1 \xrightarrow{(1,\psi_2,\psi_3)} E_1 \times E_2 \times E_3 \xrightarrow{(1,\delta_2^{-1}\psi_2^\vee,\delta_3^{-1}\psi_3^\vee)} E_1 \times E_1 \times E_1$$

induces an injective $\mathbb{Q}$-algebra homomorphism

$$\mathrm{End}^0(E_1 \times E_2 \times E_3) \hookrightarrow \mathrm{End}^0(E_1 \times E_1 \times E_1) = M_3(\mathrm{End}^0 E_1). \tag{6.1}$$

Let $\Phi$ denote the composite map

$$\Phi : K \hookrightarrow \mathrm{End}^0(E_1 \times E_2 \times E_3) \hookrightarrow M_3(\mathrm{End}^0 E_1).$$

It is easily seen that

$$\begin{pmatrix} 1 & 0 & 0 \\ 0 & \delta_2 & 0 \\ 0 & 0 & \delta_3 \end{pmatrix} \Phi(\mathcal{O}) \subset M_3(\operatorname{End} E_1).$$

Under the assumptions made in Section 6.1, we may assume that the elliptic curves $E_i$ in the formulation of the embedding problem are supersingular (Proposition 6.5). We therefore recall some well-known facts on the endomorphism ring of a supersingular elliptic curve.

Let $p \in \mathbb{Z}_{>0}$ be the rational prime lying below $\mathfrak{p}$.

**Proposition 6.6.** *Let $E$ be a supersingular elliptic curve defined over a field of characteristic $p$. Then* $\operatorname{End}^0 E$ *is a quaternion algebra over $\mathbb{Q}$ ramified at precisely the places $\{p, \infty\}$. This quaternion algebra is non-canonically isomorphic to the algebra $B_{p,\infty}$, where $B_{p,\infty} = (\frac{-1,-1}{\mathbb{Q}})$ if $p = 2$ and if $p$ is odd, $B_{p,\infty} = (\frac{-\varepsilon,-p}{\mathbb{Q}})$ where*

$$\varepsilon = \begin{cases} 1 & \text{if } p \equiv 3 \pmod 4, \\ 2 & \text{if } p \equiv 5 \pmod 8, \\ \ell & \text{if } p \equiv 1 \pmod 8. \end{cases}$$

*In the case that $p \equiv 1 \pmod 8$, $\ell \in \mathbb{Z}_{>0}$ is a prime such that $\ell \equiv 3 \pmod 4$ and $\ell$ is not a square modulo $p$. Any isomorphism sends* $\operatorname{End} E$ *to an order of $B_{p,\infty}$ and the involution given by taking the dual isogeny corresponds to the canonical involution on $B_{p,\infty}$.*

*Proof.* The fact that the endomorphism algebra $\operatorname{End}^0(E)$ of a supersingular elliptic curve is a quaternion algebra over $\mathbb{Q}$ ramified precisely at $\{p, \infty\}$ is proved, for example, in Section 21 of Mumford (1970). The statement on the Rosati involution is also proved in loc. cit. The uniqueness of the quaternion algebra is proved, for example, in Theorem III.3.1 of Vignéras (1980).

For $p = 2$, let $Q = (\frac{-1,-1}{\mathbb{Q}})$. For every odd prime $p$, let $\varepsilon$ be as in the statement of the proposition and let $Q = (\frac{-\varepsilon,-p}{\mathbb{Q}})$ be the corresponding quaternion algebra. The statement that $Q$ is exactly ramified at the places $\{p, \infty\}$ follows easily from the properties of the Hilbert symbol (page 37 of Vignéras 1980). □

For $b \in B_{p,\infty}$, we write $\operatorname{Nrd}(b) = bb^*$ (where $b^*$ represents the involution on the quaternion algebra) for the reduced norm of $b$. The reduced norm corresponds to the degree of an endomorphism under the identification in Proposition 6.6.

**Lemma 6.7 (Elements of Small Norm Commute, Goren and Lauter (2007, Corollary 2.1.2)).** *Let $R$ be a maximal order of $B_{p,\infty}$. If $k_1, k_2 \in R$ and* $\operatorname{Nrd}(k_1), \operatorname{Nrd}(k_2) < \sqrt{p}/2$, *then $k_1 k_2 = k_2 k_1$.*

## 6.3 Bounding the Primes of Bad Reduction for C

Recall that $J = \text{Jac}(C)$ is the Jacobian of a genus 3 curve $C$ which has complex multiplication by an order $\mathcal{O}$ in a sextic CM-field $K$ which does not contain an imaginary quadratic field (Assumption 6.1). Let $K^+$ denote the totally real cubic subfield of $K$. The main result of this section is Theorem 6.8 which gives an upper bound on the primes of bad reduction for $C$ satisfying Assumption 6.2.

**Theorem 6.8.** *Suppose that $K$ does not contain an imaginary quadratic subfield. Let $\mathfrak{p} \mid p$ be a prime of bad reduction for $C$ satisfying Assumption 6.2. Write $K = \mathbb{Q}(\sqrt{\alpha})$ for some totally negative element $\alpha \in K^+ \setminus \mathbb{Z}$ with $\sqrt{\alpha} \in \mathcal{O} = \text{End}(J)$. Then $p \leq 4\,\mathbf{Tr}_{K^+/\mathbb{Q}}(\alpha)^6/3^6$.*

The existence of such $\alpha$ is guaranteed because the sextic CM-field $K$ contains no imaginary quadratic subfield. By Proposition 6.5, the following result implies Theorem 6.8.

**Theorem 6.9.** *Suppose that $K$ does not contain an imaginary quadratic subfield. Let $p$ be a prime such that there exists a solution to the embedding problem (Problem 6.3) for some order $\mathcal{O}$ of $K$. Write $K = \mathbb{Q}(\sqrt{\alpha})$ for some totally negative element $\alpha \in K^+ \setminus \mathbb{Z}$ with $\sqrt{\alpha} \in \mathcal{O}$. Then $p \leq 4\,\mathbf{Tr}_{K^+/\mathbb{Q}}(\alpha)^6/3^6$.*

We break down the proof of Theorem 6.9 into several lemmas.

Let

$$Q = \begin{pmatrix} r & s & t \\ u & v & w \\ x & y & z \end{pmatrix}$$

be the image of $\sqrt{\alpha}$ in $\text{End}(E_1 \times E_2 \times E_3)$. By Proposition 4.8, the Rosati involution corresponds to complex conjugation on $K$, so we have

$$\begin{pmatrix} r^\vee & u^\vee & x^\vee \\ s^\vee & v^\vee & y^\vee \\ t^\vee & w^\vee & z^\vee \end{pmatrix} = \begin{pmatrix} -r & -s & -t \\ -u & -v & -w \\ -x & -y & -z \end{pmatrix}. \tag{6.2}$$

**Lemma 6.10.** *We may assume that the homomorphisms $s : E_2 \to E_1$ and $t : E_3 \to E_1$ are both nonzero.*

*Proof.* Suppose for contradiction that both $s$ and $t$ are zero. Then the image of $\alpha$ in $\text{End}(E_1 \times E_2 \times E_3)$ is

$$Q^2 = \begin{pmatrix} -rr^\vee & 0 & 0 \\ 0 & -vv^\vee - ww^\vee & vw + wz \\ 0 & -w^\vee v - zw^\vee & -w^\vee w - zz^\vee \end{pmatrix}.$$

For $i = 2, 3$, let $\psi_i : E_1 \to E_i$ be an isogeny of degree $\delta_i$. As seen in (6.1), the $\psi_i$ induce an injective $\mathbb{Q}$-algebra homomorphism $\operatorname{End}^0(E_1 \times E_2 \times E_3) \to \operatorname{End}^0(E_1 \times E_1 \times E_1) = M_3(\operatorname{End}^0 E_1)$ sending $Q^2$ to

$$S = \begin{pmatrix} -rr^\vee & 0 & 0 \\ 0 & -vv^\vee - ww^\vee & \delta_2^{-1}\psi_2^\vee(vw + wz)\psi_3 \\ 0 & \delta_3^{-1}\psi_3^\vee(-w^\vee v - zw^\vee)\psi_2 & -w^\vee w - zz^\vee \end{pmatrix}.$$

Since $(vw + wz)^\vee = -w^\vee v - zw^\vee$, the entries of $S$ commute and therefore form a subfield $L$ of $\operatorname{End}^0 E_1$. Since $S$ is the image of $\alpha$ under an injective $\mathbb{Q}$-algebra homomorphism, the minimal polynomial of $S$ over $L$ divides the minimal polynomial of $\alpha$ over $\mathbb{Q}$. Recall that $rr^\vee \in \mathbb{Z}$ is the degree of $r$. Now $-rr^\vee$ is an eigenvalue of $S$ and therefore a root of its minimal polynomial. But this means that the minimal polynomial of $\alpha$ over $\mathbb{Q}$ has a root in $\mathbb{Z}$, contradicting its irreducibility.

Therefore, at least one of $s, t$ is nonzero. Using $E_2$ in place of $E_1$, we see that at least one of $s, w$ is nonzero. Using $E_3$ in place of $E_1$, we see that at least one of $t, w$ is nonzero. Putting all these conditions together and reordering the elliptic curves $E_1, E_2, E_3$ if necessary, we may assume that $s$ and $t$ are both nonzero. □

Henceforth, we assume that $s$ and $t$ are nonzero. Therefore, we can use $s^\vee$ and $t^\vee$ to give an injective $\mathbb{Q}$-homomorphism $\operatorname{End}^0(E_1 \times E_2 \times E_3) \hookrightarrow \operatorname{End}^0(E_1 \times E_1 \times E_1)$ as in (6.1). The image of $\sqrt{\alpha}$ in $M_3(\operatorname{End}^0 E_1)$ is

$$T = \begin{pmatrix} r & \delta_2 & \delta_3 \\ -1 & svs^\vee/\delta_2 & swt^\vee/\delta_2 \\ -1 & -tw^\vee s^\vee/\delta_3 & tzt^\vee/\delta_3 \end{pmatrix}, \tag{6.3}$$

where $\delta_2 = \deg(s)$ and $\delta_3 = \deg(t)$.

Since $K$ contains no imaginary quadratic subfield, Lemma 6.4 shows that the elliptic curves $E_1, E_2$, and $E_3$ are supersingular. By Proposition 6.6, we may choose an isomorphism $\operatorname{End}^0 E_1 \to B_{p,\infty}$. The isomorphism sends $\operatorname{End} E_1$ to a order of $B_{p,\infty}$ and the Rosati involution on $\operatorname{End} E_1$ corresponds to the usual involution on $B_{p,\infty}$. We abuse notation slightly by continuing to write $T$ for the image of $\sqrt{\alpha}$ in $M_3(B_{p,\infty})$.

**Lemma 6.11.** *Suppose that $K$ contains no imaginary quadratic subfield. Let $T$ denote the image of $\sqrt{\alpha}$ in $M_3(B_{p,\infty})$. Then the entries of the matrix $T$ do not all commute with each other.*

*Proof.* Suppose for contradiction that the entries of $T$ commute. Let $K_1$ denote the subfield of $B_{p,\infty}$ generated by the entries of $T$. A subfield of $B_{p,\infty}$ is either $\mathbb{Q}$ or a quadratic subfield which splits $B_{p,\infty}$. But $B_{p,\infty}$ is ramified at the infinite place, so it is not split by any real field. Thus, $K_1$ is either $\mathbb{Q}$ or an imaginary quadratic field. By assumption, $K$ contains no imaginary quadratic subfield. Thus, the minimal polynomial of $\sqrt{\alpha}$ over $\mathbb{Q}$ remains irreducible over $K_1$.

Let $g$ denote the minimal polynomial of $T$ over $K_1$. The degree of $g$ is at most 3. Since $T$ is the image of $\sqrt{\alpha}$ under an injective $\mathbb{Q}$-algebra homomorphism, $g$ divides the minimal polynomial of $\sqrt{\alpha}$ over $\mathbb{Q}$, which has degree 6. Thus, the minimal polynomial of $\sqrt{\alpha}$ over $\mathbb{Q}$ factorizes over $K_1$, giving the required contradiction. □

We restrict to the case where $p$ is odd; the case $p = 2$ is very similar. By Proposition 6.6, $B_{p,\infty}$ has a $\mathbb{Q}$-basis $1, i, j, k$ where $i^2 = -\varepsilon$, $j^2 = -p$, $ij = k$, $ji = -ij$ and $\varepsilon$ is as in Proposition 6.6. We embed $B_{p,\infty}$ into $M_4(\mathbb{Q})$ via

$$1 \mapsto \begin{pmatrix} 1&0&0&0\\ 0&1&0&0\\ 0&0&1&0\\ 0&0&0&1 \end{pmatrix}, \; i \mapsto \begin{pmatrix} 0&-\varepsilon&0&0\\ 1&0&0&0\\ 0&0&0&-\varepsilon\\ 0&0&1&0 \end{pmatrix}, \; j \mapsto \begin{pmatrix} 0&0&-p&0\\ 0&0&0&p\\ 1&0&0&0\\ 0&-1&0&0 \end{pmatrix}, \; k \mapsto \begin{pmatrix} 0&0&0&-\varepsilon p\\ 0&0&-p&0\\ 0&\varepsilon&0&0\\ 1&0&0&0 \end{pmatrix}.$$

This induces an embedding $M_3(B_{p,\infty}) \hookrightarrow M_{12}(\mathbb{Q})$. Let $U$ denote the image of $\alpha$ in $M_{12}(\mathbb{Q})$. Write $\mathbf{Tr}(T^2)$ for the sum of the elements on the diagonal of $T^2$. Define $\mathbf{Tr}(Q^2)$ in the same way. It is easily checked that $\mathbf{Tr}(T^2) = \mathbf{Tr}(Q^2)$. By the construction of the embedding $B_{p,\infty} \hookrightarrow M_4(\mathbb{Q})$, we have

$$\mathbf{Tr}(U) = 4\,\mathbf{Tr}(T^2). \tag{6.4}$$

**Lemma 6.12.** *Let $T$ denote the image of $\sqrt{\alpha}$ in $M_3(B_{p,\infty})$. Then* $\mathbf{Tr}(T^2) = \mathbf{Tr}_{K^+/\mathbb{Q}}(\alpha)$.

*Proof.* Let $\alpha = \alpha_1, \alpha_2, \alpha_3$ denote the conjugates of $\alpha$. The characteristic polynomial of $U$ is $(X-\alpha_1)^{m_1}(X-\alpha_2)^{m_2}(X-\alpha_3)^{m_3}$ for some $m_1, m_2, m_3 \in \mathbb{Z}_{>0}$ with $m_1 + m_2 + m_3 = 12$. The trace of $U$ is $m_1\alpha_1 + m_2\alpha_2 + m_3\alpha_3 \in \mathbb{Q}$. If we can show that $m_1 = m_2 = m_3 = 4$, then Equation (6.4) gives

$$4\,\mathbf{Tr}(T^2) = \mathbf{Tr}(U) = m_1\alpha_1 + m_2\alpha_2 + m_3\alpha_3 = 4(\alpha_1 + \alpha_2 + \alpha_3) = 4\,\mathbf{Tr}_{K^+/\mathbb{Q}}(\alpha). \tag{6.5}$$

Therefore, it is enough to show that $m_1 = m_2 = m_3$. Since $\alpha \in \mathcal{O}_{K^+}$, we have $\alpha_1 + \alpha_2 + \alpha_3 \in \mathbb{Z}$ and therefore $(m_2 - m_1)\alpha_2 + (m_3 - m_1)\alpha_3 \in \mathbb{Q}$. Suppose for contradiction that we are not in the case $m_1 = m_2 = m_3$. Then, without loss of generality, $(m_2 - m_1) \neq 0$ and since $\alpha_2 \notin \mathbb{Q}$ it follows that $(m_3 - m_1) \neq 0$. Therefore, $\alpha_3 = \lambda\alpha_2$ for some $\lambda \in \mathbb{Q}$. But $\alpha_3$ is a Galois conjugate of $\alpha_2$ and the Galois group of the Galois closure of $K^+/\mathbb{Q}$ is either $C_3$ or $S_3$. Therefore, the automorphism sending $\alpha_2$ to $\alpha_3$ has order dividing 6 and hence $\lambda$ is a sixth root of unity in $\mathbb{Q}$. Therefore, $\lambda = -1$ and $\alpha_3 = -\alpha_2$. But this gives $\mathbf{Tr}_{K^+/\mathbb{Q}}(\alpha) = \alpha_1 + \alpha_2 + \alpha_3 = \alpha_1$. So $\alpha = \alpha_1 = \mathbf{Tr}_{K^+/\mathbb{Q}}(\alpha) \in \mathbb{Q}$, which is a contradiction. □

*Proof of Theorem 6.9.* Suppose for contradiction that $p > 4\,\mathbf{Tr}_{K^+/\mathbb{Q}}(\alpha)^6/3^6$. We will show that the entries of the matrix $T$ commute, contradicting Lemma 6.11. The key ingredients will be Lemma 6.7 (which states that elements of a maximal order whose reduced norms are smaller than $\sqrt{p}/2$ commute) and Equation (6.7) below.

Recall that

$$T = \begin{pmatrix} r & \delta_2 & \delta_3 \\ -1 & svs^\vee/\delta_2 & swt^\vee/\delta_2 \\ -1 & -tw^\vee s^\vee/\delta_3 & tzt^\vee/\delta_3 \end{pmatrix} \tag{6.6}$$

where $\delta_2 = \deg(s)$ and $\delta_3 = \deg(t)$. We have

$$\begin{pmatrix} 1 & 0 & 0 \\ 0 & \delta_2 & 0 \\ 0 & 0 & \delta_3 \end{pmatrix} T \in M_3(\operatorname{End} E_1).$$

We have chosen an isomorphism $\operatorname{End}^0 E_1 \to B_{p,\infty}$, sending $\operatorname{End} E_1$ to a order of $B_{p,\infty}$. The dual on $\operatorname{End} E_1$ corresponds to the usual involution on $B_{p,\infty}$. We identify $\operatorname{End}^0 E_1$ with $B_{p,\infty}$ and write $\mathrm{Nrd}(f) = \deg(f) = ff^\vee$ for $f \in \operatorname{End} E_1$.

By Lemma 6.12, we have $\mathbf{Tr}(T^2) = \mathbf{Tr}_{K^+/\mathbb{Q}}(\alpha)$. Writing out the entries on the diagonal of $T^2$ gives

$$\begin{aligned} 0 &< \deg(r) + 2\deg(s) + 2\deg(t) + \deg(v) + 2\deg(w) + \deg(z) \\ &= -\mathbf{Tr}_{K^+/\mathbb{Q}}(\alpha) < 3\sqrt[6]{p/4}. \end{aligned} \tag{6.7}$$

Note that the sum of degrees is a sum of non-negative integers. We want to use (6.7) to bound the reduced norms of the non-scalar entries of $\begin{pmatrix} 1 & 0 & 0 \\ 0 & \delta_2 & 0 \\ 0 & 0 & \delta_3 \end{pmatrix} T$. Recall that, in light of Lemma 6.10, we are assuming that $s$ and $t$ are nonzero. Therefore, $\deg(s), \deg(t) \geq 1$ and (6.7) gives

i) $\mathrm{Nrd}(r) = \deg(r) < 3\sqrt[6]{p/4} - 4 < \sqrt{p}/2$,
ii) $2\deg(s) + \deg(v) < 3\sqrt[6]{p/4}$,
iii) $2(\deg(s) + \deg(t) + \deg(w)) < 3\sqrt[6]{p/4}$,
iv) $2\deg(t) + \deg(z) < 3\sqrt[6]{p/4}$.

Observe that $\mathrm{Nrd}(swt^\vee) = \deg(s)\deg(w)\deg(t) = \mathrm{Nrd}(-tw^\vee s^\vee)$. So it remains to bound the reduced norms of $svs^\vee$, $swt^\vee$ and $tzt^\vee$. Let $a \in \mathbb{R}_{>0}$. The maximum of the function $f(x) = x^2(a - 2x)$ for $x \geq 0$ is achieved at $x = a/3$ and we have $f(a/3) = (a/3)^3$. Applying this to ii) with $a = 3\sqrt[6]{p/4}$, we see that

$$\mathrm{Nrd}(svs^\vee) = \deg(s)^2\deg(v) < (\sqrt[6]{p/4})^3 = \sqrt{p}/2.$$

Similarly, using iv) we get

$$\mathrm{Nrd}(tzt^\vee) = \deg(t)^2\deg(z) < (\sqrt[6]{p/4})^3 = \sqrt{p}/2.$$

Using iii), we get

$$\begin{aligned}\mathrm{Nrd}(swt^{\vee}) &= \deg(s)\deg(w)\deg(t) \leq (\deg(s)+\deg(w))^2 2\deg(t) < (\sqrt[6]{p/4})^3 \\ &= \sqrt{p}/2.\end{aligned}$$

Therefore, by Lemma 6.7, the entries of $\begin{pmatrix} 1 & 0 & 0 \\ 0 & \delta_2 & 0 \\ 0 & 0 & \delta_3 \end{pmatrix} T$ commute. Since the entries of $\begin{pmatrix} 1 & 0 & 0 \\ 0 & \delta_2 & 0 \\ 0 & 0 & \delta_3 \end{pmatrix} T$ are just scalar multiples of the entries of $T$, this means that the entries of $T$ commute. But this contradicts Lemma 6.11. Therefore, the assumption $p > 4\,\mathbf{Tr}_{K^+/\mathbb{Q}}(\alpha)^6/3^6$ does not hold. □

## 6.4 *Solutions to the Embedding Problem in the Case that K Contains an Imaginary Quadratic Subfield*

In this section, we consider the case where the sextic CM-field $K$ contains an imaginary quadratic subfield $K_1$. We show that the embedding problem 6.3 has solutions for every prime $p$ (Corollary 6.15). The solutions are constructed via the reduction at $p$ of a CM-abelian variety $A = E^3$ in characteristic zero, where $E$ is an elliptic curve. In particular, the CM-type of $A$ is imprimitive (Theorem 3.2). The solutions we construct may therefore be called *degenerate solutions* to the embedding problem.

The point is that if $K$ is a CM-field which contains an imaginary quadratic subfield then there always exist imprimitive CM-types for $K$. This is what allows for the existence of degenerate solutions to the embedding problem. Recall from Corollary 3.4 that there do not exist imprimitive CM-types $(K, \varphi)$ for CM-fields that do not contain a proper CM-subfield.

The proof of Theorem 6.8 relied on showing non-existence of solutions of the embedding problem for sufficiently large primes (Theorem 6.9) in the case where the sextic CM-field contains no proper CM-subfield. In contrast, if $C$ is a curve whose Jacobian has CM by a sextic CM-field $K$ which contains an imaginary quadratic field, then this strategy breaks down because there the embedding problem has degenerate solutions for all primes $p$ (Corollary 6.15). The embedding problem, as formulated in Problem 6.3, does not take the CM-type into consideration. It may be possible to prove an analogous result to Theorem 6.8, in the case that $K$ contains a proper CM-subfield, using a more refined formulation of the embedding problem that includes the CM-type as part of the data.

**Proposition 6.13.** *Let $K$ be a sextic CM-field containing a proper CM subfield $K_1$. Let $E$ be an elliptic curve over an arbitrary field and suppose that there exists an embedding $K_1 \hookrightarrow \mathrm{End}^0(E)$. Then there exists an order $\mathcal{O}$ of $K$ and a ring embedding*

$$\mathcal{O} \hookrightarrow \mathrm{End}(E^3) = M_3(\mathrm{End}(E))$$

*such that the Rosati involution on $\mathrm{End}(E^3)$ corresponding to the product polarization on $A = E^3$ induces complex conjugation on $\mathcal{O}$.*

*Proof.* It suffices to give an injective $\mathbb{Q}$-algebra homomorphism

$$K \hookrightarrow \mathrm{End}^0(E^3) = M_3(\mathrm{End}^0(E)). \tag{6.8}$$

This can be achieved as follows. Write $K = K^+K_1$ where $K^+/\mathbb{Q}$ is a totally real field with $[K^+ : \mathbb{Q}] = 3$. Choose a primitive element $\alpha$ of $K^+/\mathbb{Q}$, so $K^+ = \mathbb{Q}(\alpha)$. Embed $K_1$ diagonally via the fixed embedding of $K_1$ into $\mathrm{End}^0(E)$. Map $\alpha$ to a symmetric matrix $Q \in M_3(\mathbb{Q})$ which has the same minimal polynomial as $\alpha$. Since all the conjugates of $\alpha$ are real, the existence of the matrix $Q$ is proved in Theorem 4 of Bender (1968). Extend to a $\mathbb{Q}$-algebra homomorphism. □

In Remark 5.6, we reviewed the construction in characteristic 0 of genus 3 curves with CM by a sextic CM-field $K$. Similarly, when $K_1$ is an imaginary quadratic field, elliptic curves with CM by $K_1$ exist in characteristic zero. For example, we may take $E = \mathbb{C}/\mathcal{O}_{K_1}$, where we consider the maximal order $\mathcal{O}_{K_1}$ of $K_1$ as lattice in $\mathbb{C}$ (Silverman 1994, Remark II.4.1.1). Then $\mathrm{End}(E) = \mathcal{O}_{K_1}$. Moreover, $j(E)$ is an algebraic integer (Silverman 1994, Theorem II.6.1). (This can be deduced from Theorem 4.1 which states that $E$ has potentially good reduction.) In particular, $E$ can be defined over the number field $M := \mathbb{Q}(j(E))$.

We now show the existence of elliptic curves with CM by $K_1$ in positive characteristic. As above, $E/M$ is an elliptic curve defined over the number field $M$ with $\mathrm{End}(E) = \mathcal{O}_{K_1}$. We choose a rational prime $p$, and let $\mathfrak{p}$ be a prime of $M$ above $p$. After extending $M$ if necessary, we may assume that $E$ has good reduction at $\mathfrak{p}$. Write $\overline{E}_\mathfrak{p}$ for the reduction of $E$ at $\mathfrak{p}$. We obtain an embedding

$$\mathcal{O}_{K_1} = \mathrm{End}(E) \hookrightarrow \mathrm{End}(\overline{E}_\mathfrak{p}).$$

This proves the following lemma.

**Lemma 6.14.** *Let $p$ be a prime. Then there exists an elliptic curve $\overline{E}_p$ in characteristic $p$ with $\mathcal{O}_{K_1} \hookrightarrow \mathrm{End}(\overline{E}_p)$.*

The following result follows immediately from Lemma 6.14 and Proposition 6.13.

**Corollary 6.15.** *Let $K$ be a sextic CM-field containing an imaginary quadratic field $K_1$. Then there exists an order $\mathcal{O}$ of $K$ for which there exists a solution to the embedding problem for $\mathcal{O}$ and $p$ for every prime number $p$.*

Corollary 6.15 does not specify whether the elliptic curve $\overline{E}_p$ from Lemma 6.14 is ordinary or supersingular. The following proposition answers this question. Note that it follows that the set of primes where the elliptic curve $\overline{E}_p$ is supersingular has Dirichlet density $1/2$.

**Proposition 6.16 (Deuring's Theorem).** *Let $E/M$ be an elliptic curve with CM by $\mathcal{O}_{K_1}$. Let $p$ be a rational prime and $\mathfrak{p}$ be a prime of $M$ above $p$ such that $E$ has good reduction at $\mathfrak{p}$. Then the reduction $\overline{E}_{\mathfrak{p}}$ of $E$ at $\mathfrak{p}$ is supersingular if and only if $p$ is inert or ramified in $K_1$.*

Proposition 6.16 is well known, but hard to find explicitly in the literature. The statement can be proved using Theorem 10 of Section 10.4 of Lang (1987). We give the idea of the proof of the proposition. Let $\overline{E}/\mathbb{F}_q$ be an elliptic curve. Write $\pi$ for its $q$-Frobenius endomorphism. Then $\overline{E}$ is supersingular if and only if there exists integers $n, m$ such that $\pi^n = [p]^m$, where $[p]$ denotes multiplication by $p$. (See, for example, the proof of the Theorem of Deuring in Section 22 of Mumford 1970.) The theorem from Lang (1987) shows that this happens if and only if $p$ is inert or ramified in $K_1$.

## Appendix: Equations

In this section, we list the equations obtained from a possible solution to the embedding problem. We start by setting some notation.

Let $K^+$ be the maximal real subfield of the sextic CM-field $K = K^+(\eta)$. Take an integral basis of $\mathcal{O}_{K^+}$, so $\mathcal{O}_{K^+} = \alpha_1\mathbb{Z} \oplus \alpha_2\mathbb{Z} \oplus \alpha_3\mathbb{Z}$. We may assume that $K^+ = \mathbb{Q}(\alpha_1)$. We fix the following notation:

- $\mathbf{Tr}_{K/K^+}(\eta) = a_1\alpha_1 + a_2\alpha_2 + a_3\alpha_3$
- $\mathbf{Nm}_{K/K^+}(\eta) = b_1\alpha_1 + b_2\alpha_2 + b_3\alpha_3$
- $f_i(x) = x^3 + m_i x^2 + n_i x + s_i$ is the characteristic polynomial of $\alpha_i$ over $\mathbb{Q}$ for $i = 1, 2, 3$.

A solution to the embedding problem (Problem 6.3) gives us three elliptic curves $E_1, E_2, E_3$ and an embedding of $\iota : \mathcal{O}_K \hookrightarrow \mathrm{End}(E_1 \times E_2 \times E_3)$ such that Rosati involution on $E_1 \times E_2 \times E_3$ restricts to complex conjugation in the image of $\mathcal{O}_K$. This gives the following conditions on $\iota(\alpha_i)$ and $\iota(\eta)$:

1. Commutativity:

   (a) $\iota(\alpha_i)\iota(\eta) = \iota(\eta)\iota(\alpha_i)$ for all $i = 1, 2, 3$.
   (b) $\iota(\alpha_i)\iota(\alpha_j) = \iota(\alpha_j)\iota(\alpha_i)$ for all $i \neq j \in \{1, 2, 3\}$.

2. Characteristic polynomial: $f_i(\iota(\alpha_i)) = 0$ for all $i = 1, 2, 3$.
3. Norm: $\iota(\eta)\iota(\eta)^\dagger = b_1\iota(\alpha_1) + b_2\iota(\alpha_2) + b_3\iota(\alpha_3)$, where $\dagger$ denotes the conjugate transpose.
4. Trace: $\iota(\eta) + \iota(\eta)^\dagger = a_1\iota(\alpha_1) + a_2\iota(\alpha_2) + a_3\iota(\alpha_3)$.

5. Duality/Complex conjugation: $\iota(\alpha_i) = \iota(\alpha_i)^\dagger$ for all $i = 1, 2, 3$. Since we are interested in the case that Rosati involution induces complex multiplication and since $\eta$ can be chosen so that $\eta^2 \in K^+$ is totally negative, we have $\iota(\eta)^\dagger = -\iota(\eta)$.

In the rest of this appendix, we will only write the conditions for $i = 1$ which is enough if we have a power basis. In any case, the other relations for $i = 2, 3$ are similar. We now write the conditions above in terms of matrix coefficients. We are using the conventions and maps introduced in Section 6.2.

Let $M = \iota(\alpha_1)$ be the matrix $\begin{pmatrix} a & b & c \\ d & e & f \\ g & h & \ell \end{pmatrix}$ and $N = \iota(\eta)$ be the matrix $\begin{pmatrix} p & q & r \\ s & t & u \\ v & w & y \end{pmatrix}$.

## *Equations for Duality/Complex Conjugation Condition*

The relation $\iota(\eta)^\dagger = -\iota(\eta)$ translates into $M = M^\vee$ i.e., $\begin{pmatrix} a & b & c \\ d & e & f \\ g & h & \ell \end{pmatrix} = \begin{pmatrix} a^\vee & d^\vee & g^\vee \\ b^\vee & e^\vee & h^\vee \\ c^\vee & f^\vee & \ell^\vee \end{pmatrix}$.

This gives us the following relations.

*Remark 6.17.* Note that we name the relations with respect to the variables we intend to use later on. Our aim is to simplify the equations and write everything in terms of the upper triangular entries of our matrices which are $a, b, c, e, f, \ell$ in the case of $M$ and $p, q, r, t, u, y$ in the case of $N$.

(b-d) $d = b^\vee$
(c-g) $g = c^\vee$
(f-h) $h = f^\vee$
(int) $a, e, \ell$ are integral and in $\mathbb{Q}$, hence they are integers.

The relation $\iota(\eta)^\vee = -\iota(\eta)$ translates into: $\begin{pmatrix} p & q & r \\ s & t & u \\ v & w & y \end{pmatrix} = \begin{pmatrix} -p^\vee & -s^\vee & -v^\vee \\ -q^\vee & -t^\vee & -w^\vee \\ -r^\vee & -u^\vee & -y^\vee \end{pmatrix}$.

This gives us the following relations:

(q-s) $s = -q^\vee$
(r-v) $v = -r^\vee$
(u-w) $w = -u^\vee$
(trace) $p = -p^\vee$, $t = -t^\vee$, and $y = -y^\vee$
i.e., $p$, $t$, and $y$ have trace zero in $\mathrm{End}(E_1)$, $\mathrm{End}(E_2)$, and $\mathrm{End}(E_3)$ respectively.

## *Equations for Commutativity Condition*

Using $M$ and $N$ as above, the condition means $MN = NM$ which translates into the following equations:

(i-i) $ap + bs + cv = pa + qd + rg$. (By equation (int) in section "Equations for Duality/Complex Conjugation Condition", $a$ is an integer. Hence $ap = pa$ and $bs + cv = qd + rg$.)
(i-ii) $aq + bt + cw = pb + qe + rh$
(i-iii) $ar + bu + cy = pc + qf + r\ell$
(ii-i) $dp + es + fv = sa + td + ug$
(ii-ii) $dq + et + fw = sb + te + uh$ (By equation (int) in section "Equations for Duality/Complex Conjugation Condition", $e$ is an integer. Hence $et = te$ and $dq + fw = sb + uh$.)
(ii-iii) $dr + eu + fy = sc + tf + u\ell$
(iii-i) $gp + hs + \ell v = va + wd + yg$
(iii-ii) $gq + ht + \ell w = vb + we + yh$
(iii-iii) $gr + hu + \ell y = vc + wf + y\ell$ (By equation (int) in section "Equations for Duality/Complex Conjugation Condition", $\ell$ is an integer. Hence $\ell y = y\ell$ and $gr + hu = vc + wf$.)

## *Combining Duality and Commutativity Conditions*

Now we will plug in the equations we obtained in section "Equations for Duality/Complex Conjugation Condition" into the equations we obtained in section "Equations for Commutativity Condition". Note that our aim is to simplify the equations and write everything in terms of the upper triangular entries of our matrices which are $a, b, c, e, f, \ell$ in the case of $M$ and $p, q, r, t, u, y$ in the case of $N$.

| Relation | Obtained using: |
|---|---|
| $bq^{\vee} + cr^{\vee} + rc^{\vee} + qb^{\vee} = 0$ | (i-i), (c-g), (q-s), (v-r) |
| $pb + qe + rf^{\vee} - aq - bt + cu^{\vee} = 0$ | (i-ii), (u-w), (f-h) |
| $ar + bu + cy - pc - qf - r\ell = 0$ | (i-iii) |
| $b^{\vee}p - eq^{\vee} - fr^{\vee} + q^{\vee}a - tb^{\vee} - uc^{\vee} = 0$ | (ii-i), (b-d), (q-s), (r-v), (q-s) |
| $b^{\vee}q - fu^{\vee} + q^{\vee}b - uf^{\vee} = 0$ | (ii-ii), (b-d), (u-w), (q-s), (f-h) |
| $dr + eu + fy + q^{\vee}c - tf - u\ell = 0$ | (ii-iii), (q-s) |
| $c^{\vee}p - f^{\vee}q^{\vee} + (a-\ell)r^{\vee} + u^{\vee}b - yc^{\vee} = 0$ | (iii-i), (c-g), (f-h), (s-q), (r-v), (u-w), (b-d), (int) |
| $c^{\vee}q + f^{\vee}t + (e-\ell)u^{\vee} + r^{\vee}b - yf^{\vee} = 0$ | (iii-i), (c-g), (f-h), (u-w), (r-v), (int) |
| $c^{\vee}r + f^{\vee}u + r^{\vee}c + u^{\vee}f = 0$ | (iii-i), (f-h), (u-w), (r-v) |

## *Equations for Characteristic Polynomial Condition*

The characteristic polynomial condition for $i = 1$ translates into $0 = M^3 + m_1 M^2 + n_1 M + s_1$. Combining this equality with Equation (int) of section "Equations for Commutativity Condition" gives the following equations. For instance, for the top left corner of the matrix sum we get

$$0 = a^3 + abd + acg + bda + bed + bfg + cga + chd + c\ell g + m_1(a^2 + bd + cg) + n_1 a + s_1.$$

If we apply Condition (int) this turns into

$$(2a + e + m_1)bd + (2a + \ell + m_1)cg + bfg + chd + a^3 + m_1 a^2 + n_1 a + s_1 = 0.$$

The following is the list of equations coming from all nine entries.

(i) $(2a + e + m_1)bd + (2a + \ell + m_1)cg + bfg + chd + a^3 + m_1 a^2 + n_1 a + s_1 = 0$
(ii) $(a^2 + ae + e^2 + m_1 a + m_1 e + n_1)b + (e + \ell + m_1 + a)ch + bdb + bfh + cgb = 0$
(iii) $(a^2 + a\ell + \ell^2 + m_1 a + m_1 \ell + n_1)c + (a + e + \ell + m_1)bf + bdc + cgc + chf = 0$
(iv) $(a^2 + ea + e^2 + m_1 a + m_1 e + n_1)d + (e + a + \ell + m_1)fg + dbd + dcg + fhd = 0$
(v) $(a + 2e + m_1)db + (2e + \ell + m_1)fh + dch + fgb + e^3 + m_1 e^2 + n_1 + s_1 = 0$
(vi) $(a + \ell + e + m_1)dc + (e^2 + e\ell + \ell^2 + m_1 e + m_1 \ell + n_1)f + dbf + fgc + fhf = 0$
(vii) $(a^2 + \ell a + \ell^2 + m_1 a + m_1 \ell + n_1)g + (a + e + \ell + m_1)hd + gbd + gcg + hfg = 0$
(viii) $(e^2 + \ell e + \ell^2 + m_1 e + m_1 \ell + n_1)h + (a + e + \ell + m_1)gb + gch + hdb + hfh = 0$
(ix) $(a + 2\ell + m_1)gc + (e + 2\ell + m_1)hf + gbf + hdc + \ell^3 + m_1 \ell^2 + n_1 \ell + s_1 = 0$

## *Combining Duality and Characteristic Polynomial Conditions*

Now we will plug in the equations we obtained in section "Equations for Duality/Complex Conjugation Condition" into the equations we obtained in section "Equations for Characteristic Polynomial Condition". Note that our aim is to simplify the equations and write everything in terms of the upper triangular entries of our matrices which are $a, b, c, e, f, \ell$ in the case of $M$ and $p, q, r, t, u, y$ in the case of $N$. Note that $\mathrm{Nrd}(x) = xx^\vee$, $\mathbf{Tr}(x) = x + x^\vee$ denote the reduced norm and trace of an element. Since the norm and trace are scalars, they commute with everything else.

We start with the relations coming from $M$:

(I) $(2a + e + m_1)\,\mathrm{Nrd}(b) + (2a + \ell + m_1)\,\mathrm{Nrd}(c) + \mathbf{Tr}(bfc^\vee) + a^3 + m_1 a^2 + n_1 a + s_1 = 0$
(II) $(a^2 + ae + e^2 + m_1 a + m_1 e + n_1 + \mathrm{Nrd}(b) + \mathrm{Nrd}(c) + \mathrm{Nrd}(f))b + (a + e + \ell + m_1)cf^\vee = 0$
(III) $(a^2 + a\ell + \ell^2 + m_1 a + m_1 \ell + n_1 + \mathrm{Nrd}(b) + \mathrm{Nrd}(c) + \mathrm{Nrd}(f))c + (a + e + \ell + m_1)bf = 0$

(IV) $(a^2 + ae + e^2 + m_1 a + m_1 e + n_1 + \mathrm{Nrd}(b) + \mathrm{Nrd}(c) + \mathrm{Nrd}(f))b^\vee + (a + e + \ell + m_1)fc^\vee = 0$

(V) $(a + 2e + m_1)\,\mathrm{Nrd}(b) + (2e + \ell + m_1)\,\mathrm{Nrd}(f) + \mathbf{Tr}(b^\vee cf^\vee) + e^3 + m_1 e^2 + n_1 e + s_1 = 0$

(VI) $(e^2 + e\ell + \ell^2 + m_1 e + m_1 \ell + n_1 + \mathrm{Nrd}(b) + \mathrm{Nrd}(c) + \mathrm{Nrd}(f))f + (a + \ell + e + m_1)b^\vee c = 0$

(VII) $(a^2 + a\ell + \ell^2 + m_1 a + m_1 \ell + n_1 + \mathrm{Nrd}(b) + \mathrm{Nrd}(c) + \mathrm{Nrd}(f))c^\vee + (a + e + \ell + m_1)f^\vee b^\vee = 0$

(VIII) $(e^2 + e\ell + \ell^2 + m_1 e + m_1 \ell + n_1 + \mathrm{Nrd}(b) + \mathrm{Nrd}(c) + \mathrm{Nrd}(f))f^\vee + (a + e + \ell + m_1)c^\vee b = 0$

(IX) $(a + 2\ell + m_1)\,\mathrm{Nrd}(c) + (e + 2\ell + m_1)\,\mathrm{Nrd}(f) + \mathbf{Tr}(c^\vee bf) + \ell^3 + m_1 \ell^2 + n_1 \ell + s_1 = 0$

Write $\mathbf{Tr}(X)$ for the sum of the entries on the main diagonal of a matrix $X$. Notice that if we take $\eta = \sqrt{\alpha_1}$ like in Section 6.3, then

$$-m_1 = \mathbf{Tr}(\alpha_1) = \mathbf{Tr}(N^2) = \mathbf{Tr}(M) = a + e + \ell,$$

where the first equality follows by definition, the second equality is Lemma 6.12, the third equality holds because we took $\eta = \sqrt{\alpha_1}$, and the final equality is the definition of $\mathbf{Tr}(M)$. This implies that Equation (II) = Equation (IV), Equation (III)= Equation (VII) and Equation (VI)=Equation (VIII).

Combining $-m_1 = a + e + \ell$ with relations (I)–(IX), we deduce the following relations on the coefficients $m_1, n_1, s_1$ of the characteristic polynomial of $\alpha_1$.

(1) $m_1 = -(a + e + \ell)$
(2) $n_1 = ae + e\ell + a\ell - \mathrm{Nrd}(b) - \mathrm{Nrd}(c) - \mathrm{Nrd}(f)$ (using Equation (1) together with Equations (II), (III) and (VI)).
(3) $s_1 = a\,\mathrm{Nrd}(f) + e\,\mathrm{Nrd}(c) + l\,\mathrm{Nrd}(b) - ae\ell - \mathbf{Tr}(bfc^\vee)$ (using Equation (1) together with Equations (I), (V) and (IX)).

**Acknowledgements** The authors would like to thank the Centre International de Rencontres Mathématiques in Luminy for sponsoring the Women in Numbers - Europe (Femmes en Nombre) workshop and for providing a productive and enjoyable environment for our initial work on this project. We would especially like to thank the organizers of WINE, Marie José Bertin, Alina Bucur, Brooke Feigon, and Leila Schneps for making the conference and this collaboration possible. We also thank the referee for the detailed and helpful report.

The work of MM was partially supported by NSF-DMS 1102858.

## References

Bender, E.: Characteristic polynomials of symmetric matrices. Pac. J. Math. **25**, 433–441 (1968). MR0229619 (37 #5193)

Bosch, S., Lütkebohmert, W., Raynaud, M.: Néron models. Ergebnisse der Mathematik und ihrer Grenzgebiete (3) [Results in Mathematics and Related Areas (3)], vol. 21. Springer, Berlin (1990). MR1045822 (91i:14034)

Bouw, I.: The p-rank of ramified covers of curves. Compos. Math. **126**(3), 295–322 (2001). MR1834740 (2002e:14045)

Bouw, I., Wewers, S.: Group actions on curves and the lifting problem. http://math.arizona.edu/swc/aws/2012/2012BouwWewersNotesPreliminary.pdf (2012)

Bruinier, J., Yang, T.: CM-values of Hilbert modular functions. Invent. Math. **163**(2), 229–288 (2006). MR2207018 (2008b:11053)

Chai, C.-L., Conrad, B., Oort, F.: Complex Multiplication and Lifting Problems. Mathematical Surveys and Monographs, vol. 195. American Mathematical Society, Providence (2014). MR3137398

Crew, R.: Etale p-covers in characteristic p. Compos. Math. **52**(1), 31–45 (1984). MR742696 (85f:14011)

Deligne, P., Mumford, D.: The irreducibility of the space of curves of given genus. Inst. Hautes Études Sci. Publ. Math. **36**, 75–109 (1969). MR0262240 (41 #6850)

Dodson, B.: The structure of Galois groups of CM-fields. Trans. Am. Math. Soc. **283**(1), 1–32 (1984). MR735406 (86i:11063)

Goren, E., Lauter, K.: Evil primes and superspecial moduli. Int. Math. Res. Not. **19**, 53864 (2006). MR2250004 (2007f:11061)

Goren, E., Lauter, K.: Class invariants for quartic CM fields. Ann. Inst. Fourier (Grenoble) **57**(2), 457–480 (2007). MR2310947 (2008i:11075)

Goren, E., Lauter, K.: A Gross-Zagier formula for quaternion algebras over totally real fields. Algebra Number Theory **7**(6), 1405–1450 (2013). MR3107568

Gross, B., Zagier, D.: On singular moduli. J. Reine Angew. Math. **355**, 191–220 (1985). MR772491 (86j:11041)

Kani, E., Rosen, M.: Idempotent relations and factors of Jacobians. Math. Ann. **284**(2), 307–327 (1989). MR1000113 (90h:14057)

Klüners, J., Malle, G.: A database for number fields. http://galoisdb.math.upb.de/ (2011)

Koike, K., Weng, A.: Construction of CM Picard curves. Math. Comput. **74**(249), 499–518 (electronic) (2005). MR2085904 (2005g:11103)

Lang, S.: Complex Multiplication. Grundlehren der Mathematischen Wissenschaften [Fundamental Principles of Mathematical Sciences], vol. 255. Springer, New York (1983). MR713612 (85f:11042)

Lang, S.: Elliptic Functions. Graduate Texts in Mathematics, 2nd edn., vol. 112. Springer, New York (1987). With an appendix by J. Tate. MR890960 (88c:11028)

Lauter, K., Viray, B.: An arithmetic intersection theory formula for denominators of Igusa class polynomials. http://arxiv.org/abs/1210.7841 (2012)

Milne, J.S.: Complex multiplication. http://www.jmilne.org/math/CourseNotes/cm.htm (2006)

Milne, J.S.: Abelian varieties. http://www.jmilne.org/math/CourseNotes/av.html (2008)

Mumford, D.: Abelian Varieties. Tata Institute of Fundamental Research Studies in Mathematics, No. 5. Published for the Tata Institute of Fundamental Research/Oxford University Press, Bombay/London (1970). MR0282985 (44 #219)

Obus, A., Pries, R.: Wild tame-by-cyclic extensions. J. Pure Appl. Algebra **214**(5), 565–573 (2010). MR2577662 (2011h:12011)

Oort, F., Ueno, K.: Principally polarized abelian varieties of dimension two or three are Jacobian varieties. J. Fac. Sci. Univ. Tokyo Sect. IA Math. **20**, 377–381 (1973). MR0364265 (51 #520)

Paulhus, J.: Decomposing Jacobians of curves with extra automorphisms. Acta Arith. **132**(3), 231–244 (2008). MR2403651 (2009c:14049)

Serre, J.-P., Tate, J.: Good reduction of abelian varieties. Ann. Math. (2) **88**, 492–517 (1968). MR0236190 (38 #4488)

Shimura, G.: Abelian Varieties with Complex Multiplication and Modular Functions. Princeton Mathematical Series, vol. 46. Princeton University Press, Princeton (1998). MR1492449 (99e:11076)

Silverman, J.: Advanced Topics in the Arithmetic of Elliptic Curves. Graduate Texts in Mathematics, vol. 151. Springer, New York (1994). MR1312368 (96b:11074)

van der Geer, G., Moonen, B.: Abelian varieties. Preliminary version, http://www.math.ru.nl/bmoonen/research.html (2011)
Vignéras, M.-F.: Arithmétique des algèbres de quaternions. Lecture Notes in Mathematics, vol. 800. Springer, Berlin (1980). MR580949 (82i:12016)
Yui, N.: On the Jacobian variety of the Fermat curve. J. Algebra **65**(1), 1–35 (1980). MR578793 (82m:14016)

# Symmetries of Rational Functions Arising in Ecalle's Study of Multiple Zeta Values

**Adriana Salerno, Damaris Schindler, Amanda Tucker**

**Abstract** In Ecalle's theory of multiple zeta values he makes frequent use of certain properties that express symmetries of rational functions in several variables. We focus on the properties of push-invariance, circ-neutrality, and alternality. Ecalle states and uses several implications about the relations between these symmetries. In this paper we investigate two of these implications and prove two results: first, that push-invariance and circ-neutrality imply the first alternality relation, but not the more general alternality relations, and second, that alternality does, indeed, imply circ-neutrality.

## 1 Introduction

The *multiple zeta values* are the numbers

$$\zeta(n_1, \ldots, n_r) = \sum_{0<k_1<\cdots<k_r} \frac{1}{k_1^{n_1} \cdots k_r^{n_r}},$$

for $n_i \in \mathbb{N}$ with $n_r > 1$. Here, $r$ is the *depth* and $n = n_1 + \cdots + n_r$ is the *weight* of $\zeta(n_1, \ldots, n_r)$, where $1 \leq r < n$. When the depth $r = 1$ these values are known as *single zeta values* and are nothing other than the special values of the Riemann zeta function $\zeta(n) = \sum_{0<k} \frac{1}{k^n}$. Very little is known about the algebraic nature of the

A. Salerno (✉)
Mathematics Department, Bates College, Lewiston, ME, USA
e-mail: asalerno@bates.edu

D. Schindler
Hausdorff Center for Mathematics, University of Bonn, Bonn, Germany

A. Tucker
Department of Mathematics, State University of New York College at Geneseo, Geneseo, NY, USA



M.J. Bertin et al. (eds.), *Women in Numbers Europe*, Association for Women in Mathematics Series 2, DOI 10.1007/978-3-319-17987-2_6

special values of the Riemann zeta function (let alone multiple zeta values). Euler knew already in 1735 Euler (1775) that

$$\zeta(2) = \frac{\pi^2}{6}, \zeta(4) = \frac{\pi^4}{90}$$

and that, more generally,

$$\zeta(2n) = (-1)^{n+1} \frac{B_{2n}(2\pi)^n}{2(2n)!} \in \pi^{2n}\mathbb{Q}.$$

A widely believed folklore conjecture, for example, states that the numbers $\pi$, $\zeta(3)$, $\zeta(5), \zeta(7), \ldots, \zeta(2n+1)$ are algebraically independent over $\mathbb{Q}$ for any integer $n \geq 1$.

In studying algebraic independence of zeta values one is led naturally to the study of similar questions for multiple zeta values. In this context it turns out to be useful to define what is called the $\mathbb{Q}$-algebra of formal multiple zeta values, which is generated by symbols of the form $Z(k_1, \ldots, k_r)$ modulo the standard regularized shuffle and stuffle algebraic relations. The structure of this algebra is a tantalizing and much-studied question. Several authors, particularly Hoffman (1997) and Zagier (1993) have made seminal contributions to its study. J. Ecalle has designed and implemented a vast program to study it, using his own personal language and theory, which yields beautiful and natural generalizations, restatements and proofs of some of the important facts and conjectures concerning multiple zeta values.

The key to his theory is to place the whole situation within a bigger universe, known as the theory of *moulds*; in this paper, we consider only "moulds" which are in fact rational functions of several variables and which are the most relevant moulds in the study of multiple zeta values. An essential feature of Ecalle's theory is the study of symmetry properties of moulds, such as the properties of circ-neutrality,[1] push-invariance, and alternality (defined in Section 2) that we concentrate on in this paper. Ecalle's seminal article Ecalle (2011) contains a grand survey of some of his main ideas. However, because of the length and depth of the theory, there are few proofs. The following assertion (for general moulds, but we restrict it to rational functions here) appears in section 11.9 of Ecalle (2011).

**Assertion 1.1.** *If A is a rational function that is* push-*invariant and* circ-*neutral, then A is alternal.*

This statement turns out not to be true in full generality,[2] but we prove that push-invariance and circ-neutrality do imply the first alternality relations in Section 4. Moreover, we found via Maple calculations that there is no counterexample to the more general case in two variables and low degree, but we found one in degree three with five variables, which we give in Section 4.

---

[1]Ecalle uses the terminology pus-neutrality for this property.

[2]However, the statement forms part of the proof of a result for which Ecalle gave a very different but complete proof in a subsequent paper.

In Section 2.4 of Ecalle (2011) we find the following assertion (again for general moulds):

**Assertion 1.2.** *Every alternal rational function is* circ*-neutral.*

The main goal of Section 5 is to detail the proof of this assertion. Our strategy to prove Assertion 1.2 is to first reduce it to the polynomial case, which is done in Section 3. We can then treat the polynomial case using properties of certain Lie algebras and Ecalle's multiplication operator mu, whose definition we introduce in Section 5.

## 2 Definitions

There are three properties of rational functions that are of interest: push-invariance, circ-neutrality, and alternality.

If $A(u_1, \ldots, u_r)$ is a rational function in $r$ variables, we define

$$\operatorname{circ} A(u_1, \ldots, u_r) = A(u_r, u_1, \ldots, u_{r-1})$$

and, for $u_0 = -u_1 - u_2 - \cdots - u_r$,

$$\operatorname{push} A(u_1, \ldots, u_r) = A(u_0, u_1, \ldots, u_{r-1}).$$

**Definition 2.1.** $A$ is push-invariant if

$$\operatorname{push} A = A.$$

**Definition 2.2.** $A$ is circ-neutral if

$$A + \operatorname{circ} A + \operatorname{circ}^2 A + \cdots + \operatorname{circ}^{r-1} A = 0.$$

**Definition 2.3.** $A$ is alternal if

$$\sum_{w \in sh(u_1 u_2 \ldots u_k)(u_{k+1} \ldots u_r)} A(w) = 0$$

for all $1 \leq k \leq \lceil \frac{r}{2} \rceil$, where $sh(w, w')$ is the set of all possible shuffles of the words $w$ and $w'$, that is, permutations of the letters of $w$ and $w'$ that preserve the ordering in $w$ and the ordering in $w'$.

A more technical definition of the shuffle of two words is the following. Let $\mathcal{A}$ be the set of all words in the alphabet $\{a_1, \ldots, a_{r+s}\}$ and suppose $w = a_1 a_2 \cdots a_r$ and $w' = a_{r+1} a_{r+2} \cdots a_{r+s}$. Let the symmetric group on $r + s$ letters $S_{r+s}$ act on $\mathcal{A}$ by permuting the indices. Let

$$T = \{\sigma \in S_{r+s} \mid \sigma^{-1}(1) < \sigma^{-1}(2) < \cdots < \sigma^{-1}(r) \text{ and}$$
$$\sigma^{-1}(r+1) < \sigma^{-1}(r+2) < \cdots < \sigma^{-1}(r+s)\}$$

and note that if $x \leq r$ and $y > r$ there is no relation whatsoever imposed on $\sigma^{-1}(x)$ and $\sigma^{-1}(y)$. Then

$$sh(w, w') = \{w'' \in \mathcal{A} \mid w'' = \sigma(ww') \text{ for some } \sigma \in T\}.$$

*Example 2.4.* An example of a polynomial that is push-invariant, circ-neutral, and alternal is $A(u_1, u_2) = -2u_1^3 - 3u_1^2u_2 + 3u_1u_2^2 + 2u_2^3$.

*Remark 2.5.* We note here that circ-neutrality and alternality are additive properties that respect the multi-degree of a polynomial. Hence, once we have reduced to the polynomial case in Section 3 we can further reduce to the case of monomials of fixed multi-degree.

## 3 Reduction to the Polynomial Case

In this section, we reduce the study of Assertion 1.2 and similar questions to the case of polynomials. It turns out that to show that a family of rational functions satisfy circ-neutrality, push-invariance, or alternality, it suffices to show that a corresponding family of polynomials satisfies that property.

**Lemma 3.1.** *Every rational function $A(u_1, \ldots, u_r)$ can be written in the form $A = P/Q$ where $P$ and $Q$ are polynomials in $u_1, \ldots, u_r$ such that $Q(u_1, \ldots, u_r)$ is invariant under* push *and any permutation of $u_1, \ldots, u_r$. If $A = P/Q$ is such an expression for $A$, then $A$ is alternal (resp.* push*-invariant, resp.* circ*-neutral) if and only if $P$ is alternal (resp.* push*-invariant, resp.* circ*-neutral).*

*Proof.* Consider a rational function $A \in \mathbb{Q}(u_1, \ldots, u_r)$. We note that

$$\text{circ}^r = \text{id and push}^{r+1} = \text{id}.$$

Write $A = p/q$ with $p, q \in \mathbb{Q}[u_1, \ldots, u_r]$. Set $u_0 = -u_1 - \cdots - u_r$, and let the symmetric group $S_{r+1}$ act on the $r+1$ indices $0, 1, \ldots, r$ by permutation. We define

$$Q(u_1, \ldots, u_r) = \prod_{\sigma \in S_{r+1}} q(u_{\sigma(1)}, \ldots, u_{\sigma(r)}).$$

Then the polynomial $Q$ in $r$ variables is push-invariant and invariant under any permutation of the $r$ indices $1, \ldots, r$. Then we simply set

$$P(u_1,\dots,u_r) = p(u_1,\dots,u_r) \prod_{\sigma\in S_{r+1},\sigma\neq \mathrm{id}} q(u_{\sigma(1)},\dots,u_{\sigma(r)}),$$

so that $A = P/Q$, which is of the desired form.

Finally, because $Q$ is invariant under push and any permutation of $u_1,\dots,u_r$, we have that $A$ is circ-neutral (resp. push-invariant, resp. alternal) if and only if $P$ is circ-neutral (resp. push-invariant, resp. alternal), which proves the lemma. □

## 4 push-Invariance and a Counterexample to 1.1

Checking Assertion 1.1 in Maple (2013), one finds no counterexample for polynomials in fewer than three variables in degree five (or seven variables in degree four), but in 3 variables and degree 5, a nice counterexample appears; we give it at the end of this section. First, let us show that, even though push-invariance and circ-neutrality do not imply alternality in general, they do imply the first alternality relation (shuffling one variable into the rest).

**Theorem 4.1.** *If $A$ is* circ-*neutral and* push-*invariant, then the first alternal relation holds. That is,* $\sum_{w\in sh(u_1)(u_2,\dots,u_r)} A(w) = 0.$

*Proof.* The circ-neutral relation guarantees that

$$A + \mathrm{circ}\, A + \mathrm{circ}^2 A + \cdots + \mathrm{circ}^{r-1} A = 0.$$

That is,

$$A(u_1,\dots,u_r) + A(u_r,u_1,\dots,u_{r-1}) + \cdots + A(u_2,u_3,\dots u_r,u_1) = 0.$$

Now, assuming $A$ is push-invariant, we have that $A$ under any change of variables is push-invariant. It follows that $\mathrm{circ}^k A$ is push-invariant for all $k$. Thus, we can push each term an appropriate number of times, preserving equality, to get

$$\mathrm{push}\, A + \mathrm{push}^{r-1}\mathrm{circ}\, A + \mathrm{push}^{r-2}\mathrm{circ}^2 A + \cdots + \mathrm{push}^2\mathrm{circ}^{r-1} A = 0.$$

So we see that

$$A(u_0,u_1,\dots,u_{r-1}) + A(u_1,u_0,\dots,u_{r-1}) + A(u_1,u_2,u_0,\dots,u_{r-1}) + \cdots +$$
$$A(u_1,u_2,\dots,u_0,u_{r-1}) + A(u_1,\dots,u_{r-1},u_0) = 0,$$

which is precisely the first alternal relation, written in terms of the variables $u_0,u_1,\dots,u_{r-2},u_{r-1}$. Just as push-invariance implies push-invariance under any change of variable, so does knowing the first alternal relations with one set of variables imply the first alternal relations after any change of variables. This concludes the proof. □

Next we observe that Assertion 1.1 is trivially true for the case of linear forms in any given number of variables. Indeed, in the following lemma we show that there are no nonzero linear push-invariant forms at all.

**Lemma 4.2.** *There are no non-trivial* push-*invariant linear forms.*

*Proof.* Assume that the linear form $A(u_1, \dots, u_r) = a_1u_1 + \cdots + a_ru_r$ is push-invariant. Then the equation $A = \operatorname{push} A$ implies that

$$a_1u_1 + \cdots + a_ru_r = a_1(-u_1 - \cdots - u_r) + a_2u_1 + \cdots + a_ru_{r-1}.$$

We compare coefficients on both sides and obtain the system of linear equations

$$\begin{aligned} a_1 &= -a_1 + a_2 \\ a_2 &= -a_1 + a_3 \\ &\vdots \\ a_{r-1} &= -a_1 + a_r \\ a_r &= -a_1. \end{aligned}$$

If $a_r = 0$, then all the other $a_i$ have to be zero. If $a_r$ is nonzero, then we may assume after normalization that $a_r = 1$. The last equation then gives that $a_1 = -1$; the first equation gives $a_2 = -2$; the second $a_3 = -3$, until we obtain from the penultimate equation that $a_r = -r$, which is a contradiction to $a_r = 1$. □

*Remark 4.3.* Determining the dimension of the subspace of push-invariant polynomials is non-trivial. We include here Maple calculations of the dimension of the space of push-invariant, circ-neutral polynomials for small values of $r$ (number of variables) and $n$ (degree) (Maple 2013):

| $r \backslash n$ | 1 | 2 | 3 | 4 | 5 | 6 | 7 | 8 | 9 |
|---|---|---|---|---|---|---|---|---|---|
| 2 | 0 | 0 | 1 | 0 | 1 | 1 | 1 | 1 | 2 |
| 3 | 0 | 1 | 0 | 2 | 1 | 3 | 2 | 5 | 3 |
| 4 | 0 | 0 | 1 | 1 | 3 | 3 | 6 | 7 | 11 |
| 5 | 0 | 1 | 2 | 3 | 5 | 11 | 14 | 24 | 34 |
| 6 | 0 | 0 | 1 | 3 | 8 | 14 | 28 | – | – |
| 7 | 0 | 1 | 2 | 8 | 13 | 31 | 55 | – | – |
| 8 | 0 | 0 | 3 | 5 | 19 | 43 | – | – | – |
| 9 | 0 | 1 | 2 | 11 | 29 | – | – | – | – |

Entering the rows individually into the On-line Encyclopedia of Integer Sequences (OEIS), we note that the dimensions for $r = 2$ appear to correspond to the dimensions $[(2n+2)/4] - [(2n+4)/6]$ of cusp forms of weight $2n+6$ on $\mathrm{SL}_2(\mathbb{Z})$. Apart from this the significance of these dimensions is not clear.

Despite push-invariance being a very strong property, it turns out, as observed at the beginning of this section, that push-invariance and circ-neutrality together are not enough to ensure that a given rational function is alternal in higher degree. The following polynomial $P$ is the smallest counterexample that we found to Assertion 1.1. The Maple code used can be found in Appendix.

$$\begin{aligned} P = & -u_1^2u_3 + u_1^2u_4 + 2u_1u_2u_4 - 2u_1u_2u_5 - u_1u_3^2 - 2u_1u_3u_4 + u_1u_4^2 \\ & +2u_1u_4u_5 + u_2^2u_3 - u_2^2u_5 + u_2u_3^2 + 2u_2u_3u_5 - 2u_2u_4u_5 - u_2u_5^2 - u_3^2u_4 \\ & +u_3^2u_5 - u_3u_4^2 + u_3u_5^2 \end{aligned}$$

## 5 circ-Neutrality and a Proof of Assertion 1.2

The goal of this section is to prove that any alternal rational function is, in fact, circ-neutral.

**Lemma 5.1.** *To show any alternal rational function is* circ*-neutral, it suffices to show that any alternal polynomial is* circ*-neutral.*

*Proof.* Let $A(u_1, \ldots, u_r)$ be an alternal rational function. Then by Section 3, we can write $A = P/Q$ with $P$ alternal and $Q$ invariant under any permutation of the variables $u_1, \ldots, u_r$. Thus, to show $A$ is circ-neutral, it suffices to show that $P$ is. □

The remainder of this section is devoted to proving that if a polynomial $P$ is alternal then it must also be circ-neutral (see Theorem 5).

Our approach is to first prove a proposition that gives a useful characterization of circ-neutral polynomials. We then show that an alternal polynomial must fit this characterization. In order to prove this proposition, we first introduce Ecalle's multiplication operator mu on two polynomials.

**Definition 5.2.** If $A$ and $B$ are polynomials in $r_A$ (resp. $r_B$) variables, then $\mathrm{mu}(A, B)$ is a polynomial in $r = r_A + r_B$ variables defined as

$$\mathrm{mu}(A, B) = A(u_1, \ldots, u_{r_A})B(u_{r_A+1}, \ldots, u_{r_A+r_B}).$$

Furthermore, we set $[[A, B]] = \mathrm{mu}(A, B) - \mathrm{mu}(B, A)$.

Note that $\mathrm{mu}(B, A) = B(u_1, \ldots, u_{r_B})A(u_{r_B+1}, \ldots, u_{r_A+r_B})$ is also a polynomial in $r$ variables.

**Proposition 5.3.** *Let $A$ and $B$ be monomials of degree $d_A$ (resp. $d_B$) in $r_A$ (resp. $r_B$) variables. Then $M := [[A, B]]$ is* circ*-neutral.*

*Proof.* M is a homogeneous polynomial of degree $d = d_A + d_B$ in $r = r_A + r_B$ variables given by

$$M = A(u_1, \dots, u_{r_A})B(u_{r_A+1}, \dots, u_{r_A+r_B}) - B(u_1, \dots, u_{r_B})A(u_{r_B+1}, \dots, u_{r_A+r_B}).$$

Writing $A(u_1, \dots, u_{r_A}) = u_1^{a_1} \dots u_{r_A}^{a_{r_A}}$ and $B(u_1, \dots, u_{r_B}) = u_1^{b_1} \dots u_{r_B}^{b_{r_B}}$, we have

$$M(u_1, \dots, u_r) = u_1^{a_1} \dots u_{r_A}^{a_{r_A}} u_{r_A+1}^{b_1} \dots u_r^{b_{r_B}} - u_1^{b_1} \dots u_{r_B}^{b_{r_B}} u_{r_B+1}^{a_1} \dots u_r^{a_{r_A}}.$$

If we now consider the sum

$$M + \operatorname{circ} M + \operatorname{circ}^2 M + \dots + \operatorname{circ}^{r-1} M = \sum_{i=1}^{r} M(u_i, \dots, u_r, u_1, \dots, u_{i-1}),$$

we see that the positive monomial from the $i$-th term cancels with negative monomial in the $i + r_A$-th term (with indices taken modulo $r$ in the set $\{1, \dots, r\}$). Thus, in fact,

$$\sum_{i=1}^{r} M(u_i, \dots, u_r, u_{r+1}, \dots, u_{i-1}) = 0,$$

so $M$ is indeed circ-neutral. □

*Remark 5.4.* It follows directly from Proposition 5.3 and additivity of circ-neutrality that if $A$ and $B$ are any polynomials, not necessarily monomials, then the polynomial $[[A, B]]$ is circ-neutral.

Note that both alternality and circ-neutrality are empty conditions for a polynomial in one variable, so we adopt the convention of saying alternality implies circ-neutrality in this case. If $A$ is a polynomial in $r > 1$ variables, because alternality and circ-neutrality both respect degree, we can assume that $A$ is homogenous of multi-degree $d$. Additivity and Proposition 5.3 together imply that, in order to prove Assertion 1.2, it suffices to show that any alternal $A$ is a linear combination of terms of the form $[[B, D]]$; this is the method we use to prove the following theorem.

**Theorem 5.5.** *If $A$ is a homogeneous alternal polynomial of degree $d$ in $r > 1$ variables, then $A$ is* circ*-neutral.*

*Proof.* We capitalize on properties of a certain Lie algebra to show that if $A$ is alternal then it is of the desired form $[[B, D]]$.

Let $\mathbb{Q}[u_1, \dots, u_r]$ be the ring of polynomials in $r$ commuting variables and let $R = \mathbb{Q}\langle x, y\rangle$ be the ring of polynomials in two non-commuting variables. Let $R_r$ denote the $\mathbb{Q}$ vector subspace of $R$ spanned by monomials containing exactly $r$ $y$'s. Note that $R_0$ is spanned by the monomials that are powers in $x$, including $x^0 = 1$. As a $\mathbb{Q}$ vector space, $R$ is the direct sum of the $R_r$'s for all $r$.

For $r \geq 1$, define the map of $\mathbb{Q}$ vector spaces

$$\phi_r : \mathbb{Q}[u_1, \ldots, u_r] \longrightarrow R_r \tag{1}$$

by extending linearly from the map $u_1^{a_1} \cdots u_r^{a_r} \mapsto C_{a_1+1}C_{a_2+1} \ldots C_{a_{r-1}+1}C_{a_r+1}$, where

$$C_i = \mathrm{ad}(x)^{i-1}y = [x, \cdots x, [x, [x, y]]] \cdots ],$$

and $[x, y] = xy - yx$ is the standard Lie bracket. So, for example, we have

$$\begin{aligned}
\phi_r(1) &= \phi_r(u_1^0 \cdots u_r^0) = C_1^r = y^r, \\
\phi_r(u_1) &= \phi_r(u_1^1 u_2^0 \cdots u_r^0) = C_2 C_1^{r-1} = [x, y]y^{r-1} = (xy - yx)y^{r-1}, \\
\phi_r(u_1^2) &= \phi_r(u_1^2 u_2^0 \cdots u_r^0) = C_3 C_1^{r-1} = [x, [x, y]]y^{r-1}, \\
\phi_r(u_2) &= \phi_r(u_1^0 u_2^1 u_3^0 \cdots u_r^0) = C_1 C_2 C_1^{r-1} = y[x, y]y^{r-2}, \\
\phi_r(7u_1^2 - 5u_2) &= \phi_r(7u_1^2 u_2^0 \cdots u_r^0 - 5u_1^0 u_2^1 u_3^0 \cdots u_r^0) = 7C_3 C_1^{r-1} - 5C_1 C_2 C_1^{r-2}, \text{ and} \\
\phi_r(u_2^2 u_r) &= \phi_r(u_1 u_2^2 u_r) = C_1 C_3 C_1^{r-3} C_2.
\end{aligned}$$

A key observation is that if $B$ and $D$ are two polynomials in $s$ and $t$ variables, respectively, then we have $\phi_{s+t}(\mathrm{mu}(B, D)) = \phi_s(B)\phi_t(D)$, so that

$$\begin{aligned}
\phi_{s+t}(\mathrm{mu}(B, D) - \mathrm{mu}(D, B)) &= \phi_s(B)\phi_t(D) - \phi_t(D)\phi_s(B) \\
&= [\phi_s(B), \phi_t(D)],
\end{aligned} \tag{2}$$

where mu is the operator from Definition 5.2 above.

Let $\mathrm{Lie}[x, y]$ be the free Lie algebra generated by $x$ and $y$ under the Lie bracket, $[f, g] = fg - gf$, and let $\mathrm{Lie}_r[x, y]$ be the $\mathbb{Q}$ vector subspace of $\mathrm{Lie}[x, y]$ spanned by brackets of exactly $r$ $y$'s with any number of $x$'s, i.e., the space of Lie polynomials with homogeneous degree $r$ in $y$.

**Lemma 5.6.** *If $A$ is an alternal polynomial in $r$ variables, then $\phi_r(A)$ lies in $\mathrm{Lie}_r[x, y]$.*

*Proof.* We note that if $A$ is an alternal polynomial in the variables $u_1, \ldots, u_r$, then the polynomial $\phi_r(A)$ considered as a polynomial in the variables $C_i$ satisfies the shuffle relations. By Theorem 1.4 (Section 1.5) in Rautenauer (1993) any polynomial that satisfies the shuffle relations in the variables $C_i$ is a Lie polynomial in the variables $C_i$. Now we recall that $C_i$ is defined as

$$C_i = \mathrm{ad}(x)^{i-1}(y),$$

and, hence, is a Lie polynomial in the variables $x$ and $y$. Thus, any Lie polynomial in the variables $C_i$ is also a Lie polynomial in the variables $x$ and $y$. □

Let $\mathcal{A}l_r$ denote the subspace of alternal polynomials in $\mathbb{Q}[u_1, \ldots, u_r]$.

**Lemma 5.7.** *The map $\phi_r : \mathcal{A}l_r \longrightarrow \mathrm{Lie}_r[x, y]$ is a bijection for every $r \geq 1$.*

*Proof.* The main point is that the $C_i$ are algebraically independent. To see this we first observe that by Lazard elimination we have a direct sum $\mathrm{Lie}[x, y] = \mathrm{Lie}[x] \oplus \mathrm{Lie}[C_1, C_2, \ldots]$ (see Proposition 10 a) in Bourbaki 2006). Part b) of the same proposition proves that if $C'_1, C'_2, \ldots$ are non-commutative indeterminates, then the map induced by $C'_i \mapsto C_i$ is an isomorphism from the free Lie algebra $\mathrm{Lie}[C'_1, C'_2, \ldots]$ to $\mathrm{Lie}[C_1, C_2, \ldots]$. Hence the latter is a free Lie algebra. Finally, Theorem 1 (b) in § 3, no. 1, of Bourbaki (2006) shows that the universal enveloping algebra of $\mathrm{Lie}[C'_1, C'_2, \ldots]$ is just the free non-commutative polynomial algebra $\mathbb{Q}\langle C'_1, C'_2, \ldots\rangle$, and the previous isomorphism extends uniquely to an isomorphism of universal enveloping algebras. This shows that the universal enveloping algebra of $\mathrm{Lie}[C_1, C_2, \ldots]$, namely the polynomial algebra $\mathbb{Q}\langle C_1, C_2, \ldots\rangle$ is also free on these variables, showing that they are indeed algebraically independent. Thus, any element of $\mathrm{Lie}[C_1, C_2, \ldots]$ has a unique expression as a polynomial in the $C_i$, and thus a unique preimage under the map (1). So the map $\phi_r$ is both injective and surjective, which concludes the proof. □

*Remark 5.8.* The above lemma shows that the map $\psi_r : \mathrm{Lie}_r[x, y] \longrightarrow \mathcal{A}l_r$, which is the restriction to the Lie algebra of the map on $\mathbb{Q}\langle C_1, C_2, \ldots\rangle$ defined by $\psi_r(C_{a_1} \ldots C_{a_r}) = u_1^{a_1-1} \ldots u_r^{a_r-1}$, is an explicit inverse to $\phi_r$.

**Lemma 5.9.** *Let $r > 1$. Then any element of* $\mathrm{Lie}_r[x, y]$ *can be written as a sum* $\sum_i a_i[f_i, g_i]$ *with* $f_i, g_i \in \mathrm{Lie}[C_1, C_2, \ldots]$ *for all i.*

*Proof.* Note that the assertion of the Lemma is equivalent to saying that one can decompose any element of $\mathrm{Lie}_r[x, y]$ into a sum of brackets in which none of the $f_i, g_i$ is equal to $x$.

We observe that every element in $\mathrm{Lie}_r[x, y]$ for $r > 1$ can be written as a linear combination of Lie brackets of $r$ $y$'s and any number of $x$'s. By additivity it is hence enough to prove the lemma for a single Lie bracket of $r$ $y$'s and $s$ $x$'s. For this we note that any Lie bracket is of the form $[f, g]$ with $f, g \in \mathrm{Lie}[x, y]$. If both $f$ and $g$ are themselves Lie brackets, or if $f$ or $g$ is equal to $y$, then we are already in the desired form. Hence it remains to consider the case where $f$ or $g$ is equal to $x$. Without loss of generality we may assume that $f = x$. Thus we have reduced the proof to showing that a Lie bracket of the form $[x, g]$ can be rewritten in the form $\sum_i a_i[f_i, g_i]$ with none of the $f_i, g_i$ equal to $x$.

We prove this claim by induction on the degree of the bracket, which equals $r+s$ in the notation above. Recall that we have assumed $r > 1$ and hence $r+s \geq 4$, since we are considering brackets of the form $[x, g]$ where $g$ is a Lie bracket containing at least 2 $y$'s, so of degree at least 3. The base case is thus the example $g = [y, [x, y]]]$.

To write this as a linear combination $\sum_i a_i[f_i, g_i]$ with none of the $f_i, g_i$ equal to $x$, we use the Jacobi relation

$$[a, [b, c]] + [c, [a, b]] + [b, [c, a]] = 0,$$

with $a = x$, $b = y$ and $c = [x, y]$. This yields

$$[x, [y, [x, y]] + [[x, y], [x, y]] + [y, [[x, y], x]] = 0.$$

The middle term is zero, so we can rewrite $[x, g] = [x, [y, [x, y]] = -[y, [[x, y], x]]$, which is of the desired form. This completes the base case of the induction.

Next we assume that all Lie brackets up to total degree $r+s-1$ in $x$ and $y$ can be written in the form stated in the lemma. Consider a Lie bracket of the form $[x, g]$ in degree $r+s$. Then $g$ is a Lie bracket of degree $r+s-1$, and hence by our induction hypothesis we can write $g = \sum_i a_i[f_i, g_i]$ with none of the $f_i, g_i$ equal to $x$. Now we use again the Jacobi relation to rewrite the problematic Lie bracket as

$$[x, g] = \sum_i a_i[x, [f_i, g_i]] = -\sum_i a_i[g_i, [x, f_i]] - \sum_i a_i[f_i, [g_i, x]].$$

Since none of the $f_i$ and the $g_i$ are equal to $x$, this completes the proof of the lemma. □

We can now complete the proof of Theorem 5.

Let $A$ be an alternal polynomial in $r > 1$ variables. Then $\phi_r(A)$ is in $\mathrm{Lie}_r[x, y]$ by Lemma 5.6. Thus, by Lemma 5.9, we can write

$$\phi_r(A) = \sum_i a_i[f_i, g_i]$$

where none of the $f_i$ or $g_i$ is equal to $x$. For each $i$ in the sum, we thus have $f_i \in \mathrm{Lie}_{s_i}[x, y]$ for some $s_i \geq 1$, and $g_i \in \mathrm{Lie}_{t_i}[x, y]$ for some $t_i \geq 1$. Now, by Lemma 5.7, $\phi_{s_i}$ and $\phi_{t_i}$ are surjective, so there exist polynomials $B_i$ and $D_i$ satisfying $\phi_{s_i}(B_i) = f_i$ and $\phi_{t_i}(D_i) = g_i$.

Summing this up and using (2) and the linearity of $\phi_r$, we have

$$\begin{aligned}
\phi_r(A) &= \sum_i a_i[f_i, g_i] \\
&= \sum_i a_i[\phi_{s_i}(B_i), \phi_{t_i}(D_i)] \\
&= \sum_i a_i \phi_{s_i+t_i}(\mathrm{mu}(B_i, D_i) - \mathrm{mu}(D_i, B_i)) \\
&= \phi_r\left(\sum_i a_i[[B_i, D_i]]\right).
\end{aligned}$$

Finally, by the injectivity of $\phi_r$ proved in Lemma 5.7, we have that

$$A = \sum_i a_i[[B_i, D_i]].$$

By Proposition 5.3 and additivity this is enough to guarantee that $A$ is circ-neutral, which concludes the proof of Theorem 5. □

## Appendix

This is the Maple code for creating a generic push-invariant, circ-neutral polynomial of degree n in r variables and checking whether it fails on the second alternality relation. One inputs the degree n of the polynomial and the number r of variables at the beginning of the program as can be seen below. We used this code to obtain the counterexample to Assertion 1.1 in Section 4.

```
#
#(1) Create arbitrary polynomial in r variables of
     degree n
#(2) Solve the systems for push-invariant and
     circ-neutral
#(3) Check whether polynomial satisfies second
     alternality
# relation
n:=3:
r:=5:
#procedure to compute push(P)
pushP:=proc(P,r)
 local Q:
 Q:=expand(subs({u[1]=-add(u[k],k=1..r),seq(u[k]=u[k-1],
 k=2..r)},P)):
 return Q:
end proc:
#procedure to compute circ(P)
circP:=proc(P,r)
 local Q:
 Q:=expand(subs({u[1]=u[r],seq(u[k]=u[k-1],k=2..r)},P)):
 return Q:
end proc:
U:=[seq(u[i],i=1..r+1)]:
with(combinat):
#Creation of generic polynomial of deg n in r vars with
#indeterminate coefficients
```

```
C1:=partition(n+r):
C2:=[]:
for i from 1 to nops(C1) do
  if(nops(C1[i])=r) then C2:=[op(C2),op(permute(C1[i]))]
     fi:
od:
C3:=[]:
for i from 1 to nops(C2) do
  C3:=[op(C3),[seq(C2[i][j]-1,j=1..r)]]
od:
P:=0:
for i from 1 to nops(C3) do
  P:=P+a[i]*product(u[j]^C3[i][j],j=1..r):
od:
#That's it, P is the generic polynomial
#Now make linear system for the circ-neutrality relation
#P+circ(P)+...+circ^(r-1)(P)=0
PP[0]:=P:
for i from 1 to r-1 do PP[i]:=circP(PP[i-1],r): od:
Q:=add(PP[i],i=0..r-1):
COEFFScircneut:={coeffs(Q,U)}:
#Now make linear system for push-invariance
Q:=expand(P-pushP(P,r)):
COEFFSpushinv:={coeffs(Q,U)}:
Sols:=solve(COEFFScircneut union COEFFSpushinv):
P:=expand(subs(Sols,P)):
print("generic push-invariant, circ-neutral polynomial
   of
degree",n," in", r," variables"):
print(P):
#Test whether it satisfies the second alternality
   relation
#corresponding to sh((1,2),(3,..,r))
C4:=choose(r,2):
#build shuffle permutations
for i from 1 to binomial(r,2) do
    shu[i]:=[seq(m,k=1..r)]:
    shu[i][C4[i][1]]:=1:
    shu[i][C4[i][2]]:=2:
    cc:=3:
    for k from 1 to r do
      if(shu[i][k]=m) then shu[i][k]:=cc: cc:=cc+1: fi:
    od:
od:
```

```
#build shuffle relation
Q:=0:
for i from 1 to binomial(r,2) do
    Q:=Q + subs({seq(u[k]=u[shu[i][k]],k=1..r)},P)
od:
Q:=expand(Q):
if(Q=0) then print("satisfies second shuffle relation")
    fi:
if(Q<>0) then print("fails second shuffle relation") fi:
```

**Acknowledgements** We would like to express our sincerest thanks to our group leader, Leila Schneps, for all of her help and support and for suggesting such an intriguing problem. We would also like to thank the conference organizers of "Women in Numbers—Europe 2013" for all their hard work planning, funding, running, and following up on the wonderful conference that provided us with such a stimulating work environment.

## References

Bourbaki, N.: Éléments de Mathématique. Groupes et Algèbres de Lie: Chapitres 2 et 3, 2nd ed., Springer (2006)

Ecalle, J.: The flexion structure and dimorphy: flexion units, singulators, generators, and the enumeration of multizeta irreducibles. In: Asymptotics in Dynamics, Geometry and PDEs; Generalized Borel Summation, vol. II, CRM Series, Edizioni della Normale, Scuola Normale Superiore Pisa, 27–211 (2011)

Euler, L.: Meditationes circa singulare serierum genus. Novi Commun. Acad. Sci. **15**(20), 140–186 (1775)

Hoffman, M.E.: The algebra of multiple harmonic series. J. Algebra **194**, 477–495 (1997)

Maple 17: Maplesoft, a division of Waterloo Maple Inc. (2013)

Rautenauer, C.: Free Lie Algebras. London Mathematical Society Monographs New Series (Book 7). Oxford University Press, Oxford (1993)

Zagier, D.: Periods of modular forms, traces of Hecke operators, and multiple zeta values. Sūrikaisekikenkyūsho Kōkyūroku **843**, 162–170 (1993)

# On $\tau$-Li Coefficients for Rankin–Selberg $L$-Functions

**Alina Bucur, Anne-Maria Ernvall-Hytönen, Almasa Odžak, Edva Roditty-Gershon, and Lejla Smajlović**

**Abstract** The generalized $\tau$-Li criterion for a certain zeta or $L$-function states that non-negativity of $\tau$-Li coefficients associated to this function is equivalent to non-vanishing of this function in the region $\operatorname{Re} s > \tau/2$. For $\tau \in [1, 2)$ and positive integers $n$, we define $\tau$-Li coefficients $\lambda_n(\pi \times \pi', \tau)$ associated to Rankin–Selberg $L$-functions attached to convolutions of two cuspidal, unitary automorphic representations $\pi$ and $\pi'$. We investigate their properties, including the archimedean and non-archimedean terms, and the asymptotic behavior of these terms.

## 1 Introduction

A simple positivity criterion for the Riemann hypothesis, proved by Li (1997) states that the Riemann hypothesis is equivalent to non-negativity of a sequence of numbers

$$\lambda(n) = \sum_{\rho \in Z(\zeta)}{}^{*} \left(1 - \left(1 - \frac{1}{\rho}\right)^n\right),$$

A. Bucur (✉)
Department of Mathematics, 9500 Gilman Drive #0112, La Jolla, CA 92093, USA
e-mail: alina@math.ucsd.edu

A.-M. Ernvall-Hytönen
Department of Mathematics and Statistics, University of Helsinki, 00014 Helsinki, Finland
e-mail: ernvall@mappi.helsinki.fi

A. Odžak • L. Smajlović
Department of Mathematics, University of Sarajevo, Zmaja od Bosne 35, 71000 Sarajevo, Bosnia & Herzegovina
e-mail: almasa@pmf.unsa.ba; lejlas@pmf.unsa.ba

E. Roditty-Gershon
Raymond and Beverly Sackler School of Mathematical Sciences, Tel Aviv University, Tel Aviv, Israel
e-mail: rodittye@post.tau.ac.il

 
M.J. Bertin et al. (eds.), *Women in Numbers Europe*, Association for Women in Mathematics Series 2, DOI 10.1007/978-3-319-17987-2_7

where the sum runs over the set $Z(\zeta)$ of non-trivial zeros of the Riemann zeta function and $*$ in the sum indicates that it is taken in the sense of the limit $\lim_{T\to\infty} \sum_{|\mathrm{Im}\rho|\leq T}$.

The Li criterion is generalized to many classes of functions. For Dirichlet and Hecke $L$-functions it is proved in Li (2004), for automorphic $L$-functions the Li criterion is deduced in Lagarias (2007). In Smajlović (2010), a class $\mathcal{S}^{\sharp\flat}$ that contains both the Selberg class $\mathcal{S}$ and the class of automorphic $L$-functions is introduced and the Li criterion for this class is obtained. For the Rankin–Selberg $L$-functions the Li criterion is proved in Odžak and Smajlović (2010).

The Li coefficients for various classes of zeta and $L$-functions may be generalized in different ways. Droll (2012), following Freitas (2006), defined for $\tau \in [1,2)$ generalized $\tau$-Li coefficients $\lambda_F(n,\tau)$, for $F \in \mathcal{S}^{\sharp\flat}$ as

$$\lambda_F(n,\tau) = \sum_{\rho\in Z(F)}{}^{*} \left(1 - \left(\frac{\rho}{\rho-\tau}\right)^n\right) \tag{1}$$

where the sum is taken over the set $Z(F)$ of all non-trivial zeros of $F$ and proved that non-negativity of $\tau$-Li coefficients $\lambda_F(n,\tau)$ for all positive integers $n$ is equivalent to non-vanishing of $F$ in the region $\mathrm{Re}s > \tau/2$, or, equivalently, in the region $\mathrm{Re}s < 1-\tau/2$.

We will refer to this criterion as $\tau$-Li criterion. Let us note that the coefficients $\lambda_F(n,\tau)$ are a generalization of the Li coefficients $\lambda_F(n)$, introduced in Smajlović (2010), for $F \in \mathcal{S}^{\sharp\flat}$, in the sense that $\lambda_F(n,1) = \lambda_F(-n)$.

Another existing generalization of Li coefficients is given by Sekatskii (2013), in the case of the Riemann zeta function. Namely, for an arbitrary real number $a \neq 1/2$, generalized Li coefficients are defined as

$$\lambda(n,a) = \sum_{\rho\in Z(\zeta)}{}^{*} \left(1 - \left(\frac{\rho-a}{\rho+a-1}\right)^n\right).$$

The generalized Li criterion for the Riemann zeta function, proved in Sekatskii (2013) states that the Riemann hypothesis is equivalent to non-negativity of sums $\lambda(n,a)$ for any $a \in \mathbb{R} \setminus \{1/2\}$.

Rankin–Selberg $L$-function attached to the Rankin–Selberg convolution of two cuspidal, unitary automorphic representations $\pi$ and $\pi'$ of $\mathrm{GL}_m(\mathbb{A}_F)$ and $\mathrm{GL}_{m'}(\mathbb{A}_F)$ does not belong to the class $\mathcal{S}^{\sharp\flat} \supseteq \mathcal{S}$, since it might have poles at points on the line $\mathrm{Re}s = 1$, different from $s = 1$ and its functional equation is such that, in general case (if the Ramanujan hypothesis is not assumed) it might have trivial zeros inside the critical strip. Therefore, the results of Droll (2012) do not apply in this setting.

In this paper, we define the analogue of generalized $\tau$-Li coefficients ($\tau \in [1,2)$) for the Rankin–Selberg $L$-function and prove $\tau$-Li criterion for those functions. Then, we deduce arithmetic formulas for generalized $\tau$-Li coefficients. The generalized $\tau$-Li coefficient for the Rankin–Selberg $L$-function can be written as a sum of two terms: archimedean term, coming from the archimedean part of the completed $L$-function and non-archimedean term, coming from the finite part $L$-function. We derive full asymptotic expansion of the archimedean part of the $\tau$-Li coefficient and investigate the asymptotic behavior of non-archimedean part, as $n \to \infty$, for a fixed $\tau$.

## 2 Preliminaries

Let $F$ be a number field of degree $d = [F : \mathbb{Q}]$, and let $\mathbb{A}_F$ denote the ring of adeles over $F$. For a unitary cuspidal automorphic representation $\pi$ of $GL_m(\mathbb{A}_F)$, the completed $L$-function associated to $\pi$ can be written as an absolutely convergent Euler product over all places $v$ of $F$

$$\Lambda(s,\pi) = \prod_v L(s,\pi_v) = L\left(s,\pi_f\right) L\left(s,\pi_\infty\right)$$

in the half-plane $\mathrm{Re}s > 1$, see Jacquet and Shalika (1981a, p. 555, Th. 5.3). Here, $L\left(s,\pi_f\right)$ denotes the product over all finite places of $F$, whereas $L\left(s,\pi_\infty\right)$ is the product over infinite places. At the prime ideal $\mathfrak{p}$ where $\pi_\mathfrak{p}$ is unramified there is a set of $m$ non-zero Satake parameters $\{\alpha_\pi(\mathfrak{p},j)\}$ such that

$$L(s,\pi_\mathfrak{p}) = \prod_{j=1}^{m} \left(1 - \alpha_\pi(\mathfrak{p},j)\mathbf{N}\mathfrak{p}^{-s}\right)^{-1}, \tag{2}$$

where $\mathbf{N}\mathfrak{p}$ denotes the absolute norm of the ideal $\mathfrak{p}$.

At the prime ideal $\mathfrak{p}$ where $\pi_\mathfrak{p}$ is ramified, the local $L$-function is defined in terms of the Langlands parameters of $\pi_\mathfrak{p}$. It can be expressed as $P\left(\mathbf{N}\mathfrak{p}^{-s}\right)^{-1}$, where $P$ is a polynomial of a degree at most $m$, equal to 1 at zero. Therefore, all local factors of $L\left(s,\pi_f\right)$ can be written in the form (2), with the convention that some of $\alpha_\pi(\mathfrak{p},j)$ may be zero.

The local factor at infinite places is

$$L\left(s,\pi_v\right) = \prod_{j=1}^{m} \Gamma_v(s + \mu_\pi(v,j)),$$

where $\{\mu_\pi(v,j)\}_{j=1}^{m}$ are the Langlands parameters associated to $\pi_v$ and $\Gamma_v(s) = \pi^{-s/2}\Gamma(s/2) = \Gamma_\mathbb{R}(s)$, if $v$ is real and $\Gamma_v(s) = 2\left(2\pi\right)^{-s}\Gamma(s)$, if $v$ is complex.

The Rankin–Selberg $L$-function attached to the product $\pi \times \tilde{\pi}'$ of two unitary cuspidal automorphic representations of $GL_m(\mathbb{A}_F)$ and $GL_{m'}(\mathbb{A}_F)$ is given, for $\mathrm{Re}s > 1$, by an absolutely convergent Euler product of local factors

$$L\left(s, \pi_f \times \tilde{\pi}'_f\right) = \prod_{\mathfrak{p}<\infty} L(s, \pi_{\mathfrak{p}} \times \tilde{\pi}'_{\mathfrak{p}}), \tag{3}$$

as proved in Jacquet and Shalika (1981a, Th. 5.3). Here, $\tilde{\pi}$ denotes the contragradient representation of $\pi$. For any place $v$ of $F$, $\tilde{\pi}_v$ is equivalent to the complex conjugate $\overline{\pi}_v$ (Gelfand and Kazhdan 1974), hence $L\left(s, \pi_f \times \tilde{\pi}'_f\right) = \overline{L\left(\overline{s}, \tilde{\pi}_f \times \pi'_f\right)}$.

For prime ideals $\mathfrak{p}$ at which $\pi_{\mathfrak{p}}$ and $\pi'_{\mathfrak{p}}$ are unramified, the local factors are given by

$$L(s, \pi_{\mathfrak{p}} \times \tilde{\pi}'_{\mathfrak{p}}) = \prod_{j=1}^{m} \prod_{k=1}^{m'} \left(1 - \alpha_\pi(\mathfrak{p}, j)\overline{\alpha_{\pi'}(\mathfrak{p}, k)}\mathbf{N}\mathfrak{p}^{-s}\right)^{-1}.$$

At finite places $\mathfrak{p}$ ramified for $\pi$ or $\pi'$ the local $L$-function is defined in terms of the Langlands parameters. It can be written as $Q\left(\mathbf{N}\mathfrak{p}^{-s}\right)^{-1}$, where $Q$ is a polynomial of a degree at most $mm'$, equal to 1 at zero. Therefore, we can write all local factors at finite places as

$$L(s, \pi_{\mathfrak{p}} \times \tilde{\pi}'_{\mathfrak{p}}) = \prod_{j=1}^{m} \prod_{k=1}^{m'} (1 - \alpha_{\pi\times\tilde{\pi}'}(\mathfrak{p}, j, k)\mathbf{N}\mathfrak{p}^{-s})^{-1},$$

with the convention that some $\alpha_{\pi\times\tilde{\pi}'}(\mathfrak{p}, j, k)$ might be equal to zero, where $\alpha_{\pi\times\tilde{\pi}'}(\mathfrak{p}, j, k) = \alpha_\pi(\mathfrak{p}, j)\overline{\alpha_{\pi'}(\mathfrak{p}, k)}$, for the prime ideal $\mathfrak{p}$ unramified for both $\pi$ and $\pi'$.

The logarithmic derivative of $L\left(s, \pi_f \times \tilde{\pi}'_f\right)$, for Re $s > 1$, can be written as an absolutely convergent series over all integral ideals $\mathfrak{n}$ of the ring of integers $O_F$ of $F$

$$-\frac{L'}{L}(s, \pi_f \times \tilde{\pi}'_f) = \sum_{\mathfrak{n}} \frac{\Lambda(\mathfrak{n})a_{\pi\times\tilde{\pi}'}(\mathfrak{n})}{\mathbf{N}\mathfrak{n}^s} = \sum_{\mathfrak{n}} \frac{c_{\pi,\tilde{\pi}'}(\mathfrak{n})}{\mathbf{N}\mathfrak{n}^s}, \tag{4}$$

where $\Lambda(\mathfrak{n}) = \log \mathbf{N}\mathfrak{p}$ if $\mathfrak{n} = \mathfrak{p}^k$, for some integer $k \geq 1$, and $\Lambda(\mathfrak{n}) = 0$, otherwise.

Similarly, at infinite places $v$, the archimedean local factor $L(s, \pi_v \times \tilde{\pi}'_v)$ can be written as a product

$$L(s, \pi_v \times \tilde{\pi}'_v) = \prod_{j=1}^{m} \prod_{k=1}^{m'} \Gamma_v(s + \mu_{\pi\times\tilde{\pi}'}(v, j, k)),$$

where $\mu_{\pi\times\tilde{\pi}'}(v, j, k) = \mu_\pi(v, j) + \overline{\mu_{\pi'}(v, k)}$, at infinite places $v$ unramified for both $\pi$ and $\pi'$. Complex numbers $\mu_{\pi\times\tilde{\pi}'}(v, j, k)$ satisfy the trivial bound $\mathrm{Re}\mu_{\pi\times\tilde{\pi}'}(v, j, k) > -1$ (see the calculations in Rudnick and Sarnak 1996, Appendix).

Let us put

$$L\left(s, \pi_{\infty} \times \tilde{\pi}'_{\infty}\right) = \prod_{v \in S_{\infty}} L(s, \pi_v \times \tilde{\pi}'_v),$$

where $S_{\infty}$ denotes the set of all infinite places of $F$ consisting of $r_1$ real and $2r_2$ complex places. Then, as proved in Shahidi (1981), Shahidi (1984), Shahidi (1985), Shahidi (1990), Jacquet and Shalika (1981a), Jacquet and Shalika (1981b), Moeglin and Waldspurger (1989) and Gelbart and Shahidi (2001) (see also Cogdell 2007, Ths. 9.1 and 9.2), the complete Rankin–Selberg $L$-function

$$\Lambda(s, \pi \times \tilde{\pi}') = L(s, \pi_f \times \tilde{\pi}'_f) L(s, \pi_{\infty} \times \tilde{\pi}'_{\infty})$$

extends to a meromorphic function of order 1 on the whole complex plane, bounded (away from its possible poles) in vertical strips. It has simple poles at $s = 1 + it_0$ and $s = it_0$, arising from $L(s, \pi_f \times \tilde{\pi}'_f)$ if and only if $m = m'$ and $\pi' \cong \pi \otimes |\det|^{it_0}$, for some $t_0 \in \mathbb{R}$. Otherwise, it is a holomorphic function. Finally, $\Lambda(s, \pi \times \tilde{\pi}')$ satisfies the functional equation

$$\Lambda(s, \pi \times \tilde{\pi}') = \epsilon\left(\pi \times \tilde{\pi}'\right) Q_{\pi \times \tilde{\pi}'}^{1/2-s} \Lambda(1 - s, \tilde{\pi} \times \pi'), \tag{5}$$

where $Q_{\pi \times \tilde{\pi}'} > 0$ is the arithmetic conductor and $\epsilon(\pi \times \tilde{\pi}')$ is a complex number of modulus 1.

We call the zeros of $\Lambda(s, \pi \times \tilde{\pi}')$ the non-trivial zeros of $L(s, \pi_f \times \tilde{\pi}'_f)$. The set of non-trivial zeros of $L(s, \pi_f \times \tilde{\pi}'_f)$ is denoted by $Z(L(s, \pi_f \times \tilde{\pi}'_f))$, or shortly by $Z(L)$, in the case when $\pi$ and $\pi'$ are fixed.

By the functional equation and the Euler product representation, all those zeros lie in the critical strip $0 \leq \text{Re } s \leq 1$ (actually, Shahidi has proved that $L(1 + it, \pi_f \times \tilde{\pi}'_f) \neq 0$, for all $t \in \mathbb{R}$). Other trivial zeros arise from the poles of the function $L(s, \pi_{\infty} \times \tilde{\pi}'_{\infty})$. In the critical strip, the trivial zeros of $L(s, \pi_f \times \tilde{\pi}'_f)$ are at the points $s = -\mu_{\pi \times \tilde{\pi}'}(v, j, k)$, for those $v \in S_{\infty}$, $j$ and $k$ such that $\text{Re}\mu_{\pi \times \tilde{\pi}'}(v, j, k) \leq 0$. Furthermore, the functional equation implies that $Z(L(s, \pi_f \times \tilde{\pi}'_f)) = 1 - \overline{Z(L(s, \pi_f \times \tilde{\pi}'_f))}$.

Let $N_{\pi,\pi'}(T)$ denote the number of non-trivial zeros $\rho$ of $L(s, \pi_f \times \tilde{\pi}'_f)$ such that $|\text{Im}\rho| \leq T$. Then,

$$N_{\pi,\pi'}(T) = \frac{dmm'}{\pi} T \log T + c_{\pi,\pi'} T + O_{\pi,\pi'}(\log T), \text{ as } T \to \infty, \tag{6}$$

where

$$c_{\pi,\pi'} = \frac{1}{\pi} \log Q_{\pi \times \tilde{\pi}'} - \frac{dmm'}{\pi}(1 + \log 2\pi).$$

The complete proof is given in Iwaniec and Kowalski (2004, Th. 5.8) for the counting function $N_{\pi,\pi'}(T)$ of both trivial and (finitely many) non-trivial zeros in the critical strip.

To be fully precise, in the case we consider, a slight modification of the argument from Iwaniec and Kowalski (2004) should be made to include the case when the $L$-function has poles at $it_0$ and $1 + it_0$, for some $t_0 \in \mathbb{R}\setminus\{0\}$. It is easily done by considering the entire function

$$\Lambda^c(s, \pi \times \tilde{\pi}') := (s - it_0)^{\delta_{\pi,\pi'}(t_0)} (s - 1 - it_0)^{\delta_{\pi,\pi'}(t_0)} \Lambda(s, \pi \times \tilde{\pi}') \tag{7}$$

instead of $(s(1-s))^r \Lambda(s, \pi \times \tilde{\pi}')$ (cf. Iwaniec and Kowalski 2004, (5.23)). Here, we put

$$\delta_{\pi,\pi'}(t_0) = \begin{cases} 1, \text{ if } m = m' \text{ and } \pi' \cong \pi \otimes |\det|^{it_0}, \text{ for some } t_0 \in \mathbb{R}; \\ 0, \text{ otherwise.} \end{cases}$$

Let $N^+_{\pi,\pi'}(T)$ and $N^-_{\pi,\pi'}(T)$ denote counting functions for the non-trivial zeros $\rho$ of $L(s, \pi_f \times \tilde{\pi}'_f)$ with $0 < \mathrm{Im}\rho \le T$ and $-T \le \mathrm{Im}\rho < 0$, respectively. Repeating the arguments given in Lagarias (2007, Th. 2.1) it can be easily shown that

$$N^{\pm}_{\pi,\pi'}(T) = \frac{1}{2}\left(\frac{dmm'}{\pi} T \log T + c_{\pi,\pi'} T\right) + O_{\pi,\pi'}(\log T), \tag{8}$$

as $T \to \infty$.

Function $\Lambda^c(s, \pi \times \tilde{\pi}')$, defined by (7) is an entire function of order one, non-vanishing at $s = 0$, and hence possesses a representation as a Hadamard product

$$\Lambda^c(s, \pi \times \tilde{\pi}') = e^{as+b} \prod_{\rho \in Z(L(s,\pi\times\tilde{\pi}'))} \left(1 - \frac{s}{\rho}\right) e^{s/\rho}, \tag{9}$$

where $e^b = \Lambda^c(0, \pi \times \tilde{\pi}')$. Furthermore, in Odžak and Smajlović (2010), Proposition 4.1, using an explicit formula for the Rankin–Selberg $L$-function it is proved that the sum $\sum_{\rho \in Z(L(s,\pi\times\tilde{\pi}'))} \frac{1}{\rho}$ is $*$-convergent and that

$$a = - \sum_{\rho \in Z(L(s,\pi\times\tilde{\pi}'))}^{*} \frac{1}{\rho}. \tag{10}$$

## 3 $\tau$-Li Coefficients for the Rankin–Selberg $L$-Function

Let $\pi$ and $\pi'$ be arbitrary, but fixed unitary cuspidal automorphic representations of $GL_m(\mathbb{A}_F)$ and $GL_{m'}(\mathbb{A}_F)$. In this section, we show that $\tau$-Li coefficients for the Rankin–Selberg $L$-function $L(s, \pi_f \times \tilde{\pi}'_f)$ are well defined and we deduce two formulas for evaluation of $\tau$-Li coefficients.

In the sequel, we assume that representations $\pi$ and $\pi'$ are fixed, therefore, to shorten the notation, we denote

$$L(s) = L(s, \pi_f \times \tilde{\pi}'_f), \quad G(s) = L(s, \pi_\infty \times \tilde{\pi}'_\infty), \quad \Lambda^c(s) = \Lambda^c(s, \pi \times \tilde{\pi}').$$

Furthermore, we put $Z(L(s, \pi \times \tilde{\pi}')) = Z(L)$.

Repeating arguments given in Odžak and Smajlović (2010, Section 3) it is easy to see that we may write $G(s)$ as

$$G(s) = \pi^{-dmm's/2} Q_{\pi\times\tilde{\pi}'}^{s/2} \prod_{\ell=1}^{dmm'} \Gamma\left(\frac{s+\mu(\ell)}{2}\right), \tag{11}$$

where $\mu(\ell) = \mu_{\pi\times\tilde{\pi}'}(\ell) = \mu_{\pi\times\tilde{\pi}'}(v, j, k)$ for $r_1 + r_2$ places $v \in S_\infty$ and $\mu(\ell) = \mu_{\pi\times\tilde{\pi}'}(\ell) = \mu_{\pi\times\tilde{\pi}'}(v, j, k) + 1$ for the rest of $r_2$ places $v \in S_\infty$ ($j = 1, \ldots, m$, $k = 1, \ldots, m'$).

**Definition 3.1.** Let $\tau \in [1, 2)$. For an arbitrary positive integer $n$, the $n$th $\tau$-Li coefficient associated to the Rankin–Selberg $L$-function $L(s)$ attached to the product $\pi \times \tilde{\pi}'$ of two unitary cuspidal automorphic representations of $GL_m(\mathbb{A}_F)$ and $GL_{m'}(\mathbb{A}_F)$ is defined as

$$\lambda_n(\pi \times \tilde{\pi}', \tau) = \sum_{\rho\in Z(L)} \left(1 - \left(\frac{\rho}{\rho-\tau}\right)^n\right). \tag{12}$$

The following theorem shows that the coefficients $\lambda_n(\pi\times\tilde{\pi}', \tau)$ are well defined.

**Theorem 3.2.** *The coefficients $\lambda_n(\pi \times \tilde{\pi}', \tau)$ introduced in Definition 3.1 have the following properties*

(1) *The series defining $\lambda_n(\pi \times \tilde{\pi}', \tau)$ is *-convergent for every positive integer $n$.*

(2)

$$\mathrm{Re}\lambda_n(\pi \times \tilde{\pi}', \tau) = \sum_{\rho\in Z(L)} \mathrm{Re}\left(1 - \left(\frac{\rho}{\rho-\tau}\right)^n\right),$$

*where the sum on the right is absolutely convergent.*

(3) *The series*

$$\sum_{\rho\in Z(L)} \frac{1 + \left|\mathrm{Re}\left(\frac{\rho}{\tau}\right)\right|}{\left(1 + \left|\frac{\rho}{\tau}\right|\right)^2}$$

*converges.*

*Proof.* Claim (1) for $n \geq 2$ follows from the fact that $\Lambda^c(s)$ is an entire function of order one. In the case when $n = 1$, the claim follows from the fact that the sum $\sum_{\rho \in Z(L)} \frac{1}{\rho}$ is $*$-convergent.

The claim (2) follows from the claim (3) by using Bombieri and Lagarias (1999, Lemma 1).

In order to prove claim (3), we need to prove that

$$\sum_{\rho \in Z(L)} \frac{1 + \left|\mathrm{Re}\left(\frac{\rho}{\tau}\right)\right|}{\left(1 + \left|\frac{\rho}{\tau}\right|\right)^2}$$

converges. Estimate first $\left|1 + \left|\mathrm{Re}\left(\frac{\rho}{\tau}\right)\right|\right| \leq 2$. Since $\Lambda^c(s)$ is an entire function of order one, the sum

$$\sum_{\rho \in Z(L)} \frac{1}{\left(1 + \left|\frac{\rho}{\tau}\right|\right)^2}$$

converges, and the proof is complete. □

**Theorem 3.3.** *The coefficients $\lambda_n(\pi \times \tilde{\pi}', \tau)$ can be expressed in the following form*

$$\lambda_n(\pi \times \tilde{\pi}', \tau) = \tau \frac{1}{(n-1)!} \left[ \frac{d^n}{ds^n} \left( s^{n-1} \log \Lambda^c(s) \right) \right]_{s=\tau}. \tag{13}$$

*Proof.* The proof closely follows the corresponding proof in Droll (2012, Lemma 2.1.2), so we omit details. The right-hand side of (13) may be written as

$$\tau \frac{1}{(n-1)!} \left[ \frac{d^n}{ds^n} \left( s^{n-1} \log \Lambda^c(s) \right) \right]_{s=\tau}$$
$$= \sum_{k=0}^{n-1} \binom{n}{k} \frac{\tau^{n-k}}{(n-1-k)!} \left[ \frac{d^{n-k}}{ds^{n-k}} \log \Lambda^c(s) \right]_{s=\tau}.$$

Let us next study the part

$$\left[ \frac{d^{n-k}}{ds^{n-k}} \log \Lambda^c(s) \right]_{s=\tau}.$$

The Hadamard product representation (9), together with formula (10) implies that

$$\frac{\Lambda'^c(s)}{\Lambda^c(s)} = \sum_{\rho \in Z(L)}^{*} \frac{1}{s - \rho},$$

and, analogously

$$\left[\frac{d^{n-k}}{ds^{n-k}} \log \Lambda^c(s)\right]_{s=\tau} = -\sum_{\rho \in Z(L)}^{*} \frac{(n-k-1)!}{(\rho-\tau)^{n-k}}.$$

We have now derived

$$\tau \frac{1}{(n-1)!}\left[\frac{d^n}{ds^n}\left(s^{n-1} \log \Lambda^c(s)\right)\right]_{s=\tau}$$
$$= -\sum_{k=0}^{n-1}\binom{n}{k}\frac{\tau^{n-k}}{(n-1-k)!}\sum_{\rho \in Z(L)}^{*}\frac{(n-k-1)!}{(\rho-\tau)^{n-k}}$$
$$= -\sum_{\rho \in Z(L)}^{*}\sum_{k=0}^{n-1}\binom{n}{k}\frac{\tau^{n-k}}{(\rho-\tau)^{n-k}} = \sum_{\rho \in Z(L)}^{*}\left(1-\left(\frac{\rho}{\rho-\tau}\right)^n\right),$$

which completes the proof. □

Next, we will give an alternate description of the $\tau$-Li coefficients associated to the Rankin–Selberg $L$-function.

**Theorem 3.4.** *The coefficients $\lambda_n(\pi \times \tilde{\pi}', \tau)$ can be expressed in the following form*

$$\lambda_n(\pi \times \tilde{\pi}', \tau) = \frac{1}{\tau^n} d_{n-1}\left(1-\frac{1}{\tau}, L\right), \tag{14}$$

*where $d_n(z_0, L)$ are the power series coefficients in the expansion of the logarithmic derivative of $\Lambda^c\left(\frac{1}{1-s}\right)$ around $z_0 \neq 1$ which is not a zero of $\Lambda^c(\frac{1}{1-s})$, i.e., in a small neighborhood of $z_0$ we have*

$$\frac{d}{ds} \log \Lambda^c\left(\frac{1}{1-s}\right) = \sum_{n=0}^{\infty} d_n(z_0, L)(s-z_0)^n.$$

*Proof.* This proof also follows the corresponding proof in Droll's thesis (2012, Lemma 2.1.2). We start by noting that if $\frac{1}{1-s} \notin Z(L)$ and $s \neq 1$, we can use (9) to get

$$\frac{d}{ds} \log \Lambda^c\left(\frac{1}{1-s}\right) = \frac{1}{(1-s)^2}\left(a + \sum_{\rho \in Z(L)}\left(\frac{1}{\frac{1}{1-s}-\rho} + \frac{1}{\rho}\right)\right)$$
$$= \frac{a}{(1-s)^2} + \sum_{\rho \in Z(L)}\left(\frac{1}{1-s}\frac{1}{1-\rho+s\rho} + \frac{1}{\rho(1-s)^2}\right) \tag{15}$$

Let $z_0 = 1 - 1/\tau$. Using the following two identities (see Droll 2012, Lemma 2.1.2)

$$\frac{1}{1-\rho+s\rho} = -\frac{\tau}{\rho-\tau}\frac{1}{1-\frac{\rho\tau(s-z_0)}{\rho-\tau}} \quad \text{and} \quad \frac{1}{1-s} = \frac{\tau}{1-\tau(s-z_0)}$$

we may write (15) in the following form

$$\frac{a}{(1-s)^2} - \sum_{\rho\in Z(L)} \left( \frac{\tau^2}{\rho-\tau}\frac{1}{1-\frac{\rho\tau(s-z_0)}{\rho-\tau}}\frac{1}{1-\tau(s-z_0)} - \frac{1}{\rho(1-s)^2} \right).$$

For $s$ close enough to $z_0$ we may write $\frac{1}{1-\frac{\rho\tau(s-z_0)}{\rho-\tau}}$ and $\frac{1}{1-\tau(s-z_0)}$ as geometric series, hence we have (see Droll 2012, p. 59)

$$\begin{aligned}
&\frac{a}{(1-s)^2} - \sum_{\rho\in Z(L)} \left( \frac{\tau^2}{\rho-\tau}\sum_{n=0}^{\infty}\left(\frac{\rho\tau}{\rho-\tau}\right)^n (s-z_0)^n \sum_{m=0}^{\infty}\tau^m(s-z_0)^m - \frac{1}{\rho(1-s)^2} \right) \\
&= \sum_{n=0}^{\infty}\sum_{\rho\in Z(L)} \left( \tau^{n+1}\left(1-\left(\frac{\rho}{\rho-\tau}\right)^{n+1}\right)(s-z_0)^n - \frac{1}{2^{n+1}}\frac{1}{\rho(1-s)^2} \right) \\
&+ \frac{a}{(1-s)^2}.
\end{aligned}$$

We can now split the sum over $Z(L)$ into two $*$-convergent sums, using equation (10), to get

$$\begin{aligned}
&\frac{a}{(1-s)^2} - \frac{a}{2(1-s)^2}\sum_{n=0}^{\infty}\frac{1}{2^n} \\
&+ \sum_{n=0}^{\infty}\tau^{n+1}\left( \sum_{\rho\in Z(L)}{}^{*}\left(1-\left(\frac{\rho}{\rho-\tau}\right)^{n+1}\right)\right)(s-z_0)^n \\
&= \sum_{n=0}^{\infty}\tau^{n+1}\lambda_{n+1}(\pi\times\tilde{\pi}',\tau)(s-z_0)^n = \sum_{n=0}^{\infty}d_n(z_0,L)(s-z_0)^n,
\end{aligned}$$

which by uniqueness of series expansions concludes the proof. □

## 4 Li-Type Criterion for the Zero-Free Regions

We are now ready to prove the Li-type criterion for zero-free regions in the case of the Rankin–Selberg $L$-function.

**Theorem 4.1.** *The function* $\Lambda^c(s)$ *will have all its zeros in the strip* $1-\frac{\tau}{2} \leq \mathrm{Re}s \leq \frac{\tau}{2}$ *if and only if* $\mathrm{Re}\lambda_n(\pi \times \tilde{\pi}', \tau) \geq 0$ *for every positive integer* $n$.

*Proof.* The proof will also closely follow the proof of Droll (2012, Th. 2.1.3). We have

$$\mathrm{Re}\lambda_n(\pi \times \tilde{\pi}', \tau) = \sum_{\rho \in Z(L)} \mathrm{Re}\left(1 - \left(\frac{\rho}{\rho - \tau}\right)^n\right).$$

By Bombieri and Lagarias (1999, Th. 1), we know that if $R$ is a multiset of complex numbers with $1 \notin R$ and

$$\sum_{\rho \in R} \frac{1 + |\mathrm{Re}(\rho)|}{(1 + |\rho|)^2} < \infty,$$

then the conditions $\mathrm{Re}\rho \leq \frac{1}{2}$ for every $\rho \in R$ and

$$\sum_{\rho \in R} \mathrm{Re}\left(1 - \left(1 - \frac{1}{\rho}\right)^{-n}\right) \geq 0$$

are equivalent. Let us now put

$$R = \left\{\frac{\rho}{\tau} : \rho \in Z(L)\right\}.$$

Then $1 \notin R$. Furthermore, by Theorem 3.2, we have

$$\sum_{\rho \in Z(L)} \frac{1 + \left|\mathrm{Re}\frac{\rho}{\tau}\right|}{\left(1 + \left|\frac{\rho}{\tau}\right|\right)^2} < \infty,$$

so we may use Bombieri and Lagarias (1999, Th. 1). Hence, the conditions $\mathrm{Re}\frac{\rho}{\tau} \leq \frac{1}{2}$, i.e. $\mathrm{Re}\rho \leq \frac{\tau}{2}$ and

$$\sum_{\rho \in R} \mathrm{Re}\left(1 - \left(1 - \frac{1}{\rho}\right)^{-n}\right) \geq 0$$

are equivalent. Recalling the definition of our set $R$, we see that the condition $\mathrm{Re}\rho \leq \frac{\tau}{2}$ is equivalent to

$$\sum_{\rho \in Z(L)} \mathrm{Re}\left(1 - \left(1 - \frac{\tau}{\rho}\right)^{-n}\right) = \sum_{\rho \in Z(L)} \mathrm{Re}\left(1 - \left(\frac{\rho}{\rho - \tau}\right)^n\right) \geq 0.$$

Recalling that $Z(L) = 1 - \overline{Z(L)}$ completes the proof, since now bounds $\mathrm{Re}(\rho) \leq \frac{\tau}{2}$ and $\mathrm{Re}(\rho) \geq 1 - \frac{\tau}{2}$ are equivalent for all zeros $\rho$ of $\Lambda^c(s)$ (i.e., all non-trivial zeros of $L(s)$). □

## 5 Arithmetic Formulas for $\tau$-Li Coefficients

From the expression (13) for the $n$th $\tau$ Li coefficient we easily get

$$\lambda_n(\pi \times \tilde{\pi}', \tau) = \tau \sum_{k=1}^{n} \binom{n}{k} \frac{\tau^{k-1}}{(k-1)!} \left[ \frac{d^k}{ds^k} \log \Lambda^c(s) \right]_{s=\tau}. \tag{16}$$

Now we wish to analyze this expression, and to separate the archimedean and non-archimedean parts. We will treat the case $\tau \in (1,2)$, since the case $\tau = 1$ was thoroughly investigated in Odžak and Smajlović (2010).

**Lemma 5.1.** *The following expressions hold true*

$$\frac{d^k}{ds^k} \log G(s) = \begin{cases} -\frac{dmm'}{2} \log \pi + \frac{1}{2} \log Q_{\pi\times\tilde{\pi}'} + \sum\limits_{\ell=1}^{dmm'} \frac{1}{2}\psi\left(\frac{s+\mu(\ell)}{2}\right), & k = 1; \\ \frac{1}{2^k} \sum\limits_{\ell=1}^{dmm'} \psi^{(k-1)}\left(\frac{s+\mu(\ell)}{2}\right), & k > 1. \end{cases}$$

*where $\psi(s)$ is classical digamma function, i.e. $\psi(s) = \frac{\Gamma'}{\Gamma}(s)$.*

*Proof.* Definition (11) of $G(s)$ implies that

$$\frac{d}{ds} G(s) = -\frac{dmm'}{2} \log \pi + \frac{1}{2} \log Q_{\pi\times\tilde{\pi}'} + \sum_{\ell=1}^{dmm'} \frac{1}{2}\psi\left(\frac{s+\mu(\ell)}{2}\right).$$

Now it is clear that

$$\frac{d^k}{ds^k} \log G(s) = \sum_{\ell=1}^{dmm'} \frac{1}{2^k}\psi^{(k-1)}\left(\frac{s+\mu(\ell)}{2}\right),$$

when $k > 1$. □

We are now ready to formulate and prove the arithmetic formula for the $\tau$-Li coefficients.

**Theorem 5.2.** *For any positive integer $n$ and any $\tau \in (1,2)$ one has*

$$\lambda_n(\pi \times \tilde{\pi}', \tau) = S_\infty(n,\tau) + S_{NA}(n,\tau), \tag{17}$$

*where*

$$S_{\infty}(n,\tau) = n\frac{\tau}{2}\left(\log Q_{\pi\times\tilde{\pi}'} - dmm'\log\pi\right) + \sum_{k=1}^{n}\binom{n}{k}\frac{\tau^k}{(k-1)!}\sum_{\ell=1}^{dmm'}\frac{1}{2^k}\psi^{(k-1)}\left(\frac{\tau+\mu(\ell)}{2}\right) \tag{18}$$

*and*

$$S_{NA}(n,\tau) = \delta_{\pi,\pi'}(t_0)\left(2-\left(\frac{it_0}{it_0-\tau}\right)^n - \left(\frac{1+it_0}{1+it_0-\tau}\right)^n\right) - \sum_{k=1}^{n}\binom{n}{k}\frac{\tau^k}{(k-1)!}(-\log \mathbf{N}\mathfrak{n})^{k-1}\sum_{\mathfrak{n}}\frac{c_{\pi,\tilde{\pi}'}(\mathfrak{n})}{\mathbf{N}\mathfrak{n}^{\tau}}, \tag{19}$$

*where the coefficients* $c_{\pi,\tilde{\pi}'}(\mathfrak{n})$ *are defined as the coefficients of the Dirichlet series* (4).

*Proof.* Let us start with formula (16). We need to find $\left[\frac{d^k}{ds^k}\log\Lambda^c(s)\right]_{s=\tau}$. We have

$$\frac{d^k}{ds^k}\log\Lambda^c(s) = \frac{d^k}{ds^k}\delta_{\pi,\pi'}(t_0)\log\left((s-it_0)(s-1-it_0)\right) + \frac{d^k}{ds^k}\log L(s) + \frac{d^k}{ds^k}\log G(s).$$

The derivative involving the function $G$ has already been evaluated in the previous lemma, and it yields the total contribution

$$\frac{\tau n}{2}\left(-dmm'\log\pi + \log Q_{\pi\times\tilde{\pi}'}\right) + \sum_{k=1}^{n}\binom{n}{k}\frac{\tau^k}{(k-1)!}\sum_{\ell=1}^{dmm'}\frac{1}{2^k}\psi^{(k-1)}\left(\frac{\tau+\mu(\ell)}{2}\right)$$

This is exactly the archimedean contribution $S_{\infty}(n,\tau)$.

It thus suffices to concentrate on the other terms that yield the contribution of the non-archimedean term $S_{NA}(n,\tau)$. Let us start with evaluating

$$\left[\frac{d^k}{ds^k}\delta_{\pi,\pi'}(t_0)\left(\log(s-it_0)+\log(s-1-it_0)\right)\right]_{s=\tau} = \delta_{\pi,\pi'}(t_0)(k-1)!(-1)^{k-1}\left((\tau-it_0)^{-k} + (\tau-1-it_0)^{-k}\right).$$

Let us now calculate the total contribution of these terms. We have

$$\sum_{k=1}^{n}\binom{n}{k}\frac{\tau^k}{(k-1)!}\delta_{\pi,\pi'}(t_0)(k-1)!(-1)^{k-1}\left(\frac{1}{(\tau-it_0)^k}+\frac{1}{(\tau-1-it_0)^k}\right)$$

$$=\delta_{\pi,\pi'}(t_0)\sum_{k=1}^{n}\binom{n}{k}\left(-\left(\frac{-\tau}{\tau-it_0}\right)^k-\left(\frac{-\tau}{\tau-1-it_0}\right)^k\right)$$

$$=\delta_{\pi,\pi'}(t_0)\left(2-\left(\frac{it_0}{it_0-\tau}\right)^n-\left(\frac{1+it_0}{1+it_0-\tau}\right)^n\right). \tag{20}$$

The only term that remains is $\frac{d^k}{ds^k}\log L(s)$. Differentiating (4) we get

$$\begin{aligned}\left[\frac{d^k}{ds^k}\log L(s)\right]_{s=\tau} &= -\left[\frac{d^{k-1}}{ds^{k-1}}\sum_{\mathfrak{n}}\frac{c_{\pi,\tilde{\pi}'}(\mathfrak{n})}{N\mathfrak{n}^s}\right]_{s=\tau}\\ &= -\left[(-\log N\mathfrak{n})^{k-1}\sum_{\mathfrak{n}}\frac{c_{\pi,\tilde{\pi}'}(\mathfrak{n})}{N\mathfrak{n}^s}\right]_{s=\tau}\\ &= -(-\log N\mathfrak{n})^{k-1}\sum_{\mathfrak{n}}\frac{c_{\pi,\tilde{\pi}'}(\mathfrak{n})}{N\mathfrak{n}^\tau}\end{aligned}$$

Substituting this to the sum defining $\lambda_n(\pi\times\tilde{\pi}',\tau)$, together with (20) yields the expression for $S_{NA}(n,\tau)$. The proof is complete. □

## 6 Asymptotic Behavior of $\tau$-Li Coefficients

In this section we investigate the asymptotic behavior of the $n$th $\tau$-Li coefficient attached to a Rankin–Selberg $L$-function, as $n\to\infty$, while $\tau\in[1,2)$ is fixed. We will separately investigate the asymptotic behavior of archimedean and non-archimedean contribution in the arithmetic formula (17).

### 6.1 *Evaluation of the Archimedean Part of $\tau$-Li Coefficient*

For the evaluation of the archimedean part (18) it is useful to write it in terms of the Hurwitz zeta function. First, we use a recurrence relation for the digamma function (Abramowitz and Stegun 1964, 6.4.6)

$$\psi^{(n)}(z+1)=\psi^{(n)}(z)+(-1)^n n! z^{-n-1},$$

for $n\geq 0$, to make a variable shift and then the fact that (Abramowitz and Stegun 1964, 6.4.10)

$$\psi^{(n)}(z)=(-1)^{n+1}n!\zeta(n+1,z),$$

for $z\neq 0,-1,-2,\dots$ to get

$$\begin{aligned}
&\sum_{k=1}^{n}\binom{n}{k}\frac{\tau^k}{(k-1)!}\sum_{\ell=1}^{dmm'}\frac{1}{2^k}\psi^{(k-1)}\left(\frac{\tau+\mu(\ell)}{2}\right)\\
&=\sum_{k=1}^{n}\binom{n}{k}\frac{\tau^k}{(k-1)!}\sum_{\ell=1}^{dmm'}\frac{1}{2^k}\psi^{(k-1)}\left(\frac{\tau+\mu(\ell)+2}{2}\right)\\
&\quad+\sum_{\ell=1}^{dmm'}\left(\frac{\mu(\ell)}{\tau+\mu(\ell)}\right)^n-dmm'\\
&=n\frac{\tau}{2}\sum_{\ell=1}^{dmm'}\frac{\Gamma'}{\Gamma}\left(\frac{\tau+\mu(\ell)+2}{2}\right)+\sum_{k=2}^{n}\binom{n}{k}\left(-\frac{\tau}{2}\right)^k\sum_{\ell=1}^{dmm'}\zeta\left(k,\frac{\tau+\mu(\ell)+2}{2}\right)\\
&\quad+\sum_{\ell=1}^{dmm'}\left(\frac{\mu(\ell)}{\tau+\mu(\ell)}\right)^n-dmm',
\end{aligned}$$

thus the archimedean contribution is

$$\begin{aligned}
S_\infty(n,\tau)=&\frac{n\tau}{2}\left(\log Q_{\pi\times\tilde{\pi}'}-dmm'\log\pi\right)+\frac{n\tau}{2}\sum_{\ell=1}^{dmm'}\frac{\Gamma'}{\Gamma}\left(\frac{\tau+\mu(\ell)+2}{2}\right)\\
&+\sum_{k=2}^{n}\binom{n}{k}\left(\frac{-\tau}{2}\right)^k\sum_{\ell=1}^{dmm'}\zeta\left(k,\frac{\tau+\mu(\ell)+2}{2}\right)\\
&+\sum_{\ell=1}^{dmm'}\left(\frac{\mu(\ell)}{\tau+\mu(\ell)}\right)^n-dmm'. \qquad (21)
\end{aligned}$$

The following theorem gives us the full asymptotic expansion of the archimedean contribution to the $n$th $\tau$-Li coefficient. The proof follows the lines of the analogous theorem proved in Odžak and Smajlović (2010).

**Theorem 6.1.** *Let $\pi$ and $\pi'$ be two automorphic unitary cuspidal representations of $GL_m(\mathbb{A}_F)$ and $GL_{m'}(\mathbb{A}_F)$, respectively. Then, for $\tau\in(1,2)$ and an arbitrary $K\in\mathbb{N}$*

$$S_\infty(n,\tau) = dmm'\frac{\tau}{2}n\log n$$
$$+\frac{n\tau}{2}\left(\log Q_{\pi\times\tilde{\pi}'} + dmm'\left(\log\frac{\tau}{2\pi}+\gamma-1\right)\right)$$
$$+\frac{\tau-2}{4}dmm' + \sum_{\ell=1}^{dmm'}\left(\frac{\mu(\ell)}{\tau+\mu(\ell)}\right)^n + \frac{1}{2}\sum_{\ell=1}^{dmm'}\mu(\ell)$$
$$-dmm'\frac{\tau}{2}\sum_{k=1}^{K}\frac{B_{2k}}{2k}n^{-2k+1} + O_K(n^{-2K}),$$

*as $n \to \infty$, where $B_{2k}$ are the Bernoulli numbers.*

*Proof.* Let us first investigate the sum

$$\sum_{\ell=1}^{dmm'}\sum_{k=2}^{n}\binom{n}{k}\left(\frac{-\tau}{2}\right)^k\zeta\left(k,\frac{\tau+\mu(\ell)+2}{2}\right) = \sum_{\ell=1}^{dmm'}S_\infty(n,\ell),$$

appearing in (21).

Calculus of residues implies that

$$S_\infty(n,\ell) = \frac{(-1)^n}{2\pi i}n!\int_R f_\ell(s)ds, \tag{22}$$

where $R$ is positively oriented rectangle with vertices at points $\frac{3}{2}\pm i$ and $n+\frac{1}{2}\pm i$ and

$$f_\ell(s) = \frac{\Gamma(s-n)}{\Gamma(s+1)}\left(\frac{\tau}{2}\right)^s\zeta\left(s,\frac{\tau+\mu(\ell)+2}{2}\right).$$

Namely, poles of the function $f_\ell(s)$ inside $R$ are simple poles at $s = 2,3,\ldots,n$ which come from the gamma function $\Gamma(s-n)$, since functions $\left(\frac{\tau}{2}\right)^s$ and $\zeta\left(k,\frac{\tau+\mu(\ell)+2}{2}\right)$ are holomorphic in $R$. Residues are easily found using the fact that $\mathrm{Res}_{s=-n}\Gamma(s) = \frac{(-1)^n}{n!}$, thus

$$\mathrm{Res}_{s=k}f_\ell(s) = \frac{1}{k!}\left(\frac{\tau}{2}\right)^k\zeta\left(k,\frac{\tau+\mu(\ell)+2}{2}\right)\frac{(-1)^{n-k}}{(n-k)!},$$

for $k = 2,3,\ldots,n$, and hence (22) holds true.

The function $f_\ell(s)$ is uniformly bounded on the real segment joining $n+\frac{1}{2}$ and $e^n$, hence, the rectangle $R$ can be deformed to the line $(e^n - i\infty, e^n + i\infty)$. Further singularities of the function $f_\ell(s)$ are a simple pole at $s = 0$ and a pole $s = 1$ of order 2. Therefore, we get

$$\mathcal{S}_\infty(n,\ell) = (-1)^{n-1}n!\left(\mathrm{Res}_{s=0}f_\ell(s) + \mathrm{Res}_{s=1}f_\ell(s)\right) \\ +O\left(n!\int_{-\infty}^{\infty}\left|\frac{\Gamma(e^n+it-n)}{\Gamma(e^n+it+1)}\right|\left(\frac{\tau}{2}\right)^{e^n}\zeta\left(e^n,\left|\frac{2+\tau+\mu(\ell)}{2}\right|\right)dt\right) \tag{23}$$

Residue at $s=0$ is simple and given by

$$(-1)^{n-1}n!\mathrm{Res}_{s=0}f_\ell(s) = -\zeta\left(0,\frac{\tau+\mu(\ell)+2}{2}\right) = \frac{\tau+\mu(\ell)+1}{2},$$

since $\zeta(0,x)=\frac{1}{2}-x$.

Residue at $s=1$ can be found using Laurent series representations of the factors appearing in $f_\ell(a)$. Namely,

$$\zeta\left(s,\frac{2+\tau+\mu(\ell)}{2}\right) = \frac{1}{s-1} - \frac{\Gamma'}{\Gamma}\left(\frac{2+\tau+\mu(\ell)}{2}\right) + \cdots$$

$$\left(\frac{\tau}{2}\right)^s = \frac{\tau}{2} + \left(\frac{\tau}{2}\log\frac{\tau}{2}\right)(s-1) + \cdots$$

$$\frac{1}{\Gamma(s+1)} = 1 + (\gamma-1)(s-1) + \cdots$$

$$\Gamma(s-n) = \frac{(-1)^{n-1}}{(n-1)!}\frac{1}{s-1} + \frac{(-1)^{n-1}}{(n-1)!}\frac{\Gamma'}{\Gamma}(n) + \cdots.$$

and thus

$$(-1)^{n-1}n!\mathrm{Res}_{s=1}f_\ell(s) = \frac{n\tau}{2}\left(\frac{\Gamma'}{\Gamma}(n) - \frac{\Gamma'}{\Gamma}\left(\frac{2+\tau+\mu(\ell)}{2}\right) + \log\frac{\tau}{2} + \gamma - 1\right).$$

Asymptotic expansion of the digamma function appearing in the above residue is given by the Stirling formula for the digamma function (Abramowitz and Stegun 1964, 6.3.18)

$$\frac{\Gamma'}{\Gamma}(n) = \log n - \frac{1}{2n} - \sum_{k=1}^{K}\frac{B_{2k}}{2k}n^{-2k} + O_K(n^{-1-2K}),$$

as $n\to+\infty$, thus

$$(-1)^{n-1}n!\mathrm{Res}_{s=1}f_\ell(s) = \frac{\tau}{2}n\log n + \frac{n\tau}{2}\left(-\frac{\Gamma'}{\Gamma}\left(\frac{2+\tau+\mu(\ell)}{2}\right)\right. \\ \left.+\log\frac{\tau}{2} + \gamma - 1\right) - \frac{\tau}{4} - \frac{\tau}{2}\sum_{k=1}^{K}\frac{B_{2k}}{2k}n^{-2k+1} + O_K(n^{-2K}),$$

as $n\to\infty$.

For the estimation of the integral appearing in (23), let us notice that

$$\left|\left(\frac{\tau}{2}\right)^{e^n}\zeta\left(e^n, \left|\frac{2+\tau+\mu(\ell)}{2}\right|\right)\right| = o(1),$$

as $n \to \infty$, since $\tau \in [1, 2)$ and $\mathrm{Re}\mu(\ell) > -1$ for all $\ell = 1, \dots, dmm'$.

The part containing the gamma functions can be modified using the reflection formula and a functional equation for the gamma function

$$\left|\frac{\Gamma(e^n + it - n)}{\Gamma(e^n + it + 1)}\right| = \frac{\left|(-1)^{n-1}\right|}{\prod_{j=0}^{n}\left|e^n + it - n + j\right|}$$

and it decays rapidly enough since

$$\begin{aligned} n!\int_{-\infty}^{\infty}\left|\frac{\Gamma(e^n + it - n)}{\Gamma(e^n + it + 1)}\right| dt &= 2n!\int_0^{\infty}\frac{dt}{\prod\limits_{j=0}^{n}\left((e^n + j - n)^2 + t^2\right)^{1/2}} \qquad (24) \\ &\le 2n!\int_0^{\infty}\frac{dt}{\left((e^n - n)^2 + t^2\right)^{n/2}} \\ &= \frac{2n!}{(e^n - n)^{n-1}}\int_0^{\infty}\frac{dt}{(1 + t^2)^{n/2}} \\ &\ll \frac{(n+1)^{n+1/2}e^{-(n+1)}}{(e^n - n)^{n-1}} \ll e^{-n} \ll n^{-2K}, \end{aligned}$$

as $n \to \infty$, for all $K \in \mathbb{N}$, by the Stirling approximation formula for the gamma function (Abramowitz and Stegun 1964, 6.1.37). Now,

$$\begin{aligned} \sum_{\ell=1}^{dmm'} S_{\infty}(n, \ell) &= dmm'\frac{\tau}{2}n\log n + dmm'\frac{n\tau}{2}\left(\log\frac{\tau}{2} + \gamma - 1\right) - \\ &-\frac{n\tau}{2}\sum_{\ell=1}^{dmm'}\frac{\Gamma'}{\Gamma}\left(\frac{2+\tau+\mu(\ell)}{2}\right) + dmm'\frac{\tau+2}{4} + \\ &+\frac{1}{2}\sum_{\ell=1}^{dmm'}\mu(\ell) - dmm'\frac{\tau}{2}\sum_{k=1}^{K}\frac{B_{2k}}{2k}n^{-2k+1} + O_K(n^{-2K}), \end{aligned}$$

as $n \to \infty$. Combining obtained result with (21) completes the proof. □

## 6.2 Evaluation of the Non-archimedean Part of the Rankin–Selberg $\tau$-Li Coefficient

For the evaluation of the non-archimedean part of a $\tau$-Li coefficient attached to a Rankin–Selberg $L$-function let us notice that its main part comes from the logarithmic derivative of the Rankin–Selberg $L$-function evaluated at $\tau$. It can be written in the form of a Dirichlet series as in (19), or in terms of coefficients appearing in the Taylor (Laurent) series representation of $-L'/L$-functions around $\tau$. Namely, for $\tau \in (1, 2)$ (case $\tau = 1$ is treated in Odžak and Smajlović 2010), $\tau$ is not a pole of the $L$-function, thus

$$\frac{L'}{L}(s) = \sum_{l=0}^{\infty} \gamma_\tau(l)(s-\tau)^l,$$

or, equivalently

$$\frac{L'}{L}(s+\tau) = \sum_{l=0}^{\infty} \gamma_\tau(l)s^l. \tag{25}$$

Differentiation, for $k \geq 1$ implies

$$\left(\frac{L'}{L}\right)^{(k-1)}(s+\tau) = \sum_{l=0}^{\infty} \gamma_\tau(l)l(l-1)\dots(l-k+2)s^{l-k+1},$$

and passing to the limit as $s \to 0^+$, we obtain

$$\left(\frac{L'}{L}\right)^{(k-1)}(\tau) = (k-1)!\gamma_\tau(k-1).$$

Now, the non-archimedean contribution can be written in the form

$$S_{NA}(n,\tau) = \sum_{k=1}^{n} \binom{n}{k} \tau^k \gamma_\tau(k-1) + \delta_{\pi,\pi'}(t_0)\left(2 - \left(\frac{it_0}{it_0-\tau}\right)^n - \left(\frac{1+it_0}{1+it_0-\tau}\right)^n\right), \tag{26}$$

which will be used for the further asymptotic expansion. Namely, we will prove that this contribution to the $n$th $\tau$-Li coefficient can be written in terms of the incomplete $\tau$-Li coefficient up to the height $\sqrt{n}$ and the error term $O(\sqrt{n}\log n)$.

Incomplete $\tau$-Li coefficient up to the height $T$ is denoted by $\lambda_n(\pi \times \tilde{\pi}', \tau, T)$ and defined by

$$\lambda_n(\pi \times \tilde{\pi}', \tau, T) = \sum_{\substack{\rho \in Z(L) \\ |\mathrm{Im}\rho| < T}} \left(1 - \left(\frac{\rho}{\rho - \tau}\right)^n\right).$$

We will skip $\pi$ and $\pi'$ in the notation of the incomplete $\tau$-Li coefficient up to the height $T$ and denote it simply by $\lambda_n(\tau, T)$.

**Theorem 6.2.** *Let $\pi$ and $\pi'$ be two automorphic unitary cuspidal representations of $GL_m(\mathbb{A}_F)$ and $GL_{m'}(\mathbb{A}_F)$, respectively, and $\tau \in (1, 2)$. Then,*

$$S_{NA}(n, \tau) = \lambda_n(\tau, \sqrt{N}) + O(\sqrt{n} \log n).$$

*Proof.* For the proof we will use contour integration along a suitably chosen rectangle and approximation of the integrals along its sides. The proof follows the lines of the proofs of corresponding theorems in Lagarias (2007) and Odžak and Smajlović (2010), with a slightly modified integrand. Let

$$k_n(s) = \left(1 + \frac{\tau}{s}\right)^n - 1 = \sum_{j=1}^{n} \binom{n}{j} \left(\frac{\tau}{s}\right)^j,$$

and let us integrate the function $g(s) = k_n(s)\frac{L'}{L}(s + \tau)$ over the rectangle $R(n)$ formed by the lines $\mathrm{Re}s = \sigma_0$ for $-3 < \sigma_0 < -2$, $\mathrm{Re}s = 2\sqrt{n}$, $\mathrm{Im}s = \pm T$ where $T = \sqrt{n} + \varepsilon_n, 0 < \varepsilon_n < 1$ is such that the horizontal lines $\mathrm{Im}s = \pm T$ do not approach closer than $O(\frac{1}{\log n})$ to any zero of $L(s)$. This is possible due to equation (6).

Poles of $g(s)$ inside $R(n)$ are at $s = 0$, at points that correspond to trivial zeros (denoted by $\eta$) and non-trivial zeros (denoted by $\rho$) of the Rankin–Selberg $L$-function, and at points corresponding to possible poles of $L(s)$ at $1 + it_0$ and $it_0$.

When $\beta = \eta$ or $\beta = \rho$, residues of $g(s)$ at simple poles $\beta - \tau$ are given by

$$\mathrm{Res}_{s=\beta-\tau} g(s) = \left(1 + \frac{\tau}{\beta - \tau}\right)^n - 1 = \left(\frac{\beta}{\beta - \tau}\right)^n - 1.$$

When $\delta_{\pi,\pi'}(t_0) \neq 0$ residues at $1 + it_0 - \tau$ and $it_0 - \tau$ are

$$\mathrm{Res}_{s=1+it_0-\tau} g(s) = 1 - \left(\frac{1 + it_0}{1 + it_0 - \tau}\right)^n, \quad \mathrm{Res}_{s=it_0-\tau} g(s) = 1 - \left(\frac{it_0}{it_0 - \tau}\right)^n.$$

Residue at $s = 0$ is easily found using (25); it is equal to

$$\mathrm{Res}_{s=0} g(s) = \sum_{k=1}^{n} \binom{n}{k} \tau^k \gamma_\tau(k - 1).$$

Cauchy residue theorem implies

$$\frac{1}{2\pi i}\int_{R(n)} g(s)ds = \sum_{\substack{\rho\in Z(L)\\ |\mathrm{Im}\rho|<T}} \left(\left(\frac{\rho}{\rho-\tau}\right)^n - 1\right) + \sum_{\eta}\left(\left(\frac{\eta}{\eta-\tau}\right)^n - 1\right)$$
$$+\delta_{\pi,\pi'}(t_0)\left(2-\left(\frac{1+it_0}{1+it_0-\tau}\right)^n - \left(\frac{it_0}{it_0-\tau}\right)^n\right)$$
$$+\sum_{k=1}^{n}\binom{n}{k}\tau^k\gamma_\tau(k-1) \tag{27}$$

Using arguments similar to ones used in Lagarias (2007) it is easy to conclude that the contribution from the trivial zeros is $O(1)$, since $\tau \in [1,2)$. Since the sum over $\rho$ is equal to an incomplete $\tau$-Li coefficient up to a height $T$, multiplied by $(-1)$, we get

$$\frac{1}{2\pi i}\int_{R(n)} g(s)ds = S_{NA}(n,\tau) - \lambda_n(\tau, T) + O(1). \tag{28}$$

It is left to estimate the integral on the left side of the above equation. It is done analogously as in Odžak and Smajlović (2011), using the approximation

$$\frac{L(s)'}{L(s)} = \sum_{\rho}{}^{*}\frac{1}{s-\rho} + O_{\pi,\pi'}(\log|s|),$$

which implies that

$$\frac{L(s)'}{L(s)} = \sum_{|T-\mathrm{Im}\rho|\le 1}\frac{1}{s-\rho} + O_{\pi,\pi'}(\log T),$$

for $s = \sigma + iT$, uniformly in $-1 \le \sigma \le 2$, as noticed in Odžak and Smajlović (2010). Last observation, together with formula (6) for the number of non-trivial zeros of the Rankin–Selberg $L$-function up to a height $T$ and the bound on the length of the sides of rectangle $R(n)$ yields the bound

$$\frac{1}{2\pi i}\int_{R(n)} g(s)ds = O(\sqrt{n}\log n).$$

Inserting the above bound into (28) completes the proof. □

## 7 Further Research

Results presented in this paper may be generalized to different classes of functions. Besides that, various numerical computations might be conducted in order to pose conjectures related to the asymptotic behavior of generalized $\tau$-Li coefficients, as $n \to \infty$, for different values of the parameter $\tau$ and for different classes of $L$-functions.

In the paper Ernvall-Hytönen et al. (2015) we define, for $\sigma_0, \sigma_1 \in \mathbb{R}$ a very broad class $\mathcal{S}^{\sharp\flat}(\sigma_0, \sigma_1)$ of functions $F$ satisfying following four conditions:

(i) (Dirichlet series) The function $F$ possesses a Dirichlet series representation that converges absolutely for $\mathrm{Re} s > \sigma_0$.

(ii) (Analytic continuation) There exist finitely many non-negative integers $m_1, \ldots, m_n$ and complex numbers $s_1, \ldots, s_n$ such that function $\prod_{j=1}^{n} (s - s_i)^{m_i} F(s)$ is entire function of finite order.

(iii) (Functional equation) The function $F$ satisfies the functional equation

$$\xi_F(s) = w\overline{\xi_F(\sigma_1 - \bar{s})},$$

where

$$\xi_F(s) = F(s) Q_F^s \prod_{j=1}^{r} \Gamma(\lambda_j s + \mu_j) \prod_{i=1}^{2M+\delta(\sigma_1)} (s - s_i)^{m_i}$$

$$\prod_{i=2M+1+\delta(\sigma_1)}^{N} (s - s_i)^{m_i} (\sigma_1 - s - \overline{s_i})^{m_i},$$

with $|w| = 1$, $Q_F > 0$, $r \geq 0$, $\lambda_j > 0$, $\mu_j \in \mathbb{C}$, $j = 1, \ldots, r$. We assume that poles of $F$ are arranged so that the first $0 \leq 2M + \delta(\sigma_1) \leq N$ poles are such that $s_{2j-1} + \overline{s}_{2j} = \sigma_1$, for $j = 1, \ldots, M$, and $\delta(\sigma_1) = 1$ if $\sigma_1/2$ is a pole of $F$ in which case $s_{2M+\delta(\sigma_1)} = \sigma_1/2$; otherwise $\delta(\sigma_1) = 0$.

(iv) (Euler sum) The logarithmic derivative of the function $F$ possesses a Dirichlet series representation

$$\frac{F'}{F}(s) = -\sum_{n=2}^{\infty} \frac{c_F(n)}{n^s},$$

converging absolutely for $\mathrm{Re} s > \sigma_0$.

The class $\mathcal{S}^{\sharp\flat}(\sigma_0, \sigma_1)$ contains the class $\mathcal{S}^{\sharp\flat}$ as well as Rankin–Selberg $L$-functions. Besides classical $L$-functions, the class $\mathcal{S}^{\sharp\flat}(\sigma_0, \sigma_1)$ also contains all finite products of complex shifts of functions from the class $\mathcal{S}^{\sharp\flat}$.

We show that for $\tau \geq 1$ the generalized $\tau$-Li coefficients are well defined for all $F \in \mathcal{S}^{\sharp\flat}(\sigma_0, \sigma_1)$ and derive an analogue of the $\tau$-Li criterion for zero-free regions. Furthermore, we derive arithmetic formulas for generalized $\tau$-Li coefficients and deduce their asymptotic behavior.

We also conduct numerical computations in order to evaluate $\tau$-Li coefficients for finite products of shifts of Riemann zeta functions and deduce certain conjectures on the asymptotic behavior of $\tau$-Li coefficients, as $n \to \infty$, for different values of $\tau \geq 1$, depending on the real parts of shifts.

In the paper Bucur et al. (2015), we define generalized $\tau$-Li coefficients for functions $F$ in the class $\mathcal{S}^{\sharp}_{\mathbb{R}}$ that consists of all functions $F$ belonging to the extended Selberg class $\mathcal{S}^{\sharp}$, introduced in Kaczorowski and Perelli (1999) with the property that $\rho$ is a zero of $F$ if and only if $1-\rho$ is. Functions from the class $\mathcal{S}^{\sharp}_{\mathbb{R}}$ do not necessarily satisfy the generalized Riemann hypothesis. For example, Davenport-Heilbronn $L$-function (introduced in Davenport and Heilbronn 1936) belongs to $\mathcal{S}^{\sharp}_{\mathbb{R}}$ and possesses infinitely many zeros in the half-plane $\mathrm{Re} s > 1$.

We prove that generalized $\tau$-Li coefficients are well defined for functions in $\mathcal{S}^{\sharp}_{\mathbb{R}}$ and derive an arithmetic formula for their computation. The asymptotic behavior of generalized $\tau$-Li coefficients as $n \to \infty$, and as $\tau \to \infty$ (since in this case we might have no zero-free regions) will be a subject of our future investigation. We intend to conduct a numerical investigation using Davenport-Heilbronn type $L$-functions (see Bombieri and Ghosh (2011)) in order to conjecture asymptotic behavior of generalized $\tau$-Li coefficients as $\tau \to \infty$, in the case when the function violates Riemann hypothesis, but still possesses a positive proportion of zeros on the critical line.

**Acknowledgements** The authors of the paper would like to thank the organizers of the WINE conference. The conference was funded by CIRM, Microsoft research, Number theory foundation, NSF and Clay Mathematics Institute. Their support is gratefully acknowledged. The research of A.-M. E.-H. was supported by the Academy of Finland, grant no. 138337. The research of A.B. was supported by Simons Foundation Award #244988.

## References

Abramowitz, M., Stegun, I.: Handbook of Mathematical Functions, With Formulas, Graphs, and Mathematical Tables. NBS Applied Mathematics Series 55, National Bureau of Standards, Washington, DC (1964)

Bombieri, E., Ghosh, A.: Around Davenport-Heilbronn function. Uspekhi Math. Nauk 66, 15-66 (2011) [translated in Russ. Math. Surv. 66, 221–270 (2011)]

Bombieri, E., Lagarias, J.C.: Complements to Li's criterion for the Riemann hypothesis. J. Number Theory **77**, 274–287 (1999)

Bucur, A., Ernvall-Hytönen, A.-M., Odžak, A., Smajlo̒vic, L.: On a Li-type criteria for-zero free regions of certain Dirichlet series with real coefficients (2015, in preparation)

Cogdell, J.W.: L-functions and converse theorems for $GL_n$, Automorphic forms and applications, IAS/Park City Math. Ser. 12, Amer. Math. Soc, Providence, RI, 2007, 97–177

Davenport, D., Heilbronn, H.: On the zeros of certain Dirichlet series II. J. Lond. Math. Soc. **11**, 307–312 (1936)
Droll, A.D.: Variations of Li's criterion for an extension of the Selberg class. Ph.D. thesis, Queen's University, Kingston (2012)
Ernvall-Hytönen, A.-M., Odžak, A., Smajlović, L., Sušić, M.: On the modified Li criterion for a certain class of L-functions. J. Number Theory **156**, 340–367 (2015)
Freitas, P.: A Li-type criterion for zero-free half-planes of Riemann's zeta function. J. Lond. Math. Soc. **73**, 399–414 (2006)
Gelbart, S., Shahidi, F.: Boundedness of automorphic $L$-functions in vertical strips. J. Amer. Math. Soc. **14**, 79–107 (2001)
Gelfand, I.M., Kazhdan, D.: Representation of the group $GL(n, K)$, where $K$ is a local field. In: I.M. Gelfand (ed.) Lie Groups and Their Representations, pp. 95–118. Wiley, New York (1974)
Iwaniec, H., Kowalski, E.: Analytic Number Theory. American Mathematical Society Colloquium Publications, vol. 53. American Mathematical Society, Providence (2004)
Jacquet, H., Shalika, J.A.: On Euler products and the classification of automorphic representations I. Am. J. Math. **103**, 499–558 (1981a)
Jacquet, H., Shalika, J.A.: On Euler products and the classification of automorphic representations II. Am. J. Math. **103**, 777–815 (1981b)
Kaczorowski, J., Perelli, A.: On the structure of the Selberg class, I: $0 \leq d \leq 1$. Acta Math. **182**, 207–241 (1999)
Lagarias, J.C.: Li's coefficients for automorphic $L$-functions. Ann. Inst. Fourier **57**, 1689–1740 (2007)
Moeglin, C., Waldspurger, J.-L.: Le spectre résiduel de $GL(n)$. Ann. Sci. École Norm. Sup. **22**, 605–674 (1989)
Li, X.-J.: The positivity of a sequence of numbers and the Riemann hypothesis. J. Number Theory **65**, 325–333 (1997)
Li, X.-J.: Explicit formulas for Dirichlet and Hecke $L$-functions. Ill. J. Math. **48**, 491–503 (2004)
Odžak, A., Smajlović, L.: On Li's coefficients for the Rankin-Selberg L-functions. Ramanujan J. **21**, 303–334 (2010)
Odžak, A., Smajlović, L.: On asymptotic behavior of generalized Li coefficients in the Selberg class. J. Number Theory **131**, 519–535 (2011)
Rudnick, Z., Sarnak, P.: Zeros of principal $L$-functions and random matrix theory. Duke Math. J. **81**, 269–322 (1996)
Sekatskii, S.K.: Generalized Bombieri-Lagarias' theorem and generalized Li's criterion (2013) [arXiv:1304.7895]
Shahidi, F.: On certain $L$-functions. Am. J. Math. **103**, 297–355 (1981)
Shahidi, F.: Fourier transforms of intertwining operators and Plancherel measures for $GL(n)$. Am. J. Math. **106**, 67–111 (1984)
Shahidi, F.: Local coefficients as Artin factors for real groups. Duke Math. J. **52**, 973–1007 (1985)
Shahidi, F.: A proof of Langlands' conjecture on Plancherel measures. Complementary series for $p$-adic groups. Ann. Math. **132**, 273–330 (1990)
Smajlović, L.: On Li's criterion for the Riemann hypothesis for the Selberg class. J. Number Theory **130**, 828–851 (2010)

# Galois Representations and Galois Groups Over $\mathbb{Q}$

**Sara Arias-de-Reyna, Cécile Armana, Valentijn Karemaker, Marusia Rebolledo, Lara Thomas, and Núria Vila**

**Abstract** In this paper we generalize results of P. Le Duff to genus $n$ hyperelliptic curves. More precisely, let $C/\mathbb{Q}$ be a hyperelliptic genus $n$ curve, let $J(C)$ be the associated Jacobian variety, and let $\bar{\rho}_\ell : G_\mathbb{Q} \to \mathrm{GSp}(J(C)[\ell])$ be the Galois representation attached to the $\ell$-torsion of $J(C)$. Assume that there exists a prime $p$ such that $J(C)$ has semistable reduction with toric dimension 1 at $p$. We provide an algorithm to compute a list of primes $\ell$ (if they exist) such that $\bar{\rho}_\ell$ is surjective. In particular we realize $\mathrm{GSp}_6(\mathbb{F}_\ell)$ as a Galois group over $\mathbb{Q}$ for all primes $\ell \in [11, 500{,}000]$.

## 1 Introduction

In this paper we present the work carried out at the conference *Women in numbers—Europe* (October 2013), by the working group *Galois representations and Galois groups over* $\mathbb{Q}$. Our aim was to study the image of Galois representations attached to the Jacobian varieties of genus $n$ curves, motivated by the applications to the inverse Galois problem over $\mathbb{Q}$. In the case of genus 2, there are several results in this direction (e.g., Le Duff 1998; Dieulefait 2002a), and we wanted to explore the scope of these results.

---

S. Arias-de-Reyna
Mathematical Research Unit, University of Luxembourg, Luxembourg, Luxembourg

C. Armana
Laboratory of Mathematics, University of Franche-Comteé, Besançon, France

V. Karemaker
Mathematical Institute, Utrecht University, Utrecht, The Netherlands

M. Rebolledo
Laboratory of Mathematics, Blaise Pascal University, Clermont-Ferrand, Aubière, France

L. Thomas
Pure and Applied Mathematics Unit, ENS Lyon, Lyon, France

N. Vila (✉)
Department of Algebra and Geometry, University of Barcelona, Barcelona, Spain
e-mail: nuriavila@ub.edu



M.J. Bertin et al. (eds.), *Women in Numbers Europe*, Association for Women in Mathematics Series 2, DOI 10.1007/978-3-319-17987-2_8

Our result is a generalization of P. Le Duff's work to the genus $n$ setting, which allows us to produce realizations of groups $\mathrm{GSp}_6(\mathbb{F}_\ell)$ as Galois groups over $\mathbb{Q}$, for infinite families of primes $\ell$ (with positive Dirichlet density). These realizations are obtained through the Galois representations $\bar{\rho}_\ell$ attached to the $\ell$-torsion points of the Jacobian of a genus 3 curve.

The first section of this paper contains a historical introduction to the inverse Galois problem and some results obtained in this direction by means of Galois representations associated with geometric objects. Section 3 presents some theoretic tools, which we collect to prove a result, valid for a class of abelian varieties $A$ of dimension $n$, that yields primes $\ell$ for which we can ensure surjectivity of the Galois representation attached to the $\ell$-torsion of $A$ (see Theorem 3.10). In Section 4, we focus on hyperelliptic curves and explain the computations that allow us to realize $\mathrm{GSp}_6(\mathbb{F}_\ell)$ as a Galois group over $\mathbb{Q}$ for all primes $\ell \in [11, 500{,}000]$.

## 2 Images of Galois Representations and the Inverse Galois Problem

One of the main objectives in algebraic number theory is to understand the absolute Galois group of the rational field, $G_\mathbb{Q} = \mathrm{Gal}(\overline{\mathbb{Q}}/\mathbb{Q})$. We believe that we would get all arithmetic information if we knew the structure of $G_\mathbb{Q}$. This is a huge group, but it is compact with respect to the profinite topology. Two problems arise in a natural way: on the one hand, the identification of the finite quotients of $G_\mathbb{Q}$, and on the other hand, the study of $G_\mathbb{Q}$ via its Galois representations.

The inverse Galois problem asks whether, for a given finite group $G$, there exists a Galois extension $L/\mathbb{Q}$ with Galois group isomorphic to $G$. In other words, whether a finite group $G$ occurs as a quotient of $G_\mathbb{Q}$. As is well known, this is an open problem. The origin of this question can be traced back to Hilbert. In 1892, he proved that the symmetric group $S_n$ and the alternating group $A_n$ are Galois groups over $\mathbb{Q}$, for all $n$. We also have an affirmative answer to the inverse Galois problem for some other families of finite groups. For instance, all finite solvable groups and all sporadic simple groups, except the Mathieu group $M_{23}$, are known to be Galois groups over $\mathbb{Q}$.

A Galois representation is a continuous homomorphism

$$\rho : G_\mathbb{Q} \to \mathrm{GL}_n(R),$$

where $R$ is a topological ring. Examples for $R$ are $\mathbb{C}$, $\mathbb{Z}/n\mathbb{Z}$ or $\mathbb{F}_q$ with the discrete topology, and $\mathbb{Q}_\ell$ with the $\ell$-adic topology. Conjectures by Artin, Serre, Fontaine-Mazur, and Langlands, which have experienced significant progress in recent years, are connected with these Galois representations.

Since $G_{\mathbb{Q}}$ is compact, the image of $\rho$ is finite when the topology of $R$ is discrete. As a consequence, images of Galois representations yield Galois realizations over $\mathbb{Q}$ of finite linear groups

$$\mathrm{Gal}(\overline{\mathbb{Q}}^{\ker\rho}/\mathbb{Q}) \simeq \rho(G_{\mathbb{Q}}) \subseteq \mathrm{GL}_n(R).$$

This gives us an interesting connection between these two questions and provides us with a strategy to address the inverse Galois problem.

Let us assume that $\rho$ is an $\ell$-adic Galois representation associated with some arithmetic-geometric object. In this case, we have additional information on the ramification behavior, like the characteristic polynomial of the image of the Frobenius elements at unramified primes or the description of the image of the inertia group at the prime $\ell$. This gives us some control on the image of mod $\ell$ Galois representations in some cases and we can obtain, along the way, families of linear groups over finite fields as Galois groups over $\mathbb{Q}$.

More precisely, let $X/\mathbb{Q}$ be a smooth projective variety and let

$$\rho_\ell : G_{\mathbb{Q}} \to \mathrm{GL}(H^k_{\text{ét}}(X_{\overline{\mathbb{Q}}}, \mathbb{Q}_\ell)),$$

be the $\ell$-adic Galois representation on the $k$-th étale cohomology. We know that:

- $\rho_\ell$ is unramified away from $\ell$ and the primes of bad reduction for $X$,
- if $p$ is a prime of good reduction and $p \neq \ell$, the characteristic polynomial of $\rho_\ell(\mathrm{Frob}_p)$ has coefficients in $\mathbb{Z}$, is independent of $\ell$ and its roots have absolute value $p^{k/2}$.

Let us consider an attached residual Galois representation

$$\overline{\sigma}_\lambda : G_{\mathbb{Q}} \to \mathrm{GL}_n(\mathbb{F}_{\ell^r}),$$

where $\lambda$ is a prime in a suitable number field, dividing $\ell$ and $r \geq 1$ an integer. To determine the image of $\overline{\sigma}_\lambda$, we usually need to know the classification of maximal subgroups of $\mathrm{GL}_n(\mathbb{F}_{\ell^r})$, as well as a description of the image of the inertia group at $\ell$ and the computation of the characteristic polynomial of $\overline{\sigma}_\lambda(\mathrm{Frob}_p)$, for some prime of good reduction $p \neq \ell$.

Let us summarize the known cases of realizations of finite linear groups as Galois groups over $\mathbb{Q}$, obtained via Galois representations.

In the case of two-dimensional Galois representations attached to an elliptic curve $E$ defined over $\mathbb{Q}$ without complex multiplication, we know, by a celebrated result of (Serre 1972), that the associated residual Galois representation is surjective, for all but finitely many primes. Moreover, it can be shown that if we take, for example, the elliptic curve $E$ defined by the Weierstrass equation $Y^2 + Y = X^3 - X$, then the attached residual Galois representation is surjective, for all primes $\ell$. Thus we obtain that the group $\mathrm{GL}_2(\mathbb{F}_\ell)$ occurs as a Galois group over $\mathbb{Q}$, for all primes $\ell$. Actually we have additional information in this case: the Galois extension $\mathbb{Q}(E[\ell])/\mathbb{Q}$ is a Galois realization of $GL_2(\mathbb{F}_\ell)$, and it is unramified away from 37 and $\ell$, since $E$ has conductor 37.

The image of two-dimensional Galois representations, attached to classical modular forms without complex multiplication, has been studied by Ribet (1975). The image of the residual Galois representations attached to a normalized cuspidal Hecke eigenform without complex multiplication is as large as possible, for all but finitely many primes $\lambda$. This gives us that the groups $PSL_2(\mathbb{F}_{\ell^r})$ or $PGL_2(\mathbb{F}_{\ell^r})$ can occur as Galois groups over $\mathbb{Q}$. Moreover, we have effective control of primes with large image for the mod $\ell$ Galois representation attached to specific modular forms. This gives us Galois realizations over $\mathbb{Q}$ of the groups $PSL_2(\mathbb{F}_{\ell^r})$, $r$ even, and $PGL_2(\mathbb{F}_{\ell^r})$, $r$ odd; $1 \leq r \leq 10$, for explicit infinite families of primes $\ell$, given by congruence conditions on $\ell$ (cf. Reverter and Vila 1995; Dieulefait and Vila 2000).

Recently, it has been proven that the groups $PSL_2(\mathbb{F}_\ell)$ are Galois groups over $\mathbb{Q}$ for all $\ell > 3$, by considering the Galois representations attached to an explicit elliptic surface (see Zywina 2014).

Results on generically large image of compatible systems of three-dimensional Galois representations associated with some smooth projective surfaces and with some cohomological modular forms are obtained in Dieulefait and Vila (2004). The effective control of primes with large image for the residual three-dimensional Galois representations attached to some explicit examples gives us that the groups $PSL_3(\mathbb{F}_\ell)$, $PSU_3(\mathbb{F}_\ell)$, $SL_3(\mathbb{F}_\ell)$, $SU_3(\mathbb{F}_\ell)$ are Galois groups over $\mathbb{Q}$, for explicit infinite families of primes $\ell$ (cf. Dieulefait and Vila 2004).

In the case of four-dimensional Galois representations, we have results on large image for compatible systems of Galois representations attached to abelian surfaces $A$ defined over $\mathbb{Q}$ such that $\mathrm{End}_{\overline{\mathbb{Q}}}(A) = \mathbb{Z}$, to Siegel modular forms of genus two and to some pure motives (cf. Le Duff 1998; Dettweiler et al. 2001; Dieulefait 2002b; Dieulefait and Vila 2011). The effective control of primes with large image in some explicit cases gives us that the groups $PGSp_4(\mathbb{F}_\ell)$, for all $\ell > 3$; and the groups $PGSp_4(\mathbb{F}_{\ell^3})$, $PSp_4(\mathbb{F}_{\ell^2})$, $PSL_4(\mathbb{F}_\ell)$ and $PSU_4(\mathbb{F}_\ell)$, for explicit infinite families of primes $\ell$, are Galois groups over $\mathbb{Q}$ (cf. Arias-de-Reyna and Vila 2011; Dettweiler et al. 2001; Dieulefait 2002b; Dieulefait and Vila 2008).

In the next section we consider the image of residual Galois representations attached to principally polarized abelian varieties of dimension $n$, which provides Galois realizations over $\mathbb{Q}$ of the general symplectic group $GSp_{2n}(\mathbb{F}_\ell)$, for almost all $\ell$.

Finally, we remark that, using these methods, we can expect to obtain realizations of the groups $PSL_2(\mathbb{F}_{\ell^r})$, $PGL_2(\mathbb{F}_{\ell^r})$, $PGSp_{2n}(\mathbb{F}_{\ell^r})$, and $PSp_{2n}(\mathbb{F}_{\ell^r})$ as Galois groups over $\mathbb{Q}$. In fact, by considering compatible systems of Galois representations attached to certain automorphic forms, we know (cf. Wiese 2008; Dieulefait and Wiese 2011; Khare et al. 2008; Arias-de-Reyna et al. 2013) that these groups are Galois groups over $\mathbb{Q}$, for infinitely many integers $r$ and infinitely many primes $\ell$. More precisely, we have:

- "Vertical direction": For every fixed prime $\ell$, there are infinitely many positive integers $r$, such that $\mathrm{PSL}_2(\mathbb{F}_{\ell^r})$ can be realized as a Galois group over $\mathbb{Q}$. Moreover, for each $n \geq 2$, there are infinitely many positive integers $r$, such that either $\mathrm{PGSp}_{2n}(\mathbb{F}_{\ell^r})$ or $\mathrm{PSp}_{2n}(\mathbb{F}_{\ell^r})$ are Galois groups over $\mathbb{Q}$ (cf. Wiese 2008; Khare et al. 2008).
- "Horizontal direction": For every fixed $r$, there is a positive density set of primes $\ell$, such that $\mathrm{PSL}_2(\mathbb{F}_{\ell^r})$ can be realized as a Galois group over $\mathbb{Q}$. Moreover, for each $n \geq 2$, there is a set of primes $\ell$ of positive density for which either $\mathrm{PGSp}_{2n}(\mathbb{F}_{\ell^r})$ or $\mathrm{PSp}_{2n}(\mathbb{F}_{\ell^r})$ are Galois groups over $\mathbb{Q}$ (cf. Dieulefait and Wiese 2011; Arias-de-Reyna et al. 2013).

# 3 Galois Representations Attached to Abelian Varieties

## *3.1 The Image of the $\ell$-Torsion Galois Representation*

Let $A$ be an abelian variety of dimension $n$ defined over $\mathbb{Q}$. The set of $\overline{\mathbb{Q}}$-points of $A$ admits a group structure. Let $\ell$ be a prime number. Then the subgroup of the $\overline{\mathbb{Q}}$-points of $A$ consisting of all $\ell$-torsion points, which is denoted by $A[\ell]$, is isomorphic to $(\mathbb{Z}/\ell\mathbb{Z})^{2n}$ and it is endowed with a natural action of $G_{\mathbb{Q}}$. Therefore, it gives rise to a (continuous) Galois representation

$$\overline{\rho}_{A,\ell} : G_{\mathbb{Q}} \to \mathrm{GL}(A[\ell]) \simeq \mathrm{GL}_{2n}(\mathbb{F}_\ell).$$

As explained in Section 2, we obtain a realization of the image of $\overline{\rho}_{A,\ell}$ as a Galois group over $\mathbb{Q}$.

In this section, we will consider principally polarized abelian varieties, i.e. we will consider pairs $(A, \lambda)$, where $A$ is an abelian variety (defined over $\mathbb{Q}$) and $\lambda : A \to A^\vee$ is an isogeny of degree 1 (that is, an isomorphism between $A$ and the dual abelian variety $A^\vee$), induced from an ample divisor on $A$. Not every abelian variety $A$ admits a principal polarization $\lambda$ and, when it does, it causes certain restrictions on the image of $\overline{\rho}_{A,\ell}$.

Let $V$ be a vector space of dimension $2n$, which is defined over $\mathbb{F}_\ell$ and endowed with a symplectic (i.e., skew-symmetric, nondegenerate) pairing $\langle \cdot, \cdot \rangle : V \times V \to \mathbb{F}_\ell$. We consider the *symplectic group*

$$\mathrm{Sp}(V, \langle \cdot, \cdot \rangle) := \{M \in \mathrm{GL}(V) : \forall v_1, v_2 \in V, \langle Mv_1, Mv_2 \rangle = \langle v_1, v_2 \rangle\}$$

and the *general symplectic group*

$$\mathrm{GSp}(V, \langle \cdot, \cdot \rangle) : = \{M \in \mathrm{GL}(V) : \exists m \in \mathbb{F}_\ell^\times \text{ such that } \forall v_1, v_2 \in V,$$
$$\langle Mv_1, Mv_2 \rangle = m \langle v_1, v_2 \rangle\}.$$

When $A$ is a principally polarized abelian variety, the image of $\overline{\rho}_{A,\ell}$ lies inside the general symplectic group of $A[\ell]$ with respect to a certain symplectic pairing. More precisely, denote by $\mu_\ell(\overline{\mathbb{Q}})$ the group of $\ell$-th roots of unity inside a fixed algebraic closure $\overline{\mathbb{Q}}$ of $\mathbb{Q}$. Recall that the Weil pairing $e_\ell$ is a perfect pairing

$$e_\ell : A[\ell] \times A^\vee[\ell] \to \mu_\ell(\overline{\mathbb{Q}}).$$

If $(A, \lambda)$ is a principally polarized abelian variety, we can consider the pairing

$$\begin{aligned} e_{\ell,\lambda} : A[\ell] \times A[\ell] &\to \mu_\ell(\overline{\mathbb{Q}}) \\ (P, Q) &\mapsto e_\ell(P, \lambda(Q)) \end{aligned}$$

which is a non-degenerate skew-symmetric pairing (i.e., a symplectic pairing), compatible with the action of $G_\mathbb{Q}$. This last condition means that, for any $\sigma \in G_\mathbb{Q}$,

$$(e_{\ell,\lambda}(P, Q))^\sigma = e_{\ell,\lambda}(P^\sigma, Q^\sigma).$$

Note that $G_\mathbb{Q}$ acts on $\mu_\ell(\overline{\mathbb{Q}})$ via the mod $\ell$ cyclotomic character $\chi_\ell$, so that $(e_{\ell,\lambda}(P, Q))^\sigma = (e_{\ell,\lambda}(P, Q))^{\chi_\ell(\sigma)}$. If we fix a primitive $\ell$-th root of unity $\zeta_\ell$, we may write the pairing $e_{\ell,\lambda}(\cdot, \cdot)$ additively, i.e. we define

$$\langle \cdot, \cdot \rangle : A[\ell] \times A[\ell] \to \mathbb{F}_\ell$$

as $\langle P, Q \rangle := a$ such that $\zeta^a = e_{\ell,\lambda}(P, Q)$.

In other words, we have a symplectic pairing on the $\mathbb{F}_\ell$-vector space $A[\ell]$ such that, for all $\sigma \in G_\mathbb{Q}$, the linear map $\overline{\rho}(\sigma) : A[\ell] \to A[\ell]$ satisfies that there exists a scalar, namely $\chi_\ell(\sigma)$, such that

$$\langle \overline{\rho}(\sigma)(P), \overline{\rho}(\sigma)(Q) \rangle = \chi_\ell(\sigma) \langle P, Q \rangle. \tag{1}$$

That is to say, the image of the representation $\overline{\rho}_{A,\ell}$ is contained in the general symplectic group $\mathrm{GSp}(A[\ell], \langle \cdot, \cdot \rangle) \simeq \mathrm{GSp}_{2n}(\mathbb{F}_\ell)$. Therefore, below we will consider $\overline{\rho}_{A,\ell}$ as a map into $\mathrm{GSp}(A[\ell], \langle \cdot, \cdot \rangle) \simeq \mathrm{GSp}_{2n}(\mathbb{F}_\ell)$ and we will say that it is surjective if $\mathrm{Im}\overline{\rho}_{A,\ell} = \mathrm{GSp}(A[\ell]) \simeq \mathrm{GSp}_{2n}(\mathbb{F}_\ell)$.

The determination of the images of the Galois representations $\overline{\rho}_{A,\ell}$ attached to the $\ell$-torsion of abelian varieties is a topic that has received a lot of attention. A remarkable result by Serre quoted in Serre (2000, n. 136, Theorem 3) is:

**Theorem 3.1 (Serre).** *Let $A$ be a principally polarized abelian variety of dimension $n$, defined over a number field $K$. Assume that $n = 2, 6$ or $n$ is odd and furthermore assume that* $\mathrm{End}_{\overline{K}}(A) = \mathbb{Z}$. *Then there exists a bound $B_{A,K}$ such that, for all $\ell > B_{A,K}$,*

$$\mathrm{Im}\overline{\rho}_{A,\ell} = \mathrm{GSp}(A[\ell]) \simeq \mathrm{GSp}_{2n}(\mathbb{F}_\ell).$$

For arbitrary dimension, the result is not true (see, e.g., Mumford 1969 for an example in dimension 4). However, one eventually obtains symplectic image by making some extra assumptions. For example, there is the following result of C. Hall (cf. Hall 2011).

**Theorem 3.2 (Hall).** *Let $A$ be a principally polarized abelian variety of dimension $n$ defined over a number field $K$, such that* $\mathrm{End}_{\overline{K}}(A) = \mathbb{Z}$, *and satisfying the following property:*

(T) *There is a finite extension $L/K$ so that the Néron model of $A/L$ over the ring of integers of $L$ has a semistable fiber with toric dimension 1.*

*Then there is an (explicit) finite constant $B_{A,K}$ such that, for all $\ell \geq B_{A,K}$,*

$$\mathrm{Im}\overline{\rho}_{A,\ell} = \mathrm{GSp}(A[\ell]) \simeq \mathrm{GSp}_{2n}(\mathbb{F}_\ell).$$

*Remark 3.3.* In the case when $A = J(C)$ is the Jacobian of a hyperelliptic curve $C$ of genus $n$, say defined by an equation $Y^2 = f(X)$ with $f(X) \in K[X]$ a polynomial of degree $2n + 1$, Hall gives a sufficient condition for Condition (T) to be satisfied at a prime $\mathfrak{p}$ of the ring of integers of $K$; namely, the coefficients of $f(X)$ should have $\mathfrak{p}$-adic valuation greater than or equal to zero and the reduction of $f(X)$ mod $\mathfrak{p}$ (which is well-defined) should have one double zero in a fixed algebraic closure of the residue field, while all the other zeroes are simple.

Applying the result of Hall with $K = \mathbb{Q}$ yields the following partial answer to the inverse Galois problem:

**Corollary 3.4.** *Let $n \in \mathbb{N}$ be any natural number. Then for* all sufficiently large *primes $\ell$, the group $\mathrm{GSp}_{2n}(\mathbb{F}_\ell)$ can be realized as a Galois group over $\mathbb{Q}$.*

*Remark 3.5.* Several people, including the anonymous referee, pointed us to the following fact: if we consider a family of genus $n$ hyperelliptic curves $C_t$ defined over $\mathbb{Q}(t)$, with big monodromy at $\ell$, then Hilbert's Irreducibility Theorem provides us with infinitely many specializations $t = t_0 \in \mathbb{Q}$ such that the Jacobian $J_{t_0}$ of the corresponding curve $C_{t_0}$ satisfies that $\mathrm{Im}\overline{\rho}_{J_{t_0},\ell} \simeq \mathrm{GSp}_{2n}(\mathbb{F}_\ell)$. Such families of curves exist for any odd $\ell$ (see, e.g., Hall 2008 or Zarhin 2014). In particular, for any $n \in \mathbb{N}$ and any odd $\ell$, the Inverse Galois problem has an affirmative answer for the group $\mathrm{GSp}_{2n}(\mathbb{F}_\ell)$. Although ensuring the existence of the desired curve, this fact does not tell us how to find such a curve explicitly.

In the case of curves of genus 2, Le Duff has studied the image of the Galois representations attached to the $\ell$-torsion of $J(C)$, when Condition (T) in Theorem 3.2 is satisfied. The main result in Le Duff (1998) is the following:

**Theorem 3.6 (Le Duff).** *Let $C$ be a genus 2 curve defined over $\mathbb{Q}$, with bad reduction of type (II) or (IV) according to the notation in Liu (1993) at a prime $p$. Let $\Phi_p$ be the group of connected components of the special fiber of the Néron model of $J(C)$ at $p$. For each prime $\ell$ and each prime $q$ of good reduction of $C$, let $P_{q,\ell}(X) = X^4 + aX^3 + bX^2 + qaX + q^2 \in \mathbb{F}_\ell[X]$ be the characteristic* *polynomial of the image under $\overline{\rho}_{J(C),\ell}$ of the Frobenius element at $q$ and let $Q_{q,\ell}(X) = X^2 + aX + b - 2q \in \mathbb{F}_\ell[X]$, with discriminants $\Delta_P$ and $\Delta_Q$, respectively.*

*Then for all primes $\ell$ not dividing $2pq|\Phi_p|$ and such that $\Delta_P$ and $\Delta_Q$ are not squares in $\mathbb{F}_\ell$, the image of $\overline{\rho}_{J(C),\ell}$ coincides with $\mathrm{GSp}_4(\mathbb{F}_\ell)$.*

Using this result, he obtains a realization of $\mathrm{GSp}_4(\mathbb{F}_\ell)$ as Galois group over $\mathbb{Q}$ for all odd primes $\ell$ smaller than 500,000.

## 3.2 Explicit Surjectivity Result

A key point in Hall's result is the fact that the image under $\overline{\rho}_{A,\ell}$ of the inertia subgroup at the place $\mathfrak{p}$ of $L$ which provides the semistable fiber with toric dimension 1 is generated by a nontrivial transvection (whenever $\ell$ does not divide $\mathfrak{p}$ nor the cardinality of the group $\Phi_\mathfrak{p}$ of connected components of the special fiber of the Néron model at $\mathfrak{p}$). A detailed proof of this fact can be found in Proposition 1.3 of Le Duff (1998).

We expand on this point. Given a finite-dimensional vector space $V$ over $\mathbb{F}_\ell$, endowed with a symplectic pairing $\langle\cdot,\cdot\rangle : V \times V \to \mathbb{F}_\ell$, a transvection is an element $T \in \mathrm{GSp}(V, \langle\cdot,\cdot\rangle)$ such that there exists a hyperplane $H \subset V$ satisfying that the restriction $T|_H$ is the identity on $H$. We say that it is a nontrivial transvection if $T$ is not the identity[1]. It turns out that the subgroups of $\mathrm{GSp}(V, \langle\cdot,\cdot\rangle)$ that contain a nontrivial transvection can be classified into three categories as follows (for a proof, see, e.g., Arias-de-Reyna et al. 2014, Theorem 1.1):

**Theorem 3.7.** *Let $\ell \geq 5$ be a prime, let $V$ be a finite-dimensional vector space over $\mathbb{F}_\ell$, endowed with a symplectic pairing $\langle\cdot,\cdot\rangle : V\times V \to \mathbb{F}_\ell$, and let $G \subset \mathrm{GSp}(V, \langle\cdot,\cdot\rangle)$ be a subgroup that contains a nontrivial transvection. Then one of the following holds:*

1. *$G$ is reducible.*
2. *There exists a proper decomposition $V = \bigoplus_{i\in I} V_i$ of $V$ into equidimensional non-singular symplectic subspaces $V_i$ such that, for each $g \in G$ and each $i \in I$, there exists some $j \in I$ with $g(V_i) \subseteq V_j$ and such that the resulting action of $G$ on $I$ is transitive.*
3. *$G$ contains $\mathrm{Sp}(V, \langle\cdot,\cdot\rangle)$.*

*Remark 3.8.* Assume that $V$ is the $\ell$-torsion group of a principally polarized abelian variety $A$ defined over $\mathbb{Q}$ and $\langle\cdot,\cdot\rangle$ is the symplectic pairing coming from the Weil pairing. If $G = \mathrm{Im}\overline{\rho}_{A,\ell}$ satisfies the third condition in Theorem 3.7, then $G = \mathrm{GSp}(V, \langle\cdot,\cdot\rangle)$. Indeed, we have the following exact sequence

[1]We adopt the convention that identity is a transvection so that the set of transvections for a given hyperplane $H$ is a group.

$$1 \to \mathrm{Sp}(V, \langle\cdot,\cdot\rangle) \to \mathrm{GSp}(V, \langle\cdot,\cdot\rangle) \to \mathbb{F}_\ell^\times \to 1,$$

where the map $m : \mathrm{GSp}(A[\ell], \langle\cdot,\cdot\rangle) \to \mathbb{F}_\ell^\times$ associates with $M$ the scalar $a$ satisfying that, for all $u, v \in V$, $\langle Mu, Mv\rangle = a\langle u, v\rangle$. By Equation (1), the restriction of $m$ to $\mathrm{Im}(\overline{\rho}_{A,\ell})$ coincides with the mod $\ell$ cyclotomic character $\chi_\ell$. We can easily conclude the result using that $\chi_\ell$ is surjective onto $\mathbb{F}_\ell^\times$.

Even in the favorable case when we know that $\mathrm{Im}(\overline{\rho}_{A,\ell})$ contains a nontrivial transvection, we still need to distinguish between the three cases in Theorem 3.7. In this paper, we will make use of the following consequence of Theorem 3.7 (cf. Corollary 2.2 of Arias-de-Reyna and Kappen 2013).

**Corollary 3.9.** *Let $\ell \geq 5$ be a prime, let $V$ be a finite-dimensional vector space over $\mathbb{F}_\ell$, endowed with a symplectic pairing $\langle\cdot,\cdot\rangle : V \times V \to \mathbb{F}_\ell$ and let $G \subset$ $\mathrm{GSp}(V, \langle\cdot,\cdot\rangle)$ be a subgroup containing a nontrivial transvection and an element whose characteristic polynomial is irreducible and which has nonzero trace. Then $G$ contains* $\mathrm{Sp}(V, \langle\cdot,\cdot\rangle)$.

In order to apply this corollary in our situation, we need some more information on the image of $\overline{\rho}_{A,\ell}$. We will obtain this by looking at the images of the Frobenius elements $\mathrm{Frob}_q$ for primes $q$ of good reduction of $A$.

More generally, let $A$ be an abelian variety defined over a field $K$ and assume that $\ell$ is a prime different from the characteristic of $K$. Any endomorphism $\alpha$ of $A$ induces an endomorphism of $A[\ell]$, in such a way that the characteristic polynomial of $\alpha$ (which is a monic polynomial in $\mathbb{Z}[X]$, see, e.g., §3, Chapter 3 of Lang 1959 for its definition) coincides, after reduction mod $\ell$, with the characteristic polynomial of the corresponding endomorphism of $A[\ell]$. In the case when $K$ is a finite field (say of cardinality $q$), we can consider the Frobenius endomorphism $\phi_q \in \mathrm{End}_K(A)$, induced by the action of the Frobenius element $\mathrm{Frob}_q \in \mathrm{Gal}(\overline{K}/K)$. Then the reduction mod $\ell$ of the characteristic polynomial of $\phi_q$ coincides with the characteristic polynomial of $\overline{\rho}_{A,\ell}(\mathrm{Frob}_q)$. This will turn out to be particularly useful in the case when $A = J(C)$ is the Jacobian of a curve $C$ of genus $n$ defined over $K$, since one can determine the characteristic polynomial of $\overline{\rho}_{J(C),\ell}(\mathrm{Frob}_q)$ by counting the $\mathbb{F}_{q^r}$-valued points of $C$, for $r = 1, \dots, n$.

As a consequence, we can state the following result, which will be used in the next section.

**Theorem 3.10.** *Let $A$ be a principally polarized $n$-dimensional abelian variety defined over $\mathbb{Q}$. Assume that there exists a prime $p$ such that the following condition holds:*

($T_p$) *The special fiber of the Néron model of $A$ over $\mathbb{Q}_p$ is semistable with toric dimension* 1.

*Denote by $\Phi_p$ the group of connected components of the special fiber of the Néron model at $p$. Let $q$ be a prime of good reduction of $A$, let $A_q$ be the special fiber of the Néron model of $A$ over $\mathbb{Q}_q$, and let $P_q(X) = X^{2n} + aX^{2n-1} + \cdots + q^n \in \mathbb{Z}[X]$ be the characteristic polynomial of the Frobenius endomorphism acting on $A_q$.*

*Then for all primes $\ell$ which do not divide $6pq|\Phi_p|a$ and such that the reduction of $P_q(X)$ mod $\ell$ is irreducible in $\mathbb{F}_\ell$, the image of $\overline{\rho}_{A,\ell}$ coincides with* $\mathrm{GSp}_{2n}(\mathbb{F}_\ell)$.

*Remark 3.11.* The condition that $\ell$ does not divide $a$ corresponds to the Frobenius element having non-zero trace modulo $\ell$. Note that the theorem is vacuous when $a = 0$.

# 4 Galois Realization of $\mathrm{GSp}_{2n}(\mathbb{F}_\ell)$ from a Hyperelliptic Curve of Genus $n$

Let $C$ be a hyperelliptic curve of genus $n$ over $\mathbb{Q}$, defined by an equation $Y^2 = f(X)$ where $f(X) \in \mathbb{Q}[X]$ is a polynomial of degree $2n + 1$. Let $A = J(C)$ be its Jacobian variety. We assume that $A$ satisfies condition $(T_p)$ for some prime $p$. In this section we present an algorithm, based on Theorem 3.10, which computes a finite set of prime numbers $\ell$ for which the Galois representation $\overline{\rho}_{A,\ell}$ has image $\mathrm{GSp}_{2n}(\mathbb{F}_\ell)$. We apply this procedure to an example of a genus 3 curve using a computer algebra system.

## 4.1 *Strategy*

First, to apply Theorem 3.10, we restrict ourselves to hyperelliptic curves of genus $n$ whose Jacobian varieties will satisfy Condition $(T_p)$ for some $p$. Namely, we fix a prime number $p$ and then choose $f(X) \in \mathbb{Z}[X]$ monic of degree $2n + 1$ such that both of the following conditions hold:

1. The polynomial $f(X)$ only has simple roots over $\overline{\mathbb{Q}}$, so that $Y^2 = f(X)$ is the equation of an hyperelliptic curve $C$ over $\mathbb{Q}$.
2. All coefficients of $f(X)$ have $p$-adic valuation greater than or equal to zero, and the reduction $f(X)$ mod $p$ has one double zero in $\overline{\mathbb{F}}_p$, and its other zeroes are simple. This ensures that $A = J(C)$ satisfies Condition $(T_p)$ (see Remark 3.3).

Any prime of good reduction for $C$ is also a prime of good reduction for its Jacobian $A$. Primes of good reduction for the hyperelliptic curve can be computed using the discriminant of Weierstrass equations for $C$ (see Lockhart 1994). In our case, it turns out that any prime not dividing the discriminant of $f(X)$ is of good reduction for $C$, hence for $A$.

We take such a prime number $q$ of good reduction for $A$. Recall that $P_q(X) \in \mathbb{Z}[X]$ is the characteristic polynomial of the Frobenius endomorphism acting on the fiber $A_q$.

Let $\mathcal{S}_q$ denote the set of prime numbers $\ell$ satisfying the following conditions:

(i) $\ell$ divides neither $6pq|\Phi_p|$ nor the coefficient of $X^{2n-1}$ in $P_q(X)$,
(ii) the reduction of $P_q(X)$ modulo $\ell$ is irreducible in $\mathbb{F}_\ell$.

Note that if the coefficient of $X^{2n-1}$ in $P_q(X)$ is nonzero, condition (i) rules out only finitely many prime numbers $\ell$, whereas if it vanishes, condition (i) rules out all prime numbers $\ell$. By Theorem 3.10, for each $\ell \in \mathcal{S}_q$ the representation $\overline{\rho}_{A,\ell}$ is surjective with image $\mathrm{GSp}_{2n}(\mathbb{F}_\ell)$. Also, primes in $\mathcal{S}_q$ can be computed effectively up to a given fixed bound.

Since we want the polynomial $P_q(X)$ (of degree $2n$) to be irreducible modulo $\ell$, its Galois group $G$ over $\mathbb{Q}$ must be a transitive subgroup of $S_{2n}$ with a $2n$-cycle. Therefore, by an application of a weaker version of the Chebotarev density theorem due to Frobenius (Stevenhagen and Lenstra 1996, "Theorem of Frobenius", p. 32), the density of $\mathcal{S}_q$ is

$$\frac{\#\{\sigma \in G \subset S_{2n}\colon \sigma \text{ is a } 2n\text{-cycle}\}}{\#G}.$$

This estimate is far from what Theorem 3.2 provides us, namely that the density of $\ell$'s with $\mathrm{Im}(\overline{\rho}_{A,\ell}) = \mathrm{GSp}_{2n}(\mathbb{F}_\ell)$ is 1.

This leads us to discuss the role of the prime $q$. First of all, we can see that

$$\bigcup_q \mathcal{S}_q = \{\ell \text{ prime}\colon \ell \nmid 6p|\Phi_p| \text{ and } \overline{\rho}_{A,\ell} \text{ surjective}\},$$

where the union is taken over all primes $q$ of good reduction for $A$. Note that the inclusion $\subset$ follows directly from Theorem 3.10. To show the other inclusion $\supset$, suppose now that $\ell \nmid 6p|\Phi_p|$ and that the representation at $\ell$ is surjective. Its image $\mathrm{GSp}_{2n}(\mathbb{F}_\ell)$ contains an element with irreducible characteristic polynomial and nonzero trace (see, for instance, Proposition A.2 of Arias-de-Reyna and Kappen 2013). This element defines a conjugacy class $C \subset \mathrm{GSp}_{2n}(\mathbb{F}_\ell)$ and the Chebotarev density theorem ensures that there exists $q$ such that $\overline{\rho}_{A,\ell}(\mathrm{Frob}_q) \in C$, hence $\ell \in \mathcal{S}_q$.

Moreover, if, for some fixed $\ell$, the events "$\ell$ belongs to $\mathcal{S}_q$" are independent as $q$ varies, the density of primes $\ell$ for which $\overline{\rho}_{A,\ell}$ is surjective will increase when we take several different primes $q$. A sufficient condition for this density to tend to 1 is that there exists an infinite family of primes $q$ for which the splitting fields of $P_q(X)$ are pairwise linearly disjoint over $\mathbb{Q}$.

Therefore, it seems reasonable to expect that computing the sets $\mathcal{S}_q$ for several values of $q$ increases the density of primes $\ell$ for which we know the surjectivity of $\overline{\rho}_{A,\ell}$. This is what we observe numerically in the next example.

## 4.2 A Numerical Example in Genus 3

We consider the hyperelliptic curve $C$ of genus $n = 3$ over $\mathbb{Q}$ defined by $Y^2 = f(X)$, where

$$f(X) = X^2(X-1)(X+1)(X-2)(X+2)(X-3) + 7(X-28) \in \mathbb{Z}[X].$$

This is a Weierstrass equation, which is minimal at all primes $\ell$ different from 2 (see Lockhart 1994, Lemma 2.3), with discriminant $-2^{12} \cdot 7 \cdot 73 \cdot 1{,}069{,}421 \cdot 11{,}735{,}871{,}491$. Thus, $C$ has good reduction away from the primes appearing in this factorization. Clearly, $p = 7$ is a prime for which the reduction of $f(X)$ modulo 7 has one double zero in $\overline{\mathbb{F}}_7$ and otherwise only simple zeroes. Therefore, its Jacobian $J(C)$ satisfies Condition $(T_7)$. As we computed with MAGMA, the order of the component group $\Phi_7$ is 2. Recall that $P_q(X)$ coincides with the characteristic polynomial of the Frobenius endomorphism of the reduced curve $C$ modulo $q$ over $\mathbb{F}_q$.

Our method provides no significant result for $q \in \{3, 5\}$ because for $q = 3$ the characteristic polynomial $P_q(X)$ is not irreducible in $\mathbb{Z}[X]$ and for $q = 5$ it has zero trace in $\mathbb{Z}$. So in this example, we first take $q = 11$. The curve has 11, 135, and 1247 points over $\mathbb{F}_{11}$, $\mathbb{F}_{11^2}$, and $\mathbb{F}_{11^3}$, respectively. The characteristic polynomial $P_{11}(X)$ is

$$P_{11}(X) = X^6 - X^5 + 7X^4 - 35X^3 + 77X^2 - 121X + 1331$$

and it is irreducible over $\mathbb{Q}$. Its Galois group $G$ has order 48 and is isomorphic to the wreath product $S_2 \wr S_3$. This group is the direct product of 3 copies of $S_2$, on which $S_3$ acts by permutation (see James and Kerber 1981, Chapter 4): An element of $S_2 \wr S_3$ can be written as $((a_1, a_2, a_3), \sigma)$, where $(a_1, a_2, a_3)$ denotes an element of the direct product $S_2 \times S_2 \times S_2$ and $\sigma$ an element of $S_3$. The group law is defined as follows:

$$((a_1, a_2, a_3), \sigma)((a'_1, a'_2, a'_3), \sigma') = ((a_1, a_2, a_3)(a'_1, a'_2, a'_3)^\sigma, \sigma\sigma'),$$

where $(a'_1, a'_2, a'_3)^\sigma = (a'_{\sigma(1)}, a'_{\sigma(2)}, a'_{\sigma(3)})$. One can also view the wreath product $S_2 \wr S_3$ as the centralizer of $(12)(34)(56)$ in $S_6$, through an embedding $\psi : S_2 \wr S_3 \to S_6$ whose image is isomorphic to the so-called Weyl group of type $B_3$ (James and Kerber 1981, 4.1.18 and 4.1.33). More precisely, under $\psi$, the image of an element $((a_1, a_2, a_3), \sigma) \in S_2 \wr S_3$ is the permutation of $S_6$ that acts on $\{1, 2, \ldots, 6\}$ as follows: it first permutes the elements of the sets $E_1 = \{1, 2\}$, $E_2 = \{3, 4\}$ and $E_3 = \{5, 6\}$ separately, according to $a_1$, $a_2$, and $a_3$ respectively (identifying $E_2, E_3$ with $\{1, 2\}$ in an obvious way) and then permutes the pairs $E_1, E_2, E_3$ according to the action of $\sigma$ on the indices. For example, denoting $S_2 = \{\mathrm{id}, \tau\}$, the image under $\psi$ of $((\tau, \mathrm{id}, \mathrm{id}), (123))$ is the 6-cycle (135246).

Let us now determine the elements of $S_2 \wr S_3$ which map to 6-cycles in $S_6$ through the embedding $\psi$. For an element in $S_2 \wr S_3$ to be of order 6, it has to be of the form $((a_1, a_2, a_3), \gamma)$ with $\gamma$ a 3-cycle in $S_3$. Now, $\psi$ sends an element $((a_1, a_2, a_3), \gamma)$ where either one or three $a_i$'s are id, to a product of two disjoint 3-cycles in $S_6$. So the elements of $S_2 \wr S_3$ which are 6-cycles in $S_6$ are among the eight elements $((\mathrm{id}, \mathrm{id}, \tau), \gamma)$, $((\mathrm{id}, \tau, \mathrm{id}), \gamma)$, $((\tau, \mathrm{id}, \mathrm{id}), \gamma)$ and $((\tau, \tau, \tau), \gamma)$ with $\gamma = (123)$ or $\gamma = (132)$. Moreover, James and Kerber (1981, Theorem 4.2.8) (see also Gramain 2008, Lemma 3.1 or Taylor 2012) ensures that these 8 elements are conjugate. Since $\psi((\tau, \mathrm{id}, \mathrm{id}), (123)) = (135{,}246)$ is a 6-cycle, we deduce that the 8 elements listed above are exactly the elements of $S_2 \wr S_3$ which are 6-cycles in $S_6$.

To conclude, the Galois group $G$, viewed as a subgroup of $S_6$, contains exactly 8 elements that are 6-cycles. Therefore, the density of $\mathcal{S}_{11}$ is $8/48 = 1/6$.

We can compute $P_q(X)$ using efficient algorithms available in MAGMA Bosma et al. (1997) or SAGE Stein (2014), which are based on $p$-adic methods. We found that there are 6891 prime numbers $11 \leq \ell \leq 500{,}000$ that belong to $\mathcal{S}_{11}$. For these $\ell$, we know that the image of $\overline{\rho}_{A,\ell}$ is $\mathrm{GSp}_6(\mathbb{F}_\ell)$, so the groups $\mathrm{GSp}_6(\mathbb{F}_\ell)$ are realized as Galois groups arising from the $\ell$-torsion of the Jacobian of the hyperelliptic curve $C$. For instance, the first ten elements of $\mathcal{S}_{11}$ are

$$47, 71, 79, 83, 101, 113, 137, 251, 269, 271.$$

Also, the proportion of prime numbers $11 \leq \ell \leq 500{,}000$ in $\mathcal{S}_{11}$ is about 0.1659, which is quite in accordance with the density obtained from the Chebotarev density theorem.

By looking at polynomials $P_q(X)$ for several primes $q$ of good reduction, we are able to significantly improve the known proportion of primes $\ell$, up to a given bound, for which the Galois representation is surjective. Namely, we computed that

$$\{\ell \text{ prime}, 11 \leq \ell \leq 500{,}000\} \subseteq \bigcup_{11 \leq q \leq 571} \mathcal{S}_q.$$

As a consequence, for any prime $11 \leq \ell \leq 500{,}000$, the group $\mathrm{GSp}_6(\mathbb{F}_\ell)$ is realized as a Galois group arising from the $\ell$-torsion of the Jacobian of the hyperelliptic curve $C$. This is reminiscent of Le Duff's numerical data for $\mathrm{GSp}_4(\mathbb{F}_l)$ (see Theorem 3.6).

Combining all of the above suggests that the single hyperelliptic curve $C$ might provide a positive answer to the inverse Galois problem for $\mathrm{GSp}_6(\mathbb{F}_\ell)$ for any prime $\ell \geq 11$.

**Acknowledgements** The authors would like to thank Marie-José Bertin, Alina Bucur, Brooke Feigon, and Leila Schneps for organizing the WIN-Europe conference which initiated this collaboration. Moreover, we are grateful to the Centre International de Rencontres Mathématiques, the Institut de Mathématiques de Jussieu, and the Institut Henri Poincaré for their hospitality during several short visits. The authors are indebted to Irene Bouw, Jean-Baptiste Gramain, Kristin Lauter, Elisa Lorenzo, Melanie Matchett Wood, Frans Oort, and Christophe Ritzenthaler for several insightful discussions. We also want to thank the anonymous referee for her/his suggestions that helped us to improve this paper.

S. Arias-de-Reyna and N. Vila are partially supported by the project MTM2012-33830 of the Ministerio de Economía y Competitividad of Spain, C. Armana by a BQR 2013 Grant from Université de Franche-Comté and M. Rebolledo by the ANR Project Régulateurs ANR-12-BS01-0002. L. Thomas thanks the Laboratoire de Mathématiques de Besançon for its support.

## References


Arias-de-Reyna, S., Dieulefait, L., Shin, S.-W., Wiese, G.: Compatible systems of symplectic Galois representations and the inverse Galois problem III. Automorphic construction of compatible systems with suitable local properties. Math. Ann. **361**(3), 909–925 (2015)

Arias-de-Reyna, S., Dieulefait, L, Wiese, G.: Classification of subgroups of symplectic groups over finite fields containing a transvection. Demonstratio Math. (2014, preprint)

Arias-de-Reyna, S., Kappen, C.: Abelian varieties over number fields, tame ramification and big Galois image. Math. Res. Lett. **20**(1), 1–17 (2013)

Arias-de-Reyna, S., Vila, N.: Tame Galois realizations of $GSp_4(\mathbb{F}_\ell)$ over $\mathbb{Q}$. Int. Math. Res. Not. IMRN **9**, 2028–2046 (2011)

Bosma, W., Cannon, J., Playoust, C.: The Magma algebra system. I. The user language. J. Symb. Comput. **24**(3–4), 235–265 (1997). Computational algebra and number theory (London, 1993)

Dieulefait, L.V.: Explicit determination of the images of the Galois representations attached to abelian surfaces with $\mathrm{End}(A) = \mathbb{Z}$. Exper. Math. **11**(4), 503–512 (2002a)

Dieulefait, L.V.: On the images of the Galois representations attached to genus 2 Siegel modular forms. J. Reine Angew. Math. **553**, 183–200 (2002b)

Dettweiler, M., Kühn, U., Reiter, S.: On Galois representations via Siegel modular forms of genus two. Math. Res. Lett. **8**(4), 577–588 (2001)

Dieulefait, L., Vila, N.: Projective linear groups as Galois groups over $\mathbb{Q}$ via modular representations. J. Symb. Comput. **30**(6), 799–810 (2000). Algorithmic methods in Galois theory

Dieulefait, L., Vila, N.: On the images of modular and geometric three-dimensional Galois representations. Am. J. Math. **126**(2), 335–361 (2004)

Dieulefait, L., Vila, N.: Geometric families of 4-dimensional Galois representations with generically large images. Math. Z. **259**(4), 879–893 (2008)

Dieulefait, L., Vila, N.: On the classification of geometric families of four-dimensional Galois representations. Math. Res. Lett. **18**(4), 805–814 (2011)

Dieulefait, L., Wiese, G.: On modular forms and the inverse Galois problem. Trans. Am. Math. Soc. **363**(9), 4569–4584 (2011)

Gramain, J.-B.: On defect groups for generalized blocks of the symmetric group. J. Lond. Math. Soc. (2) **78**(1), 155–171 (2008)

Hall, C.: Big symplectic or orthogonal monodromy modulo $l$. Duke Math. J. **141**(1), 179–203 (2008)

Hall, C.: An open-image theorem for a general class of abelian varieties. Bull. Lond. Math. Soc. **43**(4), 703–711 (2011). With an appendix by Emmanuel Kowalski

James, G., Kerber, A.: The Representation Theory of the Symmetric Group. Encyclopedia of Mathematics and its Applications, vol. 16. Addison-Wesley, Reading (1981). With a foreword by P. M. Cohn, With an introduction by Gilbert de B. Robinson

Khare, C., Larsen, M., Savin, G.: Functoriality and the inverse Galois problem. Compos. Math. **144**(3), 541–564 (2008)

Lang, S.: Abelian Varieties. Interscience Tracts in Pure and Applied Mathematics. No. 7. Interscience, New York/London (1959)

Le Duff, P.: Représentations galoisiennes associées aux points d'ordre $\ell$ des jacobiennes de certaines courbes de genre 2. Bull. Soc. Math. France **126**(4), 507–524 (1998)

Liu, Q.: Courbes stables de genre 2 et leur schéma de modules. Math. Ann. **295**(2), 201–222 (1993)

Lockhart, P.: On the discriminant of a hyperelliptic curve. Trans. Am. Math. Soc. **342**(2), 729–752 (1994)

Mumford, D.: A note of Shimura's paper "Discontinuous groups and abelian varieties". Math. Ann. **181**, 345–351 (1969)

Ribet, K.A.: On $\ell$-adic representations attached to modular forms. Invent. Math. **28**, 245–275 (1975)

Reverter, A., Vila, N.: Some projective linear groups over finite fields as Galois groups over $\mathbb{Q}$. In: Recent Developments in the Inverse Galois Problem (Seattle, WA, 1993). Contemporary Mathematics, vol. 186, pp. 51–63. American Mathematical Society, Providence (1995)

Serre, J.-P.: Œuvres. Collected papers IV. Springer, Berlin (2000). 1985–1998

Serre, J.-P.: Propriétés galoisiennes des points d'ordre fini des courbes elliptiques. Invent. Math. **15**(4), 259–331 (1972)

Stein, W.A., et al.: Sage Mathematics Software (Version 6.0). The Sage Development Team. http://www.sagemath.org (2014)

Stevenhagen, P., Lenstra, H. W.: Chebotarëv and his density theorem. Math. Intell. **18**(2), 26–37 (1996)

Taylor, J.: Families of irreducible representations of $S_2 \wr S_3$. https://documents.epfl.ch/users/j/jt/jtaylor/www/PDF/representations_of_S2wrS3.pdf (2012)

Wiese, G.: On projective linear groups over finite fields as Galois groups over the rational numbers. In: Modular Forms on Schiermonnikoog, pp. 343–350. Cambridge University Press, Cambridge (2008)

Zarhin, Y.G.: Two-dimensional families of hyperelliptic jacobians with big monodromy (preprint, 2014) [arXiv:1310.6532]

Zywina, D.: The inverse Galois problem for $\mathrm{PSL}_2(\mathbb{F}_p)$ (preprint, 2013) [arXiv:1303.3646]

The manufacturer's authorised representative in the EU is Springer Nature Customer Service Centre GmbH, Europaplatz 3, 69115 Heidelberg, Germany. If you have any concerns regarding our products, please contact ProductSafety@springernature.com

Printed and bound by CPI Group (UK) Ltd, Croydon, CR0 4YY
30/09/2026
02225643-0003